T0192851

Solid Mechanics and Its Applications

Volume 241

Series editors

J.R. Barber, Ann Arbor, USA
Anders Klarbring, Linköping, Sweden

Founding editor

G.M.L. Gladwell, Waterloo, ON, Canada

Aims and Scope of the Series

The fundamental questions arising in mechanics are: *Why?*, *How?*, and *How much?* The aim of this series is to provide lucid accounts written by authoritative researchers giving vision and insight in answering these questions on the subject of mechanics as it relates to solids.

The scope of the series covers the entire spectrum of solid mechanics. Thus it includes the foundation of mechanics; variational formulations; computational mechanics; statics, kinematics and dynamics of rigid and elastic bodies: vibrations of solids and structures; dynamical systems and chaos; the theories of elasticity, plasticity and viscoelasticity; composite materials; rods, beams, shells and membranes; structural control and stability; soils, rocks and geomechanics; fracture; tribology; experimental mechanics; biomechanics and machine design.

The median level of presentation is to the first year graduate student. Some texts are monographs defining the current state of the field; others are accessible to final year undergraduates; but essentially the emphasis is on readability and clarity.

More information about this series at http://www.springer.com/series/6557

Joshua Pelleg

Creep in Ceramics

 Springer

Joshua Pelleg
Department of Materials Engineering
Ben-Gurion University of the Negev
Beer Sheva
Israel

ISSN 0925-0042   ISSN 2214-7764   (electronic)
Solid Mechanics and Its Applications
ISBN 978-3-319-84501-2   ISBN 978-3-319-50826-9   (eBook)
DOI 10.1007/978-3-319-50826-9

Printed on acid-free paper

This Springer imprint is published by Springer Nature
The registered company is Springer International Publishing AG
The registered company address is: Gewerbestrasse 11, 6330 Cham, Switzerland

God has given two great presents to man: work and tears. Without work, there would be no progress in the world. The tears of sorrow comfort a man in his distress, while the tears of joy accompany him, soothing his passage from birth to grave.

Joshua Pelleg

To my wife Ada and my children: Deenah and her late husband Gidon Barak; Ruth and Christer Kallevag; Shlomit and Asher Pelleg and their children: Roy, Tal, Rotem and Noa Barak; Ella and Maya Kallevag; and Ofir and Ori Pelleg.

Preface

This textbook on creep in ceramics is unique since all other treatments of the subject appear only as a part of the general concept of mechanical properties in materials. A collection of papers on creep in brittle materials was published in 1989, containing a set of papers on MgO and Si_3N_4 ceramics. Also, a chapter dealing with creep can be found in the book published in 2014 by Springer: Mechanical Properties of Ceramics, providing a taste of this important subject, both in theory and in regard to its practical technological applications. Creep is an important deformation process in ceramics, as in other materials. Although no theoretical basics have yet been formulated for creep phenomena, leaving those working in the field to rely solely on experimental observations, they should be aware that physical laws govern the complex deformation mechanism in materials exposed to creep conditions.

This textbook has two parts. Part I contains 11 chapters. Chapter 1 introduces the basic concept of creep. Chapter 2 describes the general mechanism of creep. Chapter 3 presents the relation of creep to diffusion. Chapter 4 provides a general consideration of creep in ceramics. Chapter 5 discusses creep in single-crystal ceramics and creep testing methods. Chapter 6 describes creep in nanoceramics, followed by creep rupture in Chap. 7. Superplasticity is considered in Chap. 8. Chapter 9 deals with creep and recovery, while the empirical relations related to creep are discussed in Chap. 10. Part I concludes with Chap. 11 on design for creep resistance.

Part II covers creep deformation in technologically important ceramics. The six ceramics most commonly encountered in various technological applications were selected to represent creep in ceramics. These are three oxide ceramics: Al_2O_3 (alumina) in Chap. 12; MgO (magnesia) in Chap. 13; and ZrO_2 (zirconia) in Chap. 14; followed by two carbides: SiC (silicon carbide) in Chap. 15 and BC (boron carbide) in Chap. 16; and concluding with the important silicon nitride ceramic, Si_3N_4 in Chap. 17.

Practical exercises are not given in this textbook, since each lecturer tends to provide his/her own preferred problems, which the students are expected to solve. The author of this book is also not inclined to republish exercises existing in prior textbooks. Suffice it to say that those interested in creating and/or solving new problems in the field should be encouraged to do so for everyone's benefit.

My gratitude to all the publishers and authors for their permission to use and reproduce some of their illustrations and microstructures. Thanks to Ethelea Katzenell of Ben Gurion University for improving the English.

Finally, without the tireless devotion, understanding and unlimited patience of my wife Ada, it would be difficult to imagine the completion of this book, despite my decades of teaching the mechanical behaviors of materials. Her helpful attitude was instrumental in inspiring its writing.

Here, it is impossible for me not to mention my gratitude to my grandparents for the education they gave me where I spent my childhood and adolescence; they ascended to Heaven in fire, not unlike Elijah the Prophet, though not having been called by God.

Beer Sheva, Israel   Joshua Pelleg

Contents

About the Author

Joshua Pelleg received his B.Sc. in Chemical Engineering at the Technion—Institute of Technology, Haifa, Israel; a M.Sc. in Metallurgy at the Illinois Institute of Technology, Chicago, IL and a Ph.D. in Metallurgy at the University of Wisconsin, Madison, WI. He has been in the Ben-Gurion University of the Negev (BGU) Materials Engineering Department in Beer-Sheva, Israel since 1970, and was among the founders of the department, and served as its second chairman. Professor Pelleg was the recipient of the Samuel Ayrton Chair in Metallurgy. He specializes in the mechanical properties of materials and the diffusion and defects in solids. He has chaired several university committees and served four terms as the Chairman of Advanced Studies at Ben-Gurion University of the Negev. Prior to his work at BGU, Pelleg acted as Assistant Professor and then Associate Professor in the Department of Materials and Metallurgy at the University of Kansas, Lawrence, KS. Professor Pelleg was also a Visiting Professor: in the Department of Metallurgy at Iowa State University; at the Institute for Atomic Research, US Atomic Energy Commission, Ames, IA; at McGill University, Montreal, QC; at the Tokyo Institute of Technology, Applied Electronics Department, Yokohama, Japan; and in Curtin University, Department of Physics, Perth, Australia. His nonacademic research and industrial experience includes: Chief Metallurgist in Urdan Metallurgical Works Ltd., Netanyah, Israel; Research Engineer in International Harvester Manufacturing Research, Chicago, IL; Associate Research Officer for the National Research Council of Canada, Structures and Materials, National Aeronautical Establishment, Ottawa, ON; Physics Senior Research Scientist, Nuclear Research Center, Beer-Sheva, Israel; Materials Science Division, Argonne National Labs, Argonne, IL; Atomic Energy of Canada, Chalk River, ON; Visiting Scientist, CSIR, National Accelerator Centre, Van de Graaf Group Faure, South Africa; Bell Laboratories, Murray Hill, NJ; and GTE Laboratories, Waltham, MA. His current research interests are diffusion in solids, thin film deposition and properties (mostly by sputtering) and the characterization of thin films, among them various silicides.

Abbreviations

ATZ	Alumina-toughened zirconia
BCC	Body-centered cubic
CGN	Nicalon fiber (12 wt% oxygen)
CMC	Ceramic matrix composites
CRM	Creep rate mismatch
CSR	Constant strain rate
CVD	Chemical vapor deposition
DPH	Diamond pyramid hardness
EDP	Electron diffraction pattern
EELS	Electron energy loss spectroscopy
FCC	Face-centered cubic
GBS	Grain boundary sliding
HCP	Hexagonal close packed
HIP	Hot isostatic pressure
HPSN	Hot-pressed silicon nitride
HREM	High-resolution electron microscopy
HRSEM	High-resolution SEM
HRTEM	High-resolution TEM
HSRS	High strain rate superplasticity
LMP	Larson-Miller parameter/relation
MEMS	Microelectromechanical systems
MFP	Manson-Haferd parameter
MGR	Monkman–Grant relationship
MMC	Metal matrix composites
NW	Nanowires
OSD	Orr-Sherby-Dorn
PSZ	Partially stabilized zirconia
PZT	Lead zirconate titanate
RBSN	Reaction bonded silicon nitride
SCG	Subcritical crack growth

SEM	Scanning electron microscopy
SF	Stacking fault
TEM	Transmission electron microscopy
TSZ	Tetragonally stabilized zirconia
TZP	Tetragonal zirconia polycrystal
WBDF	Weak-beam dark field
XRD	X-ray diffraction
Y-CSZ	Yttria-cubic stabilized zirconia
Y-TZP	Yttria-stabilized tetragonal zirconia polycrystal
ZTA	Zirconia-toughened alumina

Part I
Basics

Chapter 1
What Is Creep?

Abstract The concept of creep as presented originally by Andrade is discussed in this chapter. The conventional three stages of creep and the relevant equations are indicated. Creep rate as a function of the important parameters shown in the equation below summarizes the effects

$$\dot{\varepsilon} = f(\sigma, t, T)$$

of stress, time, and temperature. Creep at some stress and temperature is time dependent.

1.1 General Concept of Creep

Historically, about a century ago, Andrade was among the first, if not the first, to pioneer and systematically study the concept of creep in metals. He called it a 'viscous flow in metals' and indicated: "that for a lead wire, loaded well beyond the elastic limit, the extension, after some time, becomes proportional to the time, or the flow becomes viscous in character. The rate of this viscous flow varies with the load." The method adapted to measure creep is to load a wire (or rod) axially and measure the extension with time at a constant stress. However, one must apply a constant load, maintained at constant temperature, and record the stain over a certain period of time. Modern, universal testing machines are commercially available and provide:

  (i)   Load measurement and control;
 (ii)   Extension measurement and control;
(iii)   Time measurement;
 (iv)   Temperature measurement and control facilities;
  (v)   A chamber for controlled environmental and test conditions;
 (vi)   A computer for data acquisition and control; and
(vii)   A testing apparatus equipped with grips, preventing slippage and excessive stress.

© Springer International Publishing AG 2017
J. Pelleg, *Creep in Ceramics*, Solid Mechanics and Its Applications 241,
DOI 10.1007/978-3-319-50826-9_1

Andrade also classified creep. Creep occurs in three stages: primary creep (stage I); secondary creep (stage II); and tertiary creep (stage III). Stage I occurs at the beginning of a creep test. This creep is mostly characterized as 'transient creep', which does not occur at a steady rate. Resistance to creep increases until stage II is reached. In stage II, the rate of creep becomes steady or almost steady. This stage is also known as 'steady-state creep'. In stage III, the creep rate begins to accelerate as the cross-sectional area of the specimen decreases due to necking or until internal voiding decreases the effective area of the specimen. If the test is allowed to proceed, fracture will occur, often referred to as 'rupture'.

The first relations expressing extension under constant stress were formulated based on Andrade's results from the extension of lead (Pb) wires, given as:

$$l = l_0(1 + \beta t^{1/3})\exp(\kappa t) \tag{1.1}$$

where l is the length at time t. l_0 is about the original constant length, β expresses creep at a rate diminishing rapidly with time, hence its term as 'transient' or 'β creep'. The creeping portion which proceeds at a constant rate (steady-state creep) is also known as being 'quasi-viscous creep' or 'κ creep'.

Creep curves are stress- and temperature-dependent, thus the three parameters that determine creep rate: time, temperature and stress, may be expressed by:

$$\dot{\varepsilon} = f(\sigma, t, T) \tag{1.2}$$

Temperature is an important parameter of creep in general and especially in ceramics. Since most ceramics are intended for high-temperature applications, guidelines are essential for understanding the limitations, if appreciable creep deformation is to be eliminated. Commonly, ceramics at high temperatures are ductile. For ceramics with low-temperature ductility, creep may occur at $\sim 0.5T_m$ or may even set in at lower temperatures. The homologous temperature, defined in terms of the melting point, T_m, serves as an efficient demarcation point between low-temperature and high-temperature creep, given as:

$$\text{homologous temperature} = \frac{T}{T_m} \tag{1.3}$$

T is a relevant temperature of application and expressed (together with T_m) on the absolute scale. Low-temperature creep below $\sim 0.5T_m$ is considered to be governed by a nondiffusion-controlled mechanism, whereas high-temperature creep, above $0.5T_m$, is diffusion controlled. Stress (load), time and temperature act simultaneously, determining the creep rate as expressed in Eq. (1.2). When resistance to creep in ceramics is an essential prerequisite in high-temperature applications (preventing the risk of creep failure during their lifetime) high T_m ceramics are advisable. Some ceramics with very high melting points are: MgO (2798 °C), Al_2O_3 (2050 °C) and SiC, that will be discussed later in this book.

In order to eliminate grain boundary creep and grain boundary sliding, single-crystal ceramics are suggested for practical use in many applications, such as turbine blades, etc. In the next section, some basic relations are presented.

1.2 Basic Concepts

Most of the many equations describing creep given in the literature follow in the wake of Andrade's empirical concept that one may express the variation of strain over time as:

$$\varepsilon = \varepsilon_0 \left(1 + \beta t^{1/3}\right) \exp(\kappa t) \tag{1.2a}$$

which is equivalent to Eq. (1.1). When $\kappa = 0$, β creep is obtained and Eq. (1.2a) may be expressed as:

$$\varepsilon = \varepsilon_0 \left(1 + \beta t^{1/3}\right) \tag{1.2b}$$

Equation (1.2b) represents transient creep, indicating a decreasing creep rate over time, since it is function of time. However, when $\beta = 0$ in Eq. (1.2a), one obtains:

$$\varepsilon = \varepsilon_0 \exp(\kappa t) \tag{1.4}$$

which is κ creep and represents a stationary state. Differentiating Eqs. (1.2b) and (1.4) with respect to time results in Eqs. (1.5) and (1.6), respectively:

$$\frac{d\varepsilon}{dt} = \dot{\varepsilon} = \frac{1}{3} \varepsilon_0 \beta t^{-2/3} \tag{1.5}$$

and the creep rate from Eq. (1.4) is:

$$\dot{\varepsilon} = K\varepsilon_0 \exp(\kappa t) = \kappa\varepsilon \tag{1.6}$$

Clearly, the last term in Eq. (1.6) is the consequence of expressing the value of Eq. (1.4) in the second term.

Figure 1.1 is a schematic creep curve illustrating a constant load creep curve, and the constant stress curve is shown by the dashed line extension of the secondary creep.

All three stages of the conventional creep curve are shown in Fig. 1.1a. Furthermore, an instantaneous stain on loading is also indicated. The variation of the strain rate in the three stages is illustrated in Fig. 1.1b. One can see from this illustration that, in the primary creep, the creep rate is continuously decreasing (hence its name 'transient creep'), while, in the secondary creep, the rate is constant (hence its name 'steady-state creep') and finally, in the tertiary creep, the creep rate

Fig. 1.1 **a** A schematic creep curve along the lines of Andrade showing three stages of creep and an instantaneous elongation on application of load. A constant stress curve is incorporated in the overall illustration, indicated by the *dashed line* extension; **b** Schematic strain rate plot versus time; Pelleg, *Mechanical Properties of Materials*, Springer, 2013

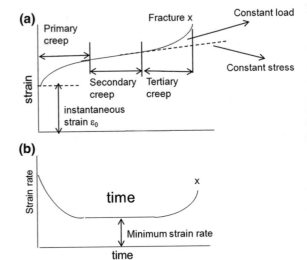

is accelerating (hence it is also known as 'accelerated creep up to fracture'. The 'constant creep rate' is the minimum creep rate, which is an important design parameter and its magnitude is temperature- and stress-dependent. Two criteria of the minimum creep rate are commonly applied to alloys: (a) the stress needed to produce a creep rate of 0.1×10^{-3}%/h (or 1% in 1×10^4 h) and (b) the stress needed to produce a creep rate of 0.1×10^{-4} %/h, namely 1% in 100×10^3 h, which is about 11.5 years. Criterion (a) is used for turbine blades, while (b) is usually applied to steam turbines.

Clearly, there is interest avoiding all forms of creep while a component is exposed to some temperature. Therefore, it is of utmost importance to evaluate the threshold stress and temperature, the additional time factors below which creep will not occur. Norton [8] attempted to determine this threshold by suggesting a relation based on the observation that a constant stress produces a constant secondary creep rate, given as:

$$\dot{\varepsilon} = A\sigma^n \tag{1.7}$$

In this equation, A and n are experimentally determined constants that are functions solely of temperature. The effect of temperature on the shape of the creep curve under constant stress is illustrated in Fig. 1.2a. The standard creep curve shown in Fig. 1.2a, may be obtained experimentally only at certain temperatures. At high and low temperatures, only segments of the curve can be observed. Note that line B is similar to the conventional creep curve often shown in many textbooks.

A similar case is one in which the temperature is kept constant while the stress is varied. σ_3, which is greater than σ_2, represent the often shown creep curve with all three stages. The reason for observing only two stages of the creep curve at low temperatures (or stress) is that creep strain did not produce the accelerated creep

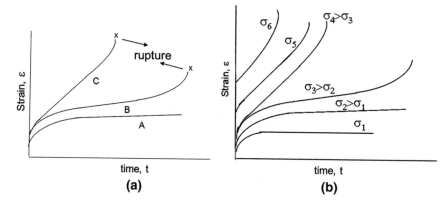

Fig. 1.2 Strain-time creep curves: **a** the shape of creep curves; *A* the standard creep curve (see Fig. 1.1a); *B* a creep curve at low temperature and stress and; *C* a high temperature and high stress curve; **b** schematic creep curves at a constant temperature with variable stress. Note that σ_3 represents the standard creep curve with all three stages. Pelleg, *Mechanical Properties of Materials*, Springer, 2013

during the time interval of the load application, a desired goal for the extended lifetime of the material since at this stage a steady-state creep rate prevails in the test specimen (or the material in use).

1.3 Additional Empirical Relations

Many formulae were originally presented for the strain time curves of creep in metallic materials, among which Andrade's formula seems to have attracted much attention, but it has not been possible to find a satisfactory explanation. Therefore, empirical equations had to be found, each expressing the behavior of a specific material of special interest. Andrade had determined the entire creep curve, as seen in Fig. 1.1, for several materials and (for lead in particular) he showed the difference obtained between constant load and constant stress, also indicated in Fig. 1.1.

Since creep is a thermally activated process, the minimum secondary-creep rate may be described by an Arrhenius equation (see [9]) as:

$$\frac{d\varepsilon}{dt} = \dot{\varepsilon} = A \exp -\left(\frac{Q_0 - \alpha\sigma}{kT}\right) \tag{1.8}$$

Here, A and α are constants and Q_0 is the activation energy for creep at zero stress. A is also known as the 'frequency' or 'pre-exponential factor,' as in the nomenclature for diffusion. An additional expression for the creep rate, where the stress and temperature terms are separated, is given as:

$$\dot{\varepsilon} = B\sigma^n \exp - \left(\frac{Q}{kT}\right) \tag{1.9}$$

In Eq. (1.9), the stress affects the frequency factor, B, while Q has the same meaning as Q_0 in Eq. 1.8.

Many experimental data indicate that the creep rate, in its early stages, may be expressed by a function suggested by Cottrell as either:

$$\frac{d\gamma}{dt} = \dot{\gamma} = At^{-n} \tag{1.10}$$

or

$$\frac{d\varepsilon}{dt} = \dot{\varepsilon} = Bt^{-n} \tag{1.11}$$

$A(B)$ and the exponent, n, are constants with $0 \leq n \leq 1$. Equation (1.10) may also be expressed in logarithmic terms and many transient regimes of creep curves may be fitted to a logarithmic law when $n = 1$. In the extreme case, when $n = 1$, which is often observed experimentally, one obtains the logarithmic creep law as:

$$\gamma = \alpha \ln t \quad (t > 1) \tag{1.12}$$

Note that Eq. (1.10) adequately describes the experimental creep data, since the creep rate.in the primary stage (transient) decreases over time, as shown in the schematic illustration in Fig. 1.1b. Various values of n, in the range 0–1, may be observed experimentally, but, very frequently, the value of 2/3 is preferred. Thus, Eq. (1.10) may be rendered as:

$$\frac{d\gamma}{dt} = \dot{\gamma} = At^{-2/3} \tag{1.13}$$

An integration of Eq. (1.10a) yields the equation for strain as:

$$\gamma = \beta t^{1/3} \tag{1.14}$$

Equation (1.13), representing transient creep, is often referred to as 'β-creep' or 'Andrade creep', since Andrade was the first to show that it applies to many materials. The creep behavior obeying Eq. (1.12) is often called 'α' or 'logarithmic creep.' Often, the instantaneous non-creep strain is also taken into account, suggesting an equation [10] in the form of:

$$\gamma = \gamma_0 + \alpha \ln(\beta t + 1) \tag{1.15}$$

with α and β being constants. Figure 1.3 schematically illustrates logarithmic creep curves at various stresses.

Fig. 1.3 Logarithmic creep. The *lines* are shown for different stresses

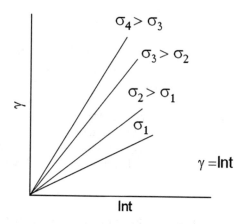

It was seen in Fig. 1.1a that stage II creep is linear, thus the function describing this region must also be linear. Much of the creep data is expressed by functions taking this linear contribution into account as:

$$\gamma = \gamma_0 + \beta t^{1/3} + \kappa t \tag{1.16}$$

Equation (1.15) is a combination of the instantaneous strain, γ_0, Eq. (1.13) and the linear contribution of second stage creep, κt [4], and it well describes many creep experiments. Usually, especially in experiments performed at high temperatures, transient and steady-state creep occur together. A graphic expression of Eq. (1.15), namely the combination of these stages, is seen in Fig. 1.4, without the instantaneous, non-creep strain, γ_0, obtained upon loading.

In tertiary creep, the strain and strain rate increase until fracture occurs. In ceramics, tertiary creep is usually not recorded, but, if the test is continued long enough, a tertiary creep may develop. In metals, entering stage III occurs when there is a reduction in the cross-sectional area due to necking or internal void formation. In ceramics, it is void formation, in the form of pores or flaws, which effectively causes a reduction in area. Thus, tertiary creep is significant in ceramic engineering, because it is often associated with the formation of structural instability, as indicated by void and/or crack formation, leading to failure-by-fracture.

Fig. 1.4 A graphic presentation of Eq. (1.15) without γ_0, obtained from the combination of transient ($\gamma_I = \beta t^{1/3}$) and steady-state ($\gamma_{II} = \kappa t$) creeps

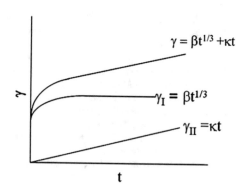

The onset of tertiary creep occurs at the end of steady-state creep. It is easier to locate the onset of tertiary creep from the $\dot{\varepsilon} - t$ relation than from $\varepsilon - t$, as seen in Fig. 1.1b, since the location of the deviation from the minimum creep rate is well defined. It is clear that the minimum creep rate parameter must limit allowable stress in practice, in order to prevent the onset of tertiary creep. In light of the minimum creep rate concept, the attention in experimental creep investigations is focused on steady-state creep, where it is constant over an extensive period of time. Generally in ceramics, tertiary creep is relatively short and sometimes even absent.

Several investigators have shown that the starting time of tertiary creep and rupture life are related in various alloys according to the relation (e.g., [6]):

$$t_2 = At_r^\alpha \tag{1.17}$$

where t_r is the rupture life, t_2 is the starting time of the tertiary creep, and A and α are constants, often ~ 1. Equation (1.16) is one of many expressions for creep, in general, and for tertiary creep, in particular, and is widely used for various materials under consideration for high-temperature applications. Other expressions are common in creep studies, such as for power, and for exponential and logarithmic functions. For example, these three functions are shown respectively as:

$$\varepsilon_{III} = \dot{\varepsilon}_{min}t + At^g \tag{1.18}$$

$$\varepsilon_{III} = \theta_3(\exp[\theta_4 t] - 1) \tag{1.19}$$

$$\varepsilon_{III} = -(\ln[1 - C\dot{\varepsilon}_{min}t])/C \tag{1.20}$$

In these expressions for tertiary creep without a primary stage, min represents the minimum creep rate and A, g, θ_3, θ_4 and C are parameters. Creep curves having higher applied stresses, with pronounced tertiary stages, may be successfully described by all three equations. Dobeš has indicated that the calculated value of g (~ 7–10) is higher than the one proposed by Graham and Walles ($g = 3$).

References

1. Andrade EN da C (1910). Proc R Soc London A 84:1
2. Andrade EN da C (1956) The concept of creep. In: Creep and Recovery. American Society for Metals, Cleveland, p 176
3. Cottrell AH (1952) J Mechan Phys Solids 1:58
4. Cottrell AH, Aytekin V (1950) J Inst Met 77:389
5. Dobeš F (1998) J Mater Sci 33:2457
6. Garofalo F (1965) Fundamentals of creep and creep-rupture in metals. Macmillan, New York
7. Graham A, Walles KFA (1955) J Iron Steel Inst 179:105
8. Norton FH (1929) The creep of steels at high temperatures. McGraw-Hill, New York
9. McLean D, Hale KF (1961) Structural processes in creep. Special Report No. 70. Iron and Steel Institute, London, p 19
10. Uchic MD, Chrzan DC, Nix WD (2001) Intermetallics 9:963

Chapter 2
General Mechanisms of Creep

Abstract The general creep mechanism is discussed in this chapter, which is classified as: (i) dislocation slip; (ii) climb; (iii) grain-boundary sliding; and (iv) diffusion flow caused by vacancies. The relevant relations and illustrations are included. These provide the basic understanding of creep.

Before discussing creep specifically in brittle materials, ductile materials, single crystals and polycrystals, it is important to consider the general creep mechanisms observed acting in materials. There are several basic mechanisms that may contribute to creep in materials (including ceramics). The various classifications of these mechanisms are not always the same and sometimes they are more detailed or combined, depending on the points being emphasized. The classification used here is somewhat arbitrary, but follows a pattern commonly found in the literature:

(i)   dislocation slip;
(ii) climb;
(iii) grain-boundary sliding; and
(iv) diffusion flow caused by vacancies.

(i) Creep by Dislocation Slip

   Actually, creep types (i) and (ii) may be combined under the general heading of 'dislocation creep', but there is merit in separating the two different types: slip (glide) and climb. Creep takes place as a result of dislocation motion in a crystalline specimen by movement known as 'slip' (glide). As a result of such dislocation motion through a crystal, one part of the dislocation moves one lattice point along a plane known as the 'slip plane', relative to the rest of the crystal. The slip plane along which the dislocation motion takes place separates both parts of the crystal. For dislocation motion to occur, the bonds between the atoms (ions, in the case of nonmetallic materials) must be broken during the deformation. Deformation by creep, which can be an important contributor to overall deformation, occurs only in certain circumstances. Creep by dislocation glide occurs over the entire range of

© Springer International Publishing AG 2017

J. Pelleg, *Creep in Ceramics*, Solid Mechanics and Its Applications 241,
DOI 10.1007/978-3-319-50826-9_2

temperatures, from low (basically absolute zero) to the melting temperature, although at low temperatures its contribution may be insignificant. The afore-mentioned factors expressed by Eq. (1.2) are important factors in creep deforma-tion. At high temperatures, however, creep can occur at stresses less than the yield stress. The stresses needed to drive dislocation glide are on the order of a tenth the theoretical shear strength of ~G/10. Glide-by-slip strengthens materials as they deform. In primary creep, stress is constant, while strain increases to a certain extent (see Fig. 1.1a) over time, but the strain rate decreases (Fig. 1.1b), until a minimum strain rate is achieved. This minimum strain rate, on a strain-time plot, represents steady-state creep.

At lower stresses, the creep rate is lower and becomes limited by the rate at which the dislocations can climb over obstacles by means of vacancy diffusion. Since a dislocation may be pinned by various obstacles, further deformation that of creep, must also occur by means of climb (discussed in the next section).

(ii) Climb

During dislocation motion, the creep rate is limited by the obstacles resisting dislocation motion. The obstacles resisting the motions of dislocations harden (strengthen) the material. High temperatures acting during deformation induce recovery processes. During steady-state creep, strain increases over time. The increased strain energy stored in the material, due to deformation, together with the high temperature, provide the driving force for the recovery process. As such, there is a balance between the processes of work hardening and recovery. Recovery involves a reduction in dislocation density and the rearrangement of dislocations into lower energy arrays, such as subgrain boundaries. For this to occur, dislocations must climb, as well as slip, and this, in turn, requires atomic movement or self-diffusion within the lattice. Hence, it is often said that the activation energies for self-diffusion and for creep are almost the same. Vacancies must be located at a site where climb is supposed to occur, to enable climb by means of a vacancy-atom exchange. As the temperature increases, the atoms gain thermal energy and the equilibrium concentrations of the vacancies in the metals increase exponentially. The number of vacancies, n, (see, for example, Damak and Dienes) is given as:

$$n = N \exp\left(-\frac{E_F}{kT}\right) \qquad (2.1)$$

This same concept of the steady-state creep-rate mechanism of dislocation climb was suggested by Mott. He assumed that the rate-controlling process is the diffusion of the vacancies. It is assumed in Mott's analysis that the rate-controlling process is the diffusion of the vacancies between the dislocations creating vacancies and those destroying them. The concentration of vacancies along a dislocation line is deter-mined by setting the change in the free energy caused by a decrease or increase in the number of vacancies equal to the change in the elastic energy occurring during dislocation climb. The creep equation that results from this analysis is:

$$\dot{\varepsilon} = C\sigma^{\alpha} \exp\left(-\frac{Q}{kT}\right), \tag{2.2}$$

where C and α are constants, Q is the activation energy (equal to the self-diffusion) of creep, σ is the stress, and kT has its usual meaning. A value for α is indicated as $\alpha \sim 3$ to 4. This same creep-rate relation was given in Eq. (1.9), reproduced here as:

$$\dot{\varepsilon} = B\sigma^{n} \exp-\left(\frac{Q}{kT}\right) \tag{1.9}$$

No theoretical treatment of creep seems to exist that leads to a creep rate as given by Eqs. (2.2) or (1.9). Mott had developed a theory leading to Eq. (2.2) in which he stated that Eq. (2.2) is valid in the stress range from the critical shear stress to a stress about equal to 108–109 dynes/cm^2. At larger stresses, the creep rate increases much more rapidly with stress. For a derivation of this equation, the reader is referred to Mott's work on the subject.

In Eq. (2.1), N is the number of lattice sites and E_F is the energy of vacancy formation. The activation energy, Q, for the jump rate, J, is given by the sum of the energy of vacancy formation and the vacancy's energy for motion, E_M, $(Q = E_F + E_M)$:

$$J = J_0 \exp\left(-\frac{Q}{kT}\right) \tag{2.3}$$

J_0 represents the respective entropies. The diffusion coefficient, D, may be given as:

$$D = D_0 \exp\left(-\frac{Q}{kT}\right) \tag{2.4}$$

D_0, the pre-exponential factor, is equivalent to J_0, and Q is the overall activation energy for self-diffusion. The rate of steady-state creep increases with temperature, as does the essential number of vacancies for effective vacancy-atom exchange for climb.

(iii) Grain-Boundary Sliding (GBS)

Different grains and grain-sizes play significant roles in the strengthening (work-hardening) mechanisms given by the Hall–Petch relation as:

$$\sigma_y = \sigma_0 + \frac{k_y}{\sqrt{d}}, \tag{2.5}$$

where σ_y is the yield stress, σ_0 represents the resistance to dislocation glide, k_y is a measure of the dislocation pile-up behind an obstacle (a grain boundary, for example) and d is the size of the grain. The various grains and their sizes are important variables characterizing the microstructure of polycrystalline materials. Grain-boundary movement plays a significant role in the characteristic behaviors of materials in regard to creep. Basically, grain-boundary sliding (GBS) is a process in which grains slide past each other along their common boundary. It has also been observed that sliding may occur in a zone immediately adjacent to the grain boundary (Wadsworth et al.).

In primary creep, the required stress increases due to work hardening (which also acts in steady-state creep, but is balanced by various recovery processes). Decreasing the grain size should strengthen a material, according to the Hall–Petch relation. Thus, for continued deformation, higher stress is required. It may be expected that materials with small grain sizes will show better creep resistance, while increasing grain size should cause an increased creep rate (for example, the secondary-creep rate). This is attributed to the decrease in boundary barriers with increasing grain size (less strengthening media exists, because there are less grain-boundary obstacles). However, this is true as long as no undesirable processes occur at the grain boundaries. For instance, large-grained materials with a small number of grain boundaries are low sources of vacancies and, therefore, dislocation climb will be reduced compared to small-grained materials. Thus, one can see that grain size in creep has a dual effect, because a small grain size strengthens the ceramics, since the large number of grains act as barriers to dislocation glide. Nonetheless, in large-grained ceramics with fewer boundaries, fewer vacancies are emitted, which are prerequisites for creep deformation by climb; therefore, this situation has reduced creep. Note that a suitable choice of grain size in ceramics is critical for achieving the best compromise regarding good creep resistance.

Major structural changes occur at the start of tertiary creep. Damage is initiated by the formation of multi-shaped cavities (in metals, either wedge-shaped or rounded cavities are observed). Wedge-shaped cavities are primarily seen at grain boundaries and their coalescence is the unmistakable sign that creep rupture will occur. It is believed that GBS is a prerequisite for the nucleation of voids and cavities and that it occurs when a sufficiently high stress concentration develops to create new surfaces. Cavity formation increases with increasing strain at high temperatures. The stresses causing GBS are the shear stresses acting on the boundaries. Whether void formation is associated with/or a consequence of GBS has not yet been completely determined, since the experiments found in the literature seem to support both concepts. In Fig. 2.1, cavities at two-grain boundary junctions may be seen in ABC-SiC. The term 'ABC-SiC' refers to SiC which has been hot-pressed with additions of Al, as well as B and C. This material has been shown to have an ambient temperature fracture toughness as high as 9 MPa m$^{1/2}$ with strengths of $\sim$650 MPa (among the highest strength property reported for SiC).

One of the concepts regarding GBS is associated with the presence of an amorphous grain-boundary film along the boundaries between the grains. More specifically, this film has often been termed a 'glassy phase' and considered responsible for GBS. This glassy film may be fully crystallized after heat treatment at high temperature for an appropriate time. Clearly, such crystallization of the

Fig. 2.1 Transmission electron micrograph of ABC-SiC showing grain-boundary cavities at two-grain junctions on the tensile edge of a specimen crept at 1400 °C for 840 h under 200 MPa. Chen et al. [5]. With kind permission of Elsevier

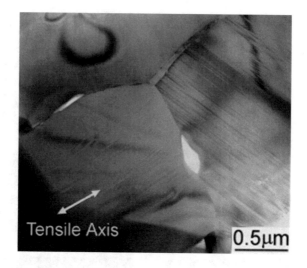

grain-boundary phase would minimize softening and GBS, which would, in turn, cause an increase in strength. As stated previously, the microstructure has a major impact on the creep properties. Creep cavitation may appear at high temperatures in grain boundaries. Cavities are observed on the tensile side, but not on the side under compression. Cavities usually form both at two-grain and multiple-grain junctions. GBS induces cavitation during creep. Thus far, there are no conclusive data proving that GBS is the driving force for the nucleation and growth of creep cavities, although a number of studies have concluded that cavity nucleation is, in fact, induced by GBS. GBS has been the subject of numerous investigations, in light of the importance of grain boundaries for many aspects of material applications. Understanding the physics of the complex behavior of grain boundaries is of great interest in regard to: grain growth, crystallization and recovery deformation, to mention just a few topics. A general review of the properties of grain boundaries may be found, for example, in the work of Valiev, et al. Here, GBS is of interest in order to gain better practical and theoretical understanding. Illuminating research results on GBS may be observed in metals. The instructive photo below (Fig. 2.2) was taken of a Mg-0.78%Al alloy strained to 2.49% at a temperature of 473 K and under an applied stress of 17.2 MPa.

The evidence of GBS is the displacement of the scratch lines during creep testing. The above figure shows scratch lines displaced across a grain boundary; transverse markings are inscribed perpendicular to the tensile axis. Clear offsets may be seen in the transverse marker line in this Mg-0.78%Al alloy. The tensile axis in this experiment is horizontal. An alternate method for evaluating GBS is by means of interferometry. An example of the offsets of the same alloy, as revealed by interferometry, is visible in Fig. 2.3. Chan and Page have developed a model describing creep-induced transient-cavity growth by assuming that cavity growth is governed by the two competing processes–transient creep and sintering. According to this model, the rate of cavity growth is described as:

Fig. 2.2 Grain-boundary sliding revealed by the boundary offsets in a transverse marker line in a Mg-0.78%Al alloy tested under creep conditions at 473 K under a stress of 17.2 MPa. From Bell and Langdon [1], reproduced from Langdon [9]. With kind permission from Springer Science and the author

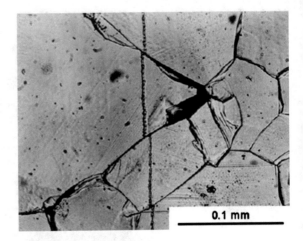

Fig. 2.3 Offset revealed by interferometry in a Mg-0.78% Al alloy pulled to an elongation of 1.5% at 473 K under a stress of 27.6 MPa. From Langdon, Mater. Sci. Eng., A166, 67 (1993), reproduced from Langdon [9]. With kind permission from Springer Science and the author

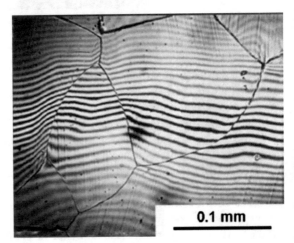

$$\dot{R} = \frac{33RG(\xi)}{4\pi^2} \left[\dot{\varepsilon}_{\mathrm{ss}}(t/t_c)^m - \frac{4\pi}{33}\left(\frac{\gamma_s}{\eta l}\right)(1/\xi - 0.9\xi) \right] \tag{2.6}$$

with

$$\xi = R/1 \tag{2.7}$$

$$G(\xi-) = \frac{2\sqrt{3} - 0.667\pi\xi^2}{0.96\xi^2 - \ln\xi - 0.23\xi^2 - 0.72} \tag{2.8}$$

In Eq. (2.6), R is the cavity radius, $\dot{\varepsilon}_{\mathrm{ss}}$ is the steady-state creep rate, t is the creep time, t_c is the characteristic time, m is an exponent ranging from −0.5 to −0.6, γ is the surface energy, η is the viscosity parameter, and 2l is the center-to-center cavity spacing. Note that the first term within the bracket in Eq. (2.6) is the transient creep rate, $\dot{\varepsilon}_{tr}$, while the second term is the sintering rate, $\dot{s}$. From Eq. (2.6), it is evident that the transient creep rate, $\dot{\varepsilon}_{tr}$, drives cavity growth, whereas the sintering rate term, $\dot{s}$, drives cavity shrinkage. In addition, imposing parameters to reach a state of equilibrium between $\dot{\varepsilon}_{tr}$ and $\dot{s}$ would result in a condition of zero cavity growth. Therefore, a critical value of $\dot{\varepsilon}_{ss}(\dot{\varepsilon}_{cr})$ may be determined by setting $R = 0$ in Eq. (2.6), which defines no-growth behavior as follows:

The viscosity parameter is given by:

$$\frac{\eta l \dot{\varepsilon}_{cr}}{\gamma_s} = \frac{4\pi}{33}(1/\xi - 0.9\xi) \tag{2.9}$$

$$\eta = \frac{1}{132} \frac{d^3 kT}{h D_b \Omega} \tag{2.10}$$

This no-growth boundary is shown in Fig. 2.4 as the solid line. In addition, cavities exhibit continuous growth in region I, where $\dot{s}_{cr} > \dot{s}_{tr}$, and the cavities will shrink when the opposite is true (region II).

Equation (4) in Fig. 2.9 is given here as Eq. 2.9.

A quantitative estimate of the contribution of GBS to overall strain, ξ, used by Tan and Tan following Langdon's proposal, is:

$$\xi = \frac{\varepsilon_{GBS}}{\varepsilon_t} \tag{2.11}$$

ε_t is the total strain at high temperatures, expressed as:

$$\varepsilon_t = \varepsilon_g + \varepsilon_{GBS} + \varepsilon_{dc} \tag{2.12}$$

Fig. 2.4 Comparison of the predicted (--) and experimentally observed conditions for zero cavity growth: (□) Lucalox, 1600 °C; (■) AD99, 1300 °C, and for cavity growth; (○) Lucalox, 1600 °C; (•) AD99, 1150 °C. Region I represents cavity growth and region II, cavity shrinkage. Blanchard and Chan [2]. With kind permission of John Wiley and Sons. (Lucalox and AD99 are aluminas)

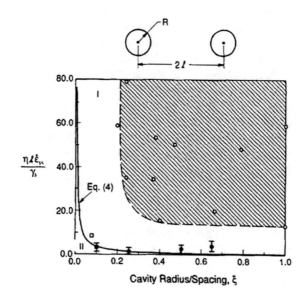

ε_g is the strain in the grain, due to processes taking place within the grain; ε_{GBS} is the strain due to GBS; and ε_{dc} is the strain due to diffusion creep. In practice, experiments are often performed with a negligible contribution of diffusion creep and, thus, Eq. (2.12) reduces to:

$$\varepsilon_t = \varepsilon_g + \varepsilon_{GBS} \tag{2.13}$$

Damage leading to failure, in the form of stress rupture, is initiated by void and crack formation. The tertiary creep, per se, is a sign that some sort of structural damage has occurred. Round or wedge-shaped voids, known as 'r-type cavities' and 'w-type cavities', are seen at first along grain boundaries and, when they coalesce, creep fracture occurs. As indicated above, the mechanism of void formation is associated with GBS and occurs due to shear stresses acting along the boundaries.

A commonly used illustration of a w-type crack initiation by GBS, its formation and growth (first presented by Chang and Grant, and found in almost every publication) is shown in Fig. 2.5. Another configuration for the initiation of intergranular cracks (somewhat more complex) is shown in Fig. 2.6.

A number of w-crack configurations have been experimentally observed at triple points. Wedge-type crack formation at triple points was initially suggested by Zener as early as 1948. According to Zener, at sufficiently high temperatures, grain boundaries behave in a viscous manner and, when near triple points under an applied tensile stress, wedge-type cracks develop due to the high stress concentration. Specifically, Zener was among the first to suggest the concept that fracturing is a consequence of plastic deformation, which is required for crack formation. His schematic illustration is shown in Fig. 2.7, where a crack can be nucleated at a dislocation site.

In Fig. 2.7b and c, the coalescence of two or three dislocations is illustrated, producing an increase in the size of the crack. The concept of crack origin at dislocation sites has been addressed and modified by various researchers. In essence, Zener suggested that cracks nucleate at dislocation pile-ups, where sufficient stress develops for the nucleation of cracks.

A dislocation model for spontaneous microcrack formation was also presented by Stroh, who calculated the elastic energy associated with wedge deformation. Stroh also determined that the nucleation of a wedge crack was due to the pile-up of dislocations on a slip plane. In Fig. 2.8, the 2D crack dislocation of a giant Burgers vector, nb, with length, c, extending to a barrier, may be seen. His expression for the elastic energy associated with wedge deformation is:

$$W_e = \frac{Gn^2b^2}{4\pi(1-v)} \ln \frac{4R}{c} \tag{2.14}$$

in which G is the shear modulus (modulus of rigidity), nb is a giant Burgers vector, with n being the number of dislocations comprising the giant vector, and R—the bounding radius in the stress field. The surface energy term, $2\gamma_s c$, may be added to obtain the total energy of the system as:

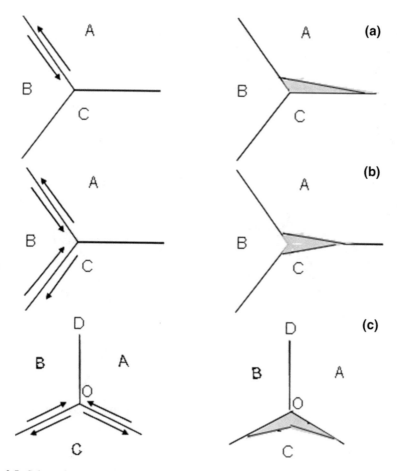

Fig. 2.5 Schematic representation of a w-type crack formation initiated by GBS. From Ref. [4]

$$W_s = \frac{Gn^2b^2}{4\pi(1-v)}\ln\frac{4R}{c} + 2\gamma_s c \qquad (2.15)$$

By differentiating Eq. (2.15) with c, the critical length, c_{min}, may be found:

$$\frac{\partial W_s}{\partial c} = 0 \qquad (2.16)$$

$$c_{min} = G\frac{n^2b^2}{4\pi(1-v)}\frac{1}{2\gamma_s} \qquad (2.17)$$

In polycrystalline solids, the typical values of b, G, v and γ_s are, respectively, (Sarfarazi and Ghosh.): $b = 2 \times 10^{-8}$ cm; $G = 1012$ dynes/cm^2; $v = 1/3$; and $\gamma_s = 103$ dynes/cm, which gives for c_{min}:

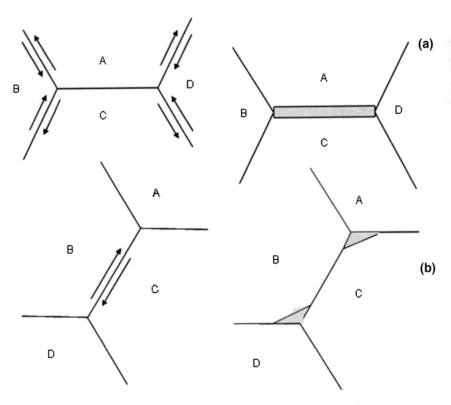

Fig. 2.6 Schematic views showing a more complex intergranular crack initiation by GBS. From Ref. [4]

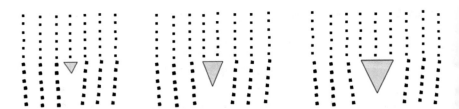

Fig. 2.7 A schematic illustration of Zener's idea, explaining how a crack of atomic dimensions can nucleate at dislocation sites; here, the growth of a crack is initiated by the coalescence of two or three dislocations (See: *Mechanical Properties of Ceramics*, Joshua Pelleg, Fig. 6.81)

$$c_{min} = 2.4 \times 10^{-8} \tag{2.17a}$$

According to the theoretical presentation of Wu et al., a wedge crack may be formed by the insertion of extra material to create the head of a crack. An extra

Fig. 2.8 Nucleation of a wedge crack due to pile-up dislocations on a slip plane (Stroh's model). Sarfarazi and Ghosh [15]. With kind permission of Elsevier

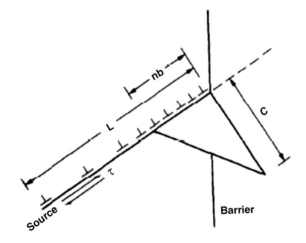

plane, present above a positive-edge dislocation, may serve as the source of a wedge crack. This is Stroh's idea, based on Zener's original concept.

GBS may be considered as a deformation mechanism above $0.5T_m$. The strain rate is important to the type of failure caused by GBS. It has been shown that r-type cavities transform into w-types with increased strain rate, leading to transgranular fracture with increasing strain rate (Gandhi and Raj).

Alloying additions may decrease the tendency for w-type cavity formation. Both cavity types are the results of GBS (Raj). GBS may produce grain-boundary (inter-granular) cracking when the grain's interior is stronger than its boundaries. GBS can be reduced by adding intergranular particles or by serrated grain boundaries. These serve as obstacles to GBS, apparently due to an increase in friction between the boundaries. Cavities have been seen to form at grain and phase boundaries prefer-entially at interfaces or triple points. The process of cavitation, associated with GBS and cavity nucleation, probably occurs at points of stress concentration in the sliding boundaries or interfaces. Creep failure occurs by the nucleation, growth and coales-cence of creep cavities at the boundaries predominantly perpendicularly oriented to the applied stress. An increase in the number of cavitated boundaries over creep-exposure time supports a mechanism of continuous cavity nucleation and growth. Some believe, on the basis of experimental observations, that there are probably preexisting cavities, voids or pores, previously introduced by the forming processes that are actually responsible for creep cavitations in engineering alloys during long-term service at low stresses and elevated temperatures. Many experiments show that GBS is a necessary condition for cavity nucleation. GBS is a key factor not only in the growth of preexisting voids, but also in nucleating voids for cavity formation.

In many polycrystalline ceramics at elevated temperatures, GBS contributes significantly to the total strain. GBS can be markedly reduced by introducing additional phases, which form precipitates (such as nitrides, carbides, borides, etc.) at the grain boundaries. Another method for improving creep resistance in materials is by the formation of serrated grain boundaries. Serrated grain boundaries are effective in improving creep-strength properties and do not permit continued creep by GBS when stress is applied at high temperatures. The effect of serration is equivalent to the 'self-locking' of the sliding process, resulting from creep deformation. Thus, materials with irregular, serrated grain boundaries have improved resistance to creep-crack growth when compared to those with smooth grain boundaries. This is explained as a consequence of the difficulty of GBS and the increase in the path of grain-boundary diffusion. The strengthening mechanisms of serrated grain boundaries are principally the result of: (1) the inhibition of GBS; (2) the retardation of grain-boundary crack initiation, caused by the decrease in stress concentration at grain-boundary triple points as a result of the decrease in GBS length and; (3) dynamic recovery at the serrated boundaries.

To summarize this section, note that GBS may account for 10–65% of the total creep strain, depending on the alloy and the conditions of its use in service (temperature, load, etc.). Its contribution to creep strain increases with rising temperature and stress and with reduced grain size. Above $\sim 0.6\ T_m$, the grain-boundary region is thought to have lower shear strength than the grains themselves, probably due to the looser atomic packing at the grain boundaries. GBS may be reduced by introducing precipitates or grain-boundary serrations, which resist GBS and significantly reduce cavity formation of the types indicated above (which is a major factor in creep failure).

(iv) Diffusion Flow Caused by Vacancies

This mechanism of creep involves diffusion and various models have been suggested for diffusion-assisted creep. As such, it will be discussed in Chap. 3.

References

1. Bell RL, Langdon TG (1967) J Mater Sci 2:313
2. Blanchard CR, Chan KS (1993) J Amer Ceram Soc 76:1651
3. Chan KS, Page KA (1992) J Mater Sci 27:1651
4. Chang HC, Grant NJ (1956) Trans AIME 206:544
5. Chen D, Sixta ME, Zhang XF, de Jonghe LC, Ritchie RO (2000) Acta Mater 48:4599
6. Damask AC, Dienes GJ (1971) Point defects in metals. Gordon and Breach, New York
7. Gandhi C, Raj R (1981) Met Trans A12:515
8. Hall EO (1951) Proc Phys Soc London 643:747
9. Langdon TG (2006) J Mater Sci 41:597
10. Mott NF (1951) Proc Phys Soc London B64:729; Phil Mag (1952): 43, 1151; Phil Mag 44, 741 (1953); Proc Roy Soc London (1953): A220, 1
11. Pelleg J (2014) Mechanical properties of ceramics. Springer, Berlin, p 193
12. Petch NJ (1953) J Iron Steel Inst 173:25
13. Raj R (1981) Met Trans A12:1089
14. Sarfarazi M, Ghosh SK (1987) Eng Fract Mech 27:257
15. Stroh AN (1957) Adv Phys 6:418
16. Stroh AN (1955) Proc R Soc London 223A:548

17. Tan JC, Tan MJ (2003) Mater Sci Eng A339:81
18. Valiev RZ, Gertsman VYu, Kaibyshev OA (1986) Phys Stat Sol (A) 97:11
19. Wadsworth J, Ruano J, Sherby OA (2002) Met Mater Trans 33A:219
20. Wu MS, Zhou H (1996) Int J Fract 78:165
21. Zener C (1948) Elasticity and anelasticity. University of Chicago Press, Chicago

Chapter 3
Creep and Its Relation to Diffusion

Abstract Creep occurs at some temperature and thus as being thermally activated is associated with diffusion. Almost every creep equation incorporates a diffusion coefficient in the relation. Creep can occur in grain boundaries also (Coble creep) and therefore lattice- or grain-boundary diffusion coefficients are indicated in the relations depending on the main creep involved in the process.

Diffusion flow by vacancies must be considered, since the mechanism of creep depends on both temperature and stress. The various methods detailed below involve some sort of diffusion occurring with vacancy-atom exchange. This may occur either by lattice diffusion or grain-boundary diffusion, or both may be involved. Bulk-diffusion-assisted creep occurs during the processes listed in (a)–(d) below, where the kinetics of atom-vacancy exchange occurs due to lattice diffusion. Afterward, creep, involving grain-boundary diffusion, will be considered (e).

(a) Nabarro-Herring creep;
(b) climb, in which the strain is actually obtained by climb;
(c) climb-assisted glide, in which climb is a mechanism allowing dislocations to bypass obstacles;
(d) thermally activated glide via cross-slip;
(e) Coble creep, involving grain-boundary diffusion.

Before entering into a detailed discussion of the above lists and based on what has been said thus far on the subject, briefly summarized: (a) creep in materials, namely time-dependent plastic deformation, may occur during mechanical stresses well below the yield stress and (b) in general, two major creep mechanisms characterize the time-dependent plastic-deformation process—dislocation creep and diffusion creep. However, it must be emphasized that even dislocation creep cannot be separated completely from diffusion phenomena, since climb, for example, is associated with the vacancies required for climb. Now, a detailed discussion of paragraphs (a)–(d) follows.

© Springer International Publishing AG 2017
J. Pelleg, *Creep in Ceramics*, Solid Mechanics and Its Applications 241,
DOI 10.1007/978-3-319-50826-9_3

3.1 Nabarro-Herring Creep

One type of creep, in which creep is diffusion controlled, is Nabarro-Herring creep. In this type of creep, atoms diffuse through a lattice, causing grains to elongate along the stress axis. Mass transport (i.e., the diffusion of atoms) takes place in regions ranging from lower to higher tensile stress. A common illustration may be seen in Fig. 3.1. This schematic figure illustrates the flow of vacancies and atomic movements as induced by tensile stress, σ. During creep deformation, vacancy-atom exchanges take place to and from the grain boundaries. One would expect that, during creep under tension, atoms would tend to diffuse from the sides of the specimen in the direction shown in Fig. 3.1 (a counterflow of vacancies), causing the sides to lengthen. Assume that local equilibrium of the vacancy concentration exists at the boundaries of the crystal when no stresses are acting on it. Also note that grain boundaries serve as vacancy sources or sinks. In this mechanism, lattice diffusion occurs within the grain and the creep rate (strain rate) is assumed to be proportional to the vacancy flux. See below that the strain rate is inversely proportional to the square of the grain size [5, 13, 14]. In Eq. (2.1), the number of vacancies is given. Equation (2. 1), in terms of vacancy concentration at equilibrium, is given as:

$$\frac{n}{N} = C_v^0 = \exp\left(-\frac{E_F}{kT}\right) \tag{3.1}$$

The energy needed to create a vacancy under acting stress is given by:

$$E_F + \sigma V \tag{3.2}$$

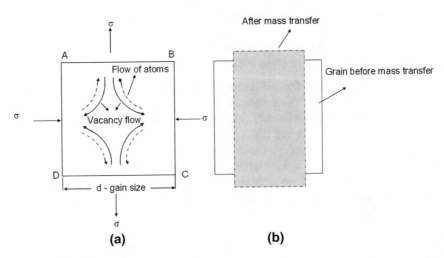

Fig. 3.1 The Nabarro-Herring concept of creep: **a** a schematic of vacancy and mass flow; **b** the elongated grain in the tensile-axis direction after mass flow

V is the atomic volume (here, it is the volume of a vacancy) and E_F is defined by Eq. (2.1). There is a small concentration difference in the vacancies between the faces of AB and BC in the above figure, where tensile and compressive stresses are acting, respectively. Denoting the vacancy concentrations at the respective faces as C_{V+} and C_{V-} and their difference as ΔC, one may write for each of them, by means of Eqs. (3.1) and (3.2), respectively:

$$C_v^+ = \exp-(\frac{E_F - \sigma V}{kT}) = C_v^0 \exp(\frac{\sigma V}{kT}) \qquad (3.3)$$

$$C_v^- = \exp-(\frac{E_F + \sigma V}{kT}) = C_v^0 \exp(-\frac{\sigma V}{kT}) \qquad (3.4)$$

$$\Delta C = C_v^+ - C_V^- = \frac{\alpha}{V}\left\{\exp(-\frac{E_F}{kT})\left[\exp(\frac{\sigma V}{kT}) - \exp(-\frac{-\sigma V}{kT})\right]\right\} \qquad (3.5)$$

Clearly, in this relation, E_F was replaced by Eq. (3.2). Equations (3.3) and (3.4) represent the local equilibrium concentrations under tension and compression (see Fig. 3.1a). Recalling that

$$\sinh x = \frac{1}{2}[\exp(x) - \exp(-x)] \qquad (3.6)$$

Equation (3.5) may be rewritten as

$$\Delta C = \frac{2\alpha}{V}\exp(-\frac{E_F}{kT}) \sinh\left(\frac{\sigma V}{kT}\right) = \frac{2\sigma}{V} C_v^0 \sinh(\frac{\sigma V}{kT}) \qquad (3.7)$$

where $C_V^0 = \exp(-E_F/kT)$ and E_F is the energy of vacancy formation in the absence of stress.

As indicated, there is a flow of atoms from the tensile to the compressed faces and an opposite flow of vacancies. When a concentration gradient exists, diffusion flux will occur. This flux of vacancies may be expressed as

$$J = -D_V \nabla C = -\frac{\alpha D_V (\Delta C)}{d} \qquad (3.8)$$

D_V is the diffusion coefficient of the vacancies and α is a geometrical factor. The corresponding transport of matter occurs in the opposite direction and produces a creep strain under the applied stress. In a unit time, Jd^2, atoms in the crystal leave the faces under compression and are added to the faces under tension. (Recall that J is the number of atoms in a unit time per unit area; thus, multiplying this value by the square of the grain size, d, one gets the number of atoms per unit time). Consequently, the grain lengthens in the tensile-axis direction and gets thinner in the transverse direction. The change in grain size may be written as:

$$\Delta d = \frac{(Jd^2)V}{d^2} = JV, \tag{3.9}$$

where V is the atomic volume (often given as Ω). The strain rate is given as:

$$\dot{\varepsilon} = \frac{\Delta d}{d} = \frac{JV}{d} \tag{3.10}$$

An expression for the strain rate, given by Eq. (3.11), is obtained by substituting the value of Δd from Eq. (3.9) into Eq. (3.10), followed by inserting J from Eq. (3.8) into Eq. (3.10) to get

$$\dot{\varepsilon} = \frac{\alpha D_V \Delta C\, V}{d} \frac{V}{d} = \frac{\alpha D_V \Delta CV}{d^2} \tag{3.11}$$

With Eq. (3.7) substituted into Eq. (3.11), it is possible to write

$$\dot{\varepsilon} = \frac{2\beta}{V} \frac{D_V V}{d^2} \exp(-\frac{E_F}{kT}) \sinh(\frac{\sigma V}{kT}) \tag{3.12}$$

For small values of stress, and since the nominator is always smaller than the denominator, the quotient is small and sinh $(\sigma V/kT) = \sigma V/kT$. Substituting this value into Eq. (3.12), one obtains

$$\dot{\varepsilon} = \frac{2\beta D_V}{V} C_V^0 \frac{\sigma V}{kT} \tag{3.13}$$

D_V is the diffusion coefficient of the vacancies and $D_V C_V^0$ is D_S, the self-diffusion coefficient. Thus, Eq. (3.13) may also be expressed as

$$\dot{\varepsilon} = \frac{2\beta D_S}{d^2} \frac{\sigma V}{kT} \tag{3.14}$$

More exact calculations, in terms of shear strain (i.e., $\gamma = 2b/d$) and macroscopic shear stress, τ, (i.e., $\sigma = \beta \tau$ and β is close to unity and recalling that the shear stress at 45° is given by $\tau = \sigma\sqrt{2}$) gives

$$\dot{\gamma}_S = \frac{32\alpha\beta D_S \tau V}{\pi d^2} \frac{1}{kT} \tag{3.15}$$

This relation defines a simple, ideal, viscous solid. One sees that increasing grain size reduces creep rate. Creep-rate change is proportional to d^{-2}. Nabarro-Herring creep is a low-stress and high-temperature process.

A somewhat alternate method for showing that $\dot{\varepsilon} \propto \frac{1}{d^2}$ follows. Based on Eq. (3.1) through (3.4), the difference in concentration may be expressed as

$$\Delta C = C_V^+ - C_V^- = \frac{\alpha}{V}\left\{\exp(-\frac{E_F}{kT}\left[\exp(\frac{\sigma V}{kT}) - \exp(-\frac{-\sigma V}{kT})\right]\right\} \qquad (3.16)$$

The flux of the vacancies, going from the tensile to the compressive regions, is

$$J_V = -D_V\frac{\Delta C}{\Delta x} \qquad (3.17)$$

where Δx is the distance in the x direction, so that $\Delta C/\Delta x$ is a gradient. Bear in mind that the atomic flux, J, is in the opposite direction to the vacancy flux, J_V, and, therefore, $D\Delta C = -D_V\Delta C_V$. In our case, the diffusion distance is l.

Stress is not constant along the grain faces, therefore, the diffusion paths are shorter near the corners. Due to stress relaxation, one may assume that $\sigma = \beta\sigma_S$ at distance $d/4$ from the boundaries (when σ_S is the macroscopic shear stress and β is nearly unity). The length of the diffusion path through this point is $1 = \pi/2(d/4)$. The atomic flux across the area of a single atom is given by

$$J = \alpha D_V\frac{\Delta C}{l} = \alpha D_V\frac{8\Delta C}{\pi d} \qquad (3.18)$$

The previous expression is the result of substituting for the value of $1 = \pi/2(d/4)$. D_V is the diffusivity of the vacancies. One may rewrite Eqs. (3.9) and (3.10) as

$$\Delta d = \frac{(Jd^2)V}{d^2} = JV \qquad (3.19)$$

$$\varepsilon = \frac{\Delta d}{d} = \frac{JV}{d} \qquad (3.20)$$

Substituting from Eq. (3.17) for J yields:

$$\dot{\varepsilon} = \alpha D_V\frac{8\Delta C}{\pi d}\frac{V}{d} \qquad (3.21)$$

and from Eq. (3.17):

$$\Delta C = \frac{2\alpha}{V}\exp(-\frac{E_F}{kT})\sinh\left(\frac{\sigma V}{kT}\right) = \frac{2\sigma}{V}C_v^0\sinh(\frac{\sigma V}{kT}) \qquad (3.7)$$

When the argument in the hyperbolic function is small, as mentioned earlier, it is equal to the argument itself; thus, for the strain rate, one may write

$$\dot{\varepsilon} = 16\alpha\frac{D_V C_V^0}{\pi d^2}\frac{\sigma V}{kT} = \frac{16\alpha D_S}{\pi d^2}\frac{\sigma V}{kT} \qquad (3.8)$$

D_S is the self-diffusion coefficient and is equal to $C_V^0 D_V$. Again, the strain rate is proportional to d^{-2}.

One sees in Eqs. (3.15) and (3.8) that the strain rate is linearly proportional to the stress and inversely proportional to the grain size. In Eq. (3.15), the expression is given in terms of shear strain and macroscopic shear stress. The above expressions explain why large-grained materials are preferential for creep applications at high temperatures.

As mentioned previously, one of the bulk-diffusion-assisted creeps occurs in the Nabarro-Herring model, though the Coble creep mechanism is also diffusion assisted. As such, the interpretation of creep results is, to a large extent, chosen by the researchers.

3.2 Climb—Dislocation Creep

Bulk-diffusion-assisted creep occurs in the processes listed above, namely in (b) climb; (c) climb-assisted glide and; (d) thermally activated glide via cross-slip. All these are obviously associated with dislocation motion. High stress, below yield stress, causes creep by conservative dislocation motion, namely by dislocation glide within its slip plane. This readily occurs at high temperatures above $0.3\ T_m$ in pure metals and at about $0.4\ T_m$ in alloys, where the dependence on the strain rate becomes quite strong. For ceramics, $T > 0.4 - 0.5\ T_m$ (K). A formulation used for such creep is

$$\dot{\gamma} \sim \left(\frac{\sigma_S}{G}\right)^n \tag{3.22}$$

where n has a value of 3–10 in high-temperature regimes. Since n is in the exponent, this creep is referred to as "power-law creep." At high temperatures, obstacle-blocked dislocations can climb, not only glide. If gliding dislocations are blocked by some obstacle, climbing may release them to move on until they meet another obstacle, where the same process is repeated. Climb is performed by the diffusion of vacancies through the lattice or along the dislocation core, diffusing into or out of the dislocation core. By climbing, dislocations change their slip planes, enabling them to bypass their obstacles. Dislocation glide is responsible for most strain, while the average dislocation density is determined by the climb step in the deformation process. This mechanism is known as "climb-controlled creep."

3.3 Climb-Controlled Creep

At relatively high stresses, beyond the elastic region or the shear moduli, creep is controlled by dislocation-glide movement and by glide in adjacent planes following climb. Real materials contain various internal obstacles (such as dislocations) or external ones (introduced intentionally, such as solutes and particles, or unintentionally by the fabrication process), which block dislocation glide in their respective slip planes. Dislocation motion is also hindered by the crystal structure itself, namely by crystal resistance (an internal obstacle). At high temperatures, obstacle-blocked dislocations may be released by dislocation climb. Creep arises as a consequence of climb, when further deformation by glide is enabled by means of vacancy-atom exchange. The creep rate is a function of several factors, usually given as

$$\sum \dot{\varepsilon} = f(\sigma, T, S, GS, P)$$

S is the structure, GS is grain size and P represents the material properties, such as the lattice parameter, atomic volume, etc. Vacancies increase with increasing temperature and are likely to diffuse into dislocations, thus, decreasing the overall free energy of the system. By the diffusion of vacancies to locations at which dislocations are blocked by obstacles, climb becomes possible, letting the dislocations bypass those obstacles. Climb allows further glide in an adjacent slip plane to occur and, by such deformation, creep strain arises.

A steady-state-based model for edge-dislocation climb [5, 13] was suggested by Weertman. He assumed that strain hardening occurs whenever dislocations are hindered in their motion by some obstacle and pileup behind it. The dislocations beyond the barrier, such as a Lomer-Cottrell lock, may escape by climbing. However, climbing beyond Lomer-Cottrell barriers leads to the generation of new dislocation loops and to a steady-state creep rate (which is applicable to face-centered cubic (FCC) and body-centered cubic (BCC) structures, but not to hexagonal close-packed (HCP) ones). Weertman also suggested that edge dislocations with opposite signs, gliding on parallel slip planes, would interact and pile up when a critical distance of $2r$ between them is not exceeded. In such a case, as in the prior case, dislocations might escape from the piled up array by means of climb. Dislocation pile-ups lead to work hardening, whereas climb is a recovery process. A steady state is reached when the hardening and recovery rates are equal. The creep rate will, therefore, be controlled by the rate at which dislocations can climb. This climb mechanism requires the creation of vacancies or their destruction at the obstacle-blocked dislocations (in this case, at the pile-up) in order to maintain the equilibrium concentration required to satisfy the climb rate. At the tip of a pile-up dislocation, a nonvanishing, hydrostatic stress, $\pm\sigma_1$, may develop, exerting a force on the dislocation in a normal direction to the slip plane and causing a positive (up) or negative (down) climb. Vacancies are absorbed where the stress is compressive and are created where the stress is tensile. A change in vacancy concentration develops in the vicinity of the dislocation line, and a vacancy flux is established between the segments of the dislocations, acting as sources or segments of sinks.

The vacancy concentration, C_e, in equilibrium with the leading dislocation in the pile-up, is given by

$$C_e = C_0 \exp\left(\frac{\pm 2L\sigma_S^2 b^2}{GkT}\right) \tag{3.23}$$

$2L$ is the length of the dislocation pile-up, and C_0 is the equilibrium concentration of the vacancies in a dislocation-free crystal. The vacancy concentration at a distance, r, from each pile-up is assumed to be equal to C_0. The rate of climb, $\dot{X}$, is given (Garofalo) as

$$\dot{X} = \frac{2C_0 D_V \sigma_S^2 L b^4}{GkT} \tag{3.24}$$

D_V is the vacancy-diffusion coefficient and $2Lb^2\sigma_S^2/GkT < 1$.

When self-diffusion occurs due to the vacancy mechanism, $C_0 D_V$ may be replaced by

$$C_0 D_V = D_S = \frac{v}{b}\exp(\frac{\Delta S}{R})\exp(-\frac{\Delta H}{RT}) \tag{3.25}$$

and $\dot{X}$ is given by

$$\dot{X} = \frac{2\sigma_S^2 L b^3}{GkT} v \exp(\frac{\Delta S}{R})\exp(-\frac{\Delta H}{RT}) \tag{3.26}$$

ΔH is the activation energy for self-diffusion, v is a frequency factor, and S is an entropy term. Equation (3.26) is obtained under the assumption that vacancies are easily destroyed or created and that an equilibrium concentration exists between pile-ups of dislocations. However, the diffusion of flux vacancies may be different in specific climb processes.

In an additional model created by Weertman, the rate of dislocation climb is also given by Eqs. (3.24) or (3.26) and the steady-state creep-rate model, in this case, becomes

$$\dot{\gamma} = NAb\frac{\dot{X}}{2r} \tag{3.27}$$

N is the density of the dislocations participating in the climb process (or the density of the sources), A is the area swept out by a loop in a pile-up, and $2r$ is the separation between those pile-ups. The stress necessary to force two groups of dislocation loops to pass each other on parallel slip planes must be greater than $\frac{Gb}{4\pi\sigma_S}$ (in terms of shear stress). When this relation is satisfied, an estimate for r may be made

$$r = \frac{Gb}{4\pi\sigma_S} \tag{3.28}$$

The probability, p, of blocking the dislocation loops generated from one source by means of loops emanating from three other sources is given as

$$p = \frac{8\pi NL^2 r}{3} = \frac{2NL^2 Gb}{3\sigma_S} \tag{3.29}$$

Using Eqs. (3.24) and (3.27)–(3.29) and setting $p = 1$ and $A = 4\pi/2$, the creep rate at low stresses becomes

$$\dot{\gamma}_S = \frac{C\pi^2 \sigma_S^{4.5} D_S}{\sqrt{bN} G^{3.5} kT} \tag{3.30}$$

C is a numerical constant in the order of 0.25 and D_S is the coefficient of self-diffusion. Equation (3.30) has been substantiated experimentally for pure metals to a greater extent than other theoretical relations. Exceptions to the exponent 4.5 were obtained, but this value is very close to the observed experimental values.

3.4 Thermally Activated Glide via Cross-Slip

Edge dislocations climb when their motion is hindered. The nonconservative motion of screw dislocations is by cross-slip, since they cannot climb. The ease of cross-slip is stacking-fault-dependent. Materials with high stacking fault (SF) energy cross-slip readily, but not so when the SF energy is low. For screw-oriented dislocations, the Burgers vector is parallel to the dislocation line and, therefore, it can move in any plane in which it lies (in isotropic materials). In real crystals (which are in most cases anisotropic), screw dislocations may favor those planes with the lowest energy. Cross-slip can occur without diffusion, but thermal activation helps cross-slip movement from the original to other slip planes. Climb and cross-slip are recovery processes. Recall that steady-state creep is a deformation process, balanced by work hardening and dynamic recovery. The temperature-dependence of creep is

$$\dot{\varepsilon} \sim \exp -\left(\frac{Q_c}{kT}\right) \tag{3.31}$$

One of the known equations for steady-state creep, indicating stress dependence [4] is

$$\dot{\varepsilon}_s = A\sigma^n \exp -\left(\frac{Q_c(\sigma)}{kT}\right) \tag{3.32}$$

Here, Q_c is the activation energy for (stress-dependent) creep and n is the stress exponent. A similar expression may be given for climb-controlled creep:

$$\dot{\varepsilon}_s = A\sigma^n \exp -\left(\frac{Q_c}{kT}\right) \tag{3.33}$$

But in this case, expressed by Eq. (3.33), Q_c is independent of applied stress [4]. At lower temperatures, cross-slips made by screw dislocations are the means by which obstacles in the slip plane may be bypassed.

Since the study of cross-slip is more informative in single crystals, many experiments have been performed on single crystals with various structures. For instance, in order to investigate the glide system in FCC metals, Al single crystals were deformed by compression parallel to [00I] at temperatures between 225 and 365 °C and at strain rates between 9×10^{-6} and 9×10^{-4}/sec [8]. (Note that Al has high SF energy and readily cross-slips). Their stress-strain curves exhibit three stages, which have been correlated with observations of slip lines and dislocation structures. The unique observation was that, after a small percentage of deformation cross-slips of $\frac{a}{2}\langle 1\bar{1}0 \rangle$, screw dislocations from the {111} to the {110} planes occurred, that might be responsible for the {110} slip. Stage I deformation occurs, as expected in FCC metals, on the {111} planes, but, after a small deformation, slip on the {110} plane sets in once the stress reaches the critical value, σ_{110}. This stress is thermally activated and decreases with temperature increase. It is not clear why dislocations cross-slip on the {110} planes, rather than on the {111} planes (as is usually the case), though several explanations have been proffered. An activation energy for the creep rate, $\dot{\varepsilon}$, of 28 kcal/mol, determined at a constant stress of σ_{110}, is close to the reported cross-slip in Al. It is likely that these observations are compatible with the mechanism of cross-slip by screw dislocations from the {111} to the {110} planes and that a SF which is stable at high temperatures stabilizes slip in the {110} plane. The possibility of a SF in the {110} plane is explained on geometrical grounds and the dislocation proposed is expressed as

$$\frac{a}{2}[110] = \frac{a}{12}[110] + \frac{a}{3}[110] + \frac{a}{12}[110]$$

SF energy, which determines the separation of the partial dislocations, improves creep resistance if it is low. Contrary to the high SF energy observed in Al (in which cross-slip or climb occurs readily), in low-energy SF materials with large separation, cross-slip by creep or climb is suppressed. This was observed by Suzuki et al. in their work on Mg-Y alloys with added zinc. The addition of small amounts of Zn has a beneficial effect on creep resistance, because it widens the separation between the partials by decreasing the SF energy. The average separation of partials in this alloy is given as

$$d_S = \frac{Gb_1b_2}{8\pi\gamma}\left(\frac{2-v}{1-v}\right)\left(1 - \frac{2v\cos(2\alpha)}{2-v}\right), \qquad (3.34)$$

where d_S is the separation width between the partials, γ is the SF energy, v is the Poisson ratio, and α is the angle between the total Burgers vector and the dislocation line. A large SF energy drop was calculated, compared with pure magnesium. Mg alloys are being used for more and more applications in which the components are subjected to elevated temperatures. Consequently, research is being focused on the development of alloys able to withstand high stresses at temperatures up to 300 °C, depending on the application. Thus, for example, in other Mg alloys, improved creep properties are produced by the addition of rare earth alloys [9]. At low temperatures, a climb mechanism for edge dislocations exists, whereas, at higher temperatures, the cross-slip mechanism of screw dislocations is believed to operate. The opinions regarding the cross-slip mechanism are not unanimous, but a majority of the researchers support it. Whether the acting mechanism is climb or cross-slip, it is most likely that the beneficial effect of alloying stems from the fact that they both widen the separation between the partials.

During the discussion on cross-slip, metals were considered as examples of its effect on creep. However, it is necessary to emphasize that those examples were provided to clarify the process that might occur at high temperatures, when creep is sometimes unavoidable. Yet, partial dislocations and SFs occur in structures other than metals, such as in ceramics. This will be considered in the next chapter devoted to creep in ceramics.

The activation energy for cross-slipping is rendered by Schoeck and Seeger as

$$\dot{\varepsilon} = C \exp - \left[\frac{\Delta H_0 - c \, \ln(\frac{\sigma}{\sigma_c})}{kT} \right] \qquad (3.35)$$

ΔH_0 is the energy for cross-slip, σ_c is the critical resolved shear stress, σ is the applied stress, and C and c are constants. A model of creep controlled by cross-slip from the $\{111\}$ to the $\{100\}$ plane in the temperature range of 530–680 °C over the stress range of 360–600 MN m^{-2} was found to be in good agreement with the experimental results. The energy for forming a restriction between the partials, namely to recombine the Shockley partials, was evaluated on the basis of Dorn's expression (Dorn) (see also [6] for the creep mechanism at intermediate temperatures in Ni$_3$Al).

Poirier pointed out that when the cross-slip and climb of dislocations operate at the same time, $\dot{\varepsilon}$ may be written as

$$\dot{\varepsilon} = \dot{\varepsilon}_{\text{cross-slip}} + \dot{\varepsilon}_{\text{clmb}} = \dot{\varepsilon}_{0_1} \left(\frac{\sigma}{\mu} \right)^{n1} \exp\left(-\frac{Q_1}{kT} \right) + \dot{\varepsilon}_{0_2} \left(\frac{\sigma}{\mu} \right)^{n2} \exp\left(-\frac{Q_2}{kT} \right) \qquad (3.36)$$

The subscripts and superscripts 1 and 2 refer to cross-slip and climb, respectively. Dislocation motion must overcome significant structural barriers or the dislocation must cross-slip or climb past obstructions. At the lower temperatures, dislocation cross-slip and climb both occur; at the higher temperatures, dislocation climb becomes a rate-controlling mechanism and classic values of the stress exponent ($n = 4.5$) are obtained. The creep-activation energy is that of diffusion.

In general, creep at temperatures below $0.5 \, T_m$ is not thought to occur by means of the lattice-diffusion-controlled mechanism.

Seldom does a lone creep mechanism operate at any given time. Creep mechanisms may operate simultaneously (in parallel) or independently. For both mechanisms, one may write

$$\dot{\varepsilon} = \sum_i \dot{\varepsilon}_i \qquad (3.37)$$

or

$$\frac{1}{\dot{\varepsilon}} = \sum_i \frac{1}{\dot{\varepsilon}_i} \qquad (3.38)$$

In the case of parallel creep mechanisms, the fastest mechanism will dominate the overall creep, whereas when they operate in sequence, the slowest process controls creep deformation.

3.5 Coble Creep, Involving Grain-Boundary Diffusion

Coble creep is also a type of diffusion creep, but involves grain-boundary diffusion. The diffusion of atoms along grain boundaries produces a change in dimensions, due to the flow of the material. Of the two kinds of self-diffusions in polycrystalline materials, the one occurring at low temperatures is grain-boundary dominated, whereas lattice diffusion occurs at high temperatures. Figure 3.2 is an illustration of ideal grain structure, showing the flow of atoms along the boundaries under the influence of a tensile stress. In a polycrystalline matrix, the grain shape is not as indicated in Fig. 3.2 (for an ideal structure), but varies in orientation, making it difficult to analyze.

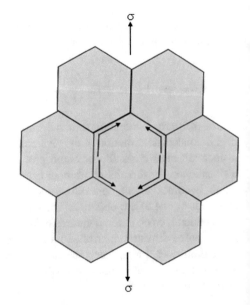

Fig. 3.2 Seven grains are shown in a two-dimensional *hexagonal* array before creep deformation. Following diffusion, the grains elongate in one direction and decrease perpendicularly to the tensile axis. A void formation develops between the grains, but GBS, which may accompany this process, removes these voids

Coble, in his original paper, used a spherical grain (apparently following the Nabarro-Herring approach for lattice-controlled-diffusion creep). In Coble's analysis of creep, a spherical grain was used once again. Based on the experimental results for Al_2O_3, where it was observed that the Al ion-diffusion coefficient is larger by orders than that of oxygen ions and, since the creep rate in lattice-controlled diffusion is limited by the least mobile species, it was expected that the O^{2-} species would determine the rate of creep. Coble suggested that grain-boundary diffusion, rather than lattice diffusion, might control creep deformation. He proposed that Al diffuses in the lattice and O^{2-} in the grain boundaries, where its diffusion coefficient is enhanced in comparison with the values in the lattice. It was assumed that the spherical grain maintains a constant volume and, thus, the areas of the vacancies at the source and the sink must also be equal (grain boundaries may act as sources or sinks for vacancies). The average gradient of the spherical grain, with a radius, R, is given as $\frac{\Delta C}{(R\pi/2)}$. The problem is to evaluate the concentration gradient at the 60° boundary, which for equal areas of rotational symmetry, lies at 60° below the pole of a hemisphere.

For steady-state creep, where Fick's law applies, the flux at the 60° boundary is

$$J_{vac \ sec^{-1}} = D_V N \left[\frac{\Delta C}{(R\pi/2)} \right] (W) 2\pi R \sin 60 \qquad (3.39)$$

Here, D_V is the diffusion coefficient of the vacancies in the boundary; N is a proportionality constant relating the average vacancy gradient, $\Delta C/(R\pi//2)$, and the maximum gradient, $1/R(dc/d\theta)_{\theta=60}$; W is the effective boundary width and; $(2\pi R \sin 60)$ is the length of the zone in which the diffusion flux is at maximum. Thus, the cross-sectional area for diffusion is $2\pi RW\sin60$. After a detailed and lengthy evaluation of the relevant parameters, Coble arrived at the final equation for creep rate, given as

$$\dot{e} = \frac{148\sigma (D_b W) a_0^3}{(GS)^3 kT} \qquad (3.40)$$

where a_0^3 $(=\Omega)$ is the atomic volume of a vacancy. For lattice diffusion, the expression [2] is

$$\dot{e} = \frac{10\sigma (D_L \Omega)}{(GS)^2 kT} \qquad (3.41)$$

Other expressions are given for Coble's creep, the difference being in the coefficient representing the assumptions in each case (in Coble and Guerard it is $\dot{e} = \frac{150\sigma (D_b W)}{(GS)^3 kT}$). Thus, by writing the coefficient as a constant, the common expression is

$$\frac{d\varepsilon_{gb}}{dt} = \dot{\varepsilon} = A\frac{\sigma\Omega D_{gb}\delta}{l^3 kT} \qquad (3.42)$$

The subscripts refer to the grain boundaries. Ω is the atomic volume (of a vacancy), δ is the grain-boundary width, and l is the grain size. (In Nabarro-Herring, grain size was denoted by d). The D_S in Eqs. (3.14) and (3.15) is replaced, in Coble's equation, by $D_{gb}\delta$. Factor $1/l$ represents the density of the cross-section of the grain boundaries per unit area. Hence, δ/l is the cross-sectional area of the grain boundaries per unit area. In a realistic structure, A depends on grain structure and on how the average grain size is determined. Creep by grain-boundary diffusion has a stronger dependence on grain size than on lattice diffusion. In terms of shear strain and shear stress (Rieth et al.) the expression is

$$\dot{\gamma} = 42 D_S \frac{\pi\delta\tau\Omega}{d^3 kT} \qquad (3.43)$$

Here, d is equivalent to l and D_S to D_{gb}. When creep deformation is influenced by both lattice- and grain-boundary diffusion, an expression may be derived as follows. Equation (3.8) may be written with the same designations used in Eq. (3.42) as

$$2\dot{\varepsilon} = \frac{16\alpha D_S \sigma\Omega}{\pi l^2 kT}\left(1 + \frac{A\pi}{16\alpha}\frac{D_{gb}\delta}{lD_S}\right) \qquad (3.44)$$

Designating that $16\alpha/2\pi = B$ and $A\pi/2 \times 16\alpha = C$ gives

$$\dot{\varepsilon} = \frac{BD_S \sigma\Omega}{l^2 kT}\left(1 + \frac{CD_{gb}\delta}{D_S l}\right) \qquad (3.45)$$

An expression for creep may be given in terms of shear-strain rate and shear stress, when both lattice- and grain-boundary diffusion are involved in the deformation. For most polycrystalline materials, diffusion in grain boundaries is more rapid than in the lattice.

To summarize this section, it may be stated that in Coble creep the atoms diffuse along the grain boundaries and elongate the grains along the stress axis. This causes Coble creep to have stronger grain-size dependence than Nabarro-Herring creep. Since the grain boundary is the controlling diffusion mechanism in Coble creep, the process occurs at lower temperatures than Nabarro-Herring creep does. Coble creep is still temperature dependent and, as the temperature increases, so does the grain-boundary diffusion. It also exhibits a linear dependence on stress, as does Nabarro-Herring creep. Coble creep and Nabarro-Herring creep can take place in parallel, so that actual creep rates may involve both components and both diffusion coefficients.

References

1. Coble RL (1963) J Appl Phys 34:1679
2. Coble RL, Guerard YH (1969) J Am Ceram Soc 46:353
3. Dorn JE (1964) Energetics in metallurgical phenomena, vol 1. Gordon and Breach, London
4. Friedel J (1964) Dislocations. Addison-Wesley, Reading, MA
5. Garofalo F (1965) Fundamentals of creep and creep-rupture in metals. Macmillan, New York
6. Hemker KJ, Mills MJ, Nix WD (1991) Acta Metall Mater 39:1901
7. Herring C J (1950) Appl Phys 21:437
8. Le Hazif R, Poirier JP (1975) Acta Met 23:865
9. Mordike BL (2002) Mater Sci Eng A324:103
10. Nabarro FRN (2001) Mater Sci Eng A309:227
11. Nabarro FRN (1948) Report on a conference on strength of solids. Physical Society, London, p 75
12. Poirier JP (1985) Creep in crystals. Cambridge University Press, Cambridge
13. Rieth M, Falkenstein A, Graf P, Heger S, Jantsch U, Klimiankou M, Materna-Morris E, Zimmermann H (2004) Forschungszentrum Karlsruhe, Wissenschaftliche Berichte, FZKA, 7065
14. Ruoff AL (1973) Materials science. Prentice-Hall, Englewood Cliffs, NJ
15. Schoeck G, Seeger A (1955) Report on the conference on defects in a crystalline solid. Physical Society, London, p 340
16. Suzuki M, Kimura T, Koike J, Maruyama K (2004) Mat Sci Eng A387:706
17. Weertman J (1955) J Appl Phys 26:1213

Chapter 4
Creep in Ceramics

Abstract Creep rate in ceramics is smaller than in metals but experimental evidence indicates similar creep behavior between them. Metals and ceramics exhibit diffusion creep with $n = 1$ at low stresses and $n \sim {>}3$ at high stresses. Creep, both in metals and ceramics in the steady state, is diffusion controlled and the homologue temperature of T/T_m, regarding the diffusion coefficient, also applies to both materials. One of the reasons for the smaller creep rate is related to the diffusion rate, which is smaller in ceramics than in metals at the same homologue temperature. The diffusion-controlled creep rate in ceramics has a higher activation energy, Q, than in metals causing a slower creep rate. There is an advantage of using single crystal due to the absence of grain boundaries. Further, one should choose

(i)   materials with high melting points;
(ii) the use of strongly bonded ceramics; and
(iii) alloying for dislocation pinning.

Modern technology needs structural materials for a wide range of high-temperature applications. Ceramic materials possess a unique combination of great strength and resistance to oxidation at high temperatures, which are important properties for engineering applications. Creep at these temperatures is very significant not only for understanding the mechanisms involved, but also for assuring product longevity during use, based on experimental evaluations of the creep performance of materials. The study of creep in ceramic materials lagged behind that of metals and alloys, mainly because of the disbelief in their practical use, due to the inherent brittleness of most ceramics and their susceptibility to thermal shock. However, their high-temperature strength and good resistance to corrosive and oxidizing atmospheres outweighed their deficiencies. Creep in individual ceramics, such as alumina magnesia, etc., will be discussed later on in separate chapters, but this chapter provides a general overview of the subject accompanied by some specific examples.

© Springer International Publishing AG 2017
J. Pelleg, *Creep in Ceramics*, Solid Mechanics and Its Applications 241,
DOI 10.1007/978-3-319-50826-9_4

Some of the key distinctions between deformation in ceramics and deformation in metals include the facts that ceramics are typically brittle and have many pre-existing flaws that control their mechanical responses; also in ceramics, viscous glassy phases can coexist with crystalline phases as part of the same microstructure. Nonetheless, ceramics are excellent materials for engineering applications in extreme environments, because they retain mechanical strength at high temperatures and in harsh chemical environments. Therefore, experimental data and theoretical information are both of utmost importance.

Creep is a deformation which occurs by some form of dislocation motion. Even climb or cross slip, associated with dislocation motion, require the presence of vacancies at sites where climb is occurring. The arrival of such vacancies involves their diffusion, in order to enable climb to happen. Thus, diffusion phenomena cannot be ignored in high-temperature creep. Unlike metallic systems, in ceramics, the presence of two atoms (or rather ionic species) is involved in the diffusion process, further complicating the analysis of the creep data. Since the cations and anions both participate in the diffusive process, it is necessary to consider ambipolar diffusion and mass transport along parallel diffusion paths. 'Ambipolar diffusion' is the diffusion of positive and negative species with opposite electrical charges. In the case of ionic crystals, the fluxes of the diffusing species are coupled.

An objective of selecting engineering materials or developing new materials is to slow down dislocation motion, as much as possible—more specifically, to retard climb as long as possible, in order to reduce creep and, thus, ensure a long lifetime of service. The techniques used to study creep in ceramics and to record the findings have been borrowed from numerous experimental researches in alloys. As such, the generally accepted method for recording the results of a creep test in ceramics is by plotting strain versus time, as shown schematically in Fig. 1. 1a. As in the alloys, in ceramics, temperature and stress both affect the shapes of the creep curves.

There is no universal creep behavior which characterizes all high-temperature structural ceramics. The dominant creep mechanism may vary from ceramic to ceramic. As observed in Fig. 1.1, transient creep decreases over time up to the point at which no more extension occurs under the effect of the load. This is a result of the equilibrium between strain hardening and thermal softening (creep). Creep curves are analyzed according to the usual, general constitutive law for high-temperature steady-state creep, as done by Bretheau et al. [3]. Here, for example, is an equation for zirconia:

$$\dot{\varepsilon} = A \frac{\mu b}{kT} \left[\frac{\sigma}{\mu} \right]^n \left[\frac{b}{d} \right]^p \left[\frac{p_{O_2}}{p_{O_2}^*} \right]^m \exp\left[-\frac{Q}{kT} \right], \tag{4.1}$$

where A is a dimensionless constant, b is the Burgers vector, d is the grain size (relevant for polycrystals), and $p_{O_2}^*$ is a reference oxygen partial pressure. Deformation and diffusion mechanisms determine the parameters n, p, m,

Fig. 4.1 Steady-state creep rate, normalized to a stress of 100 MPa, plotted as a function of reciprocal temperature. Martinez-Fernandez et al. [20]. With kind permission of John Wiley and Sons

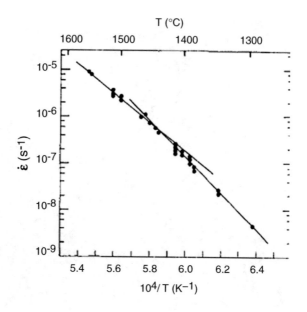

and Q. Changes in σ, T, or p_{O_2} may be determined experimentally from the variation of $\dot{\varepsilon}$. p is the exponent of the inverse grain size; for single crystal, $p = 0$. The dislocation substructures were obtained by transmission electron microscopy (TEM). Standard grain-boundary analysis was carried out to determine the Burgers vectors of isolated dislocations. The 3D nature of the dislocation arrangement was determined using stereo pairs obtained with $g = (220)$. Dislocation densities in the foils were determined by standard means. The steady-state creep rate versus $1/T$ is shown in Fig. 4.1

The above figure suggests two deformation regimes, namely two operating mechanisms with a transition in the 1400–1450 °C range. A least-square fit for both these regimes gives $Q = 6.2 \pm 0.4$ eV and $Q = 7.7 \pm 0.4$ eV at $T \leq 1400$ °C. Figures 4.2 and 4.3 show the dislocation substructures. These deformed samples exhibit different dislocation substructures at 'low' and 'high' deformation temperatures. At 1300 °C, the dislocation density, ρ, was high ($\sim 10^{13}$ m^{-2}) and showed substantial dislocation reactions and node formation (Fig. 4.2). The grain-boundary analysis (not shown in the figure) revealed that most of the dislocations had a Burgers vector $b = \frac{1}{2}[110]$ and existed along the (001) primary slip plane. However, the stereo pair in Fig. 4.2 indicates that some dislocations, belonging to the primary slip plane, changed slip planes to lie along the $(\bar{1}11)$ and $(1\bar{1}1)$ planes, indicating that a significant amount of cross-slip occurred during that deformation. In other experiments, cross-slip was also observed in samples deformed at temperatures as low as 400 °C under hydrostatic confining pressure. At 1500 °C, the dislocation density is lower ($\rho \sim 5 \times 10^{11}$ m^{-2}). A stereo pair (Fig. 4.3) shows that many dislocation segments are perpendicular to the (001) primary slip plane; these segments lie on the (100) and (010) planes. All six (110) Burgers vectors are

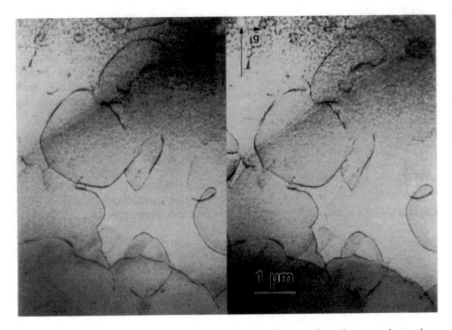

Fig. 4.2 Stereo pair of transmission electron micrographs of dislocation substructure in specimen crept at 1300 °C. Martinez-Fernandez et al. [20]. With kind permission of John Wiley and Sons

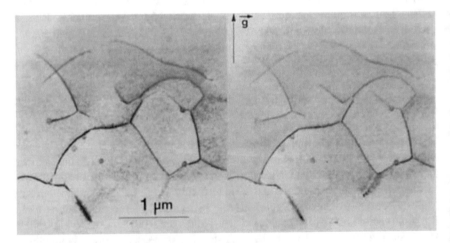

Fig. 4.3 Stereo pair of transmission electron micrographs of dislocation substructure in specimen crept at 1500 °C. Martinez-Fernandez et al. [20]. With kind permission of John Wiley and Sons

present, indicating that numerous slip systems have been activated; however, the low dislocation density indicates that significant recovery has occurred and that diffusion must be reasonably rapid at this temperature.

An equation was obtained by Cannon and Langdon for the steady-state creep rate:

$$\dot{\varepsilon} = \frac{ADGb}{kT} \left(\frac{b}{d}\right)^p \left(\frac{\sigma}{G}\right)^n. \tag{4.2}$$

The symbols which are common to Eqs. (4.1) and (4.2) have the same meaning. G is equivalent to μ, indicating free energy; A is a parameter which is a constant derived from the need to take into account the effects of structural variables, such as grain shape. The expression for D, the known form seen in earlier chapters, is reproduced here as

$$D = D_0 \exp(-Q/RT). \tag{2.4}$$

The steady-state creep rate, according to Nabarro-Herring, if the vacancies flow through the grains of the lattice, is

$$\dot{\varepsilon} = \frac{B_l \Omega D_l \sigma}{d^2 kT}, \tag{4.3}$$

where Ω is the atomic volume and B_l is a constant. The subscript l refers to the lattice. Equation (4.3) is the same as Eq. (3.13). Ω replaces V, D_l is equivalent to D_v, and B_l stands for $2\beta C_v^0$. The calculated values of B_1 range from ~ 12 to 40, for different experimental conditions, but Herring obtained $B_1 = 13.3$ for polycrystals with complete grain-boundary relaxation tested under uniaxial tension. Using the values of $B_1 = 13.3$ and $\Omega = 0.7b^3$, Eq. (4.3) may be expressed in the form of Eq. (4.2) as:

$$\dot{\varepsilon} = 9.3 \frac{D_l Gb}{kT} \left(\frac{b}{d}\right)^2 \left(\frac{\sigma}{G}\right). \tag{4.4}$$

If the vacancies flow along the grain boundaries, the process of Coble creep may be used and the steady-state creep rate is given by

$$\dot{\varepsilon} = \frac{150 \Omega \delta D_{gb} \sigma}{\pi d^3 kT}. \tag{4.5}$$

This relation is equivalent to Eq. (3.40), given in Chap. 3. Here, δ is the effective width of the grain boundary for vacancy diffusion and D_{gb} is the grain-boundary diffusion coefficient. G is the shear modulus. The other symbols have their usual meanings. With a value for the atomic volume, expressed in terms of the Burgers vector, b as $\Omega = 0.7b^3$, Eq. (4.5) may be rewritten in the form of

$$\dot{\varepsilon} = 33.4 D_{gb} \left(\frac{\delta}{b}\right) \left(\frac{b}{d}\right)^3 \left(\frac{\sigma}{G}\right). \tag{4.6}$$

Note that, in Eq. (4.6), d in the denominator is cubed, whereas in Eq. (4.4), d is squared. Thus, a comparison between them indicates that Coble creep is preferable to Nabarro-Herring creep when the grain size is very small, since p (see Eq. (4.2)) is

3 and 2 for the Coble and Nabarro creeps, respectively. Furthermore, the Cobble creep is also preferable at lower temperatures, because the respective activation energies, $Q_{gb} < Q_l$, determine the respective values of D_{gb} and D_l. Nonetheless, since the Coble and Nabarro-Herring creeps operate independently, their respective rates are additive and one may write a relation given as follows for the total creep:

$$\dot{\varepsilon} = 9.3 \frac{D_l G b}{kT} \left(\frac{b}{d}\right)^2 \left(\frac{\sigma}{b}\right) \left[1 + 3.6 \frac{D_{gb}}{D_l} \left(\frac{\delta}{d}\right)\right]. \tag{4.7}$$

The coefficient before the last term in the square brackets is clearly the result of dividing 33.4 by 9.3.

The creep rate in ceramics is more complicated than in metals, because a ceramic is composed of two species, an anion and a cation. Both must move in concert during the deformation in order to preserve the atomic ratios, namely their stoichiometric ratio in the stoichiometric composition of the ceramic. Relevant equations, such as Eq. (4.7), should be combined for the chemical species comprising each specific ceramic, as given by Cannon and Langton [7] in the wake of Gordon, and shown below:

$$\dot{\varepsilon} = 9.3 \left(\left\{ (1/\alpha) \left[D_{c(l)} + 3.6 D_{c(gb)} (\delta_c/d) \right] \right\} \Big/ \left\{ 1 + \left(\frac{\beta}{\alpha}\right) \frac{\left[D_{c(l)} + 3.6 D_{c(gb)} (\delta_c/d) \right]}{D_{a(l)} + 3.6 D_{a(gb)} (\delta_a/d)} \right\} \right)$$
$$\times \left(\frac{Gb}{kT}\right) \left(\frac{b}{d}\right)^2 \left(\frac{\sigma}{G}\right). \tag{4.8}$$

The subscripts "a" and "c" stand for anion and cation, respectively, and the symbols α and β represent their respective valences. Equation (4.8) is practically simplified, because of the large difference in the respective diffusion coefficients of the species of the ceramics. As a consequence, the observed creep rates are determined by the movement of the slower diffusing species along the faster diffusion path.

It has been observed that the slower moving species does not always control the rate of the creep process. Usually, the anion is considered as the slower diffusing species in oxide ceramics, while the cations are the faster moving species. However, in some oxide ceramics, the anion is the faster moving species and creep is controlled by the cation, such as in the case of Al_2O_3, where Al controls the rate of creep. Similar observations have been seen in MgO, BeO, etc. Thus, one cannot generalize and predict, a priori, whether the diffusion of the anion or the cation will control the creep rate in materials. Exemplary graphic presentations of creep rates for various materials appear as plots of $(\dot{\varepsilon}kT/DGb)(d/b)^2$ versus σ/G in Figs. 4.4 and 4.5. In each case, the diffusion coefficient was selected to provide the best fit with the theoretical model. Usually, the best fit was obtained for the fastest-diffusing species through the crystal lattice. The data selected for these illustrations are for materials with $n = 1$.

The various values for each material appearing in Figs. 4.4 and 4.5 and the parameters used to construct these figures are listed in Table 4.1, with additional

Fig. 4.4 Normalized creep rate plotted against normalized stress for Nabarro-Herring creep in Al$_2$O$_3$ containing no additives [47–49]. Cannon and Langdon [7]. With kind permission of Springer. References: Dixon-Stubbs and Wilshire [14], Oishi and Kingery [25], Rhodes and Carter [27], Yavari et al. [39]

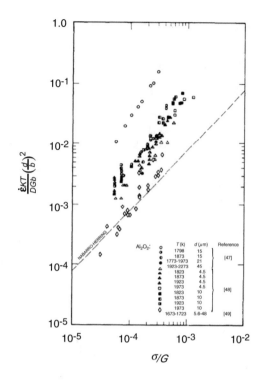

information for graph construction not considered in this book. Among the parameters used and appearing in the above equations for the materials, D, G, and b are included. G, in the above equations, was estimated using

$$G = G_0 - (\Delta G)T, \tag{4.9}$$

where G_0 was obtained by extrapolation from high temperature to absolute zero and ΔG is the variation of the shear modulus per degree Kelvin.

Note in the figures that the data differ by an order or two from the Nabarro-Herring line (presented as a dashed line in Fig. 4.5). The authors of Table 4.1 mention three reasons for this deviation from the theoretical Nabarro-Herring line: (1) the experimental testing methods were different; (2) impurities influenced the diffusion coefficients and, thus, the creep rates; and (3) there were small deviations in stoichiometry in some of the materials used in the experiments.

Some ceramics exhibit a stress exponent in the $n \sim 3$–5 range and creep rates independent of grain size. This was interpreted, as previously in the case of creep in metals, as being due to the intragranular motion of dislocations. This motion is explained in terms of the glide and climb of intragranular dislocations, occurring when they pile up and the climb process is rate-controlling. Following Weertman, Cannon and Langdon gave the steady-state creep rate as

Fig. 4.5 Normalized creep rate plotted against normalized stress for Nabarro-Herring creep in (zx) MgO [53], (O, e) BeO [54, 55], (fin) c ~ -SiC [56] and (v) Si₃N₄ [57]. Cannon and Langdon [7]. With kind permission of Springer. References: Tremper [31], Vandervoort and Barmore [32], Barmore and Vandervoort [1], Davis [13], Seltzer [28]

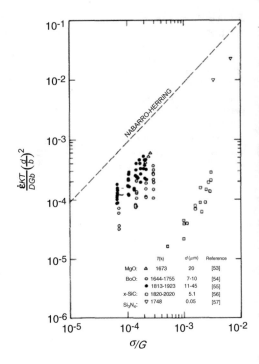

$$\dot{\varepsilon} = \frac{B_2 \Omega D_l \sigma^{4.5}}{G^{3.5} M^{0.5} b^{3.5} kT}, \tag{4.10}$$

where M is the concentration of the active dislocation and B_2 is a constant in the range of 0.015– 0.33 Weertman [36], due to the decomposition of the pile-ups into groups of dislocation dipoles. By assigning a value of $B_2 = 0.2$, Eq. (4.10) reduces to

$$\dot{\varepsilon} = \frac{0.14}{b^{1.5} M^{5.5}} \left(\frac{D_l G b}{kT}\right) \left(\frac{\sigma}{G}\right)^{4.6}. \tag{4.11}$$

In Eq. (4.11), D_l is the lattice diffusion coefficient, which applies above a temperature of $0.5T_m$, but breaks down at lower temperatures, where pipe diffusion along the dislocation cores is dominant. Therefore, experimental data were obtained at temperatures above $0.6T$ and using lattice diffusion coefficients for the slower moving species.

For $n \cong 3$, two possible mechanisms should be considered. One, mainly for metallic solid solutions, relates to the glide and climb process controlled by glide (mainly because of the solute atom atmosphere surrounding dislocations, which is dragged by their motion). For steady-state creep (Weertman; Mohamed and Langdon), this process gives

Table 4.1 Values of the material parameters. Cannon and Langdon [7]. With kind permission of Springer. For the references, check Table 1 of Cannon and Langdon

Material	D (cm^2 s^{-1})[a]	G_0 (MPa)	ΔG (MPa K^{-1})	b (cm)	References D	References G
Al$_2$O$_3$	$D_l(O^{2-}) = 2 \times 10^3 \exp(-635{,}000/RT)$ $D_l(Al^{3+})=28 \exp(-476{,}000/RT)$	1.7×10^5	23.4	4.75×10^{-8}	[18] [20]	[19]
MgO	$D_l(O^{2-}) = 2.5 \times 10^{-6} \exp(-261{,}000/RT)$ $D_l(Mg^{2+})=2.49 \times 10^{-1} \exp(-330{,}000/RT)$	1.387×10^5	26.2	2.98×10^{-8}	[21] [23]	[22]
BeO	$D_l(O^{2-}) = 2.7 \times 10^{-5} \exp(-284{,}000/RT)$ $D_l(Be^{2+}) = 1.5 \exp(-384{,}000/RT)$	1.86×10^5	20	2.7×10^{-8}	[24] [26]	[25]
SiC	$D_l(Si^{4+})(\alpha) = 2.6 \times 10^8 \exp(-840{,}000/RT)$ $D_l(Si^{4+})(\beta) = 8.4 \times 10^7 \exp(-911{,}000/RT)$	1.6×10^5	23	3×10^{-8}	[27] [29]	[28]
Si$_3$N$_4$	$D_l(N^{3-})(\alpha) = 1.2 \times 10^{-12} \exp(-233{,}000/RT)^b$	1.3×10^{5c}		3×10^{-8d}	[30]	[31]
UO$_2$	$D_l(U^{4+}) = 6.8 \times 10^{-5} \exp(-410{,}000/RT)$	8.9×10^4	21	3.87×10^{-8}	[32]	[33]
ThO$_2$	$D_l(Th^{4+}) = 1.25 \times 10^{-7} \exp(-245{,}000/RT)$	9.0×10^4	19	3.96×10^{-8}	[34]	[35, 36]
NaCl	$D_l(Cl^-) = 1.2 \times 10^2 \exp(-214{,}000/RT)$	1.79×10^4	9.6	3.99×10^{-8}	[37]	[38]
LiF	$D_l(F^-) = 64 \exp(-212{,}000/RT)$	5.52×10^4	33.2	2.85×10^{-8}	[39]	[38]
KCl	$D_l(Cl^-) = 2.1 \exp(-189{,}000/RT)$	1.225×10^4	6.57	4.45×10^{-8}	[37]	[38]
UC	$D_l(U^{4+}) = 7.5 \times 10^{-5} \exp(-339{,}000/RT)$	2.058×10^6	16.1	3.51×10^{-8}	[40]	[41]
Fe$_2$O$_3$	$D_l(O^{2-}) = 2.0 \exp(-326{,}000/RT)$	9.28×10^4	6.0	5.03×10^{-8}	[42]	[43, 44]
ZrO$_2$	$D_l(Zr^{4+}) = 3.5 \times 10^{-2} \exp(-387{,}000/RT)$	1.54×10^5	35.2	2.57×10^{-8}	[45]	[46]

[a] Activation energies in J mol^{-1} ($R = 8.31$ J mol^{-1} K^{-1})
[b] Measured by isotope exchange assuming rapid boundary diffusion of N
[c] Value of G at test temperature of 1748 K
[d] Estimated from $b = (\Omega/0.7)^{1/3}$

$$\dot{\varepsilon} = \frac{\pi(1 - \mu)kT\tilde{D}\sigma^3}{6e^2cb^3G^4}. \tag{4.12}$$

In Eq. (4.12), e is the solute–solvent size difference and c is the concentration of solute. $\tilde{D}$ is the solute interdiffusion coefficient and μ is Poisson's ratio. Using $\mu = 0.34$, Eq. (4.12) reduces to

$$\dot{\varepsilon} = \frac{0.35}{e^2c}\left(\frac{kT}{Gb^3}\right)\left(\frac{\tilde{D}Gb}{kT}\right)\left(\frac{\sigma}{G}\right)^3. \tag{4.13}$$

Now, most ceramics with $n \cong 3$ do not contain solutes; thus, solute-atmospheric dragging is irrelevant.

If creep is controlled by dislocation climb from Bardeen–Herring sources, the creep rate Nabarro [22] is given by

$$\dot{\varepsilon} = \frac{B_3\pi\Omega D_l\sigma^3}{G^2b^2kT}, \tag{4.14}$$

where B_3 is a constant having a value Weertman [37] of ~ 0.1. With this value, Eq. (4.14) reduces to

$$\dot{\varepsilon} = 0.22\frac{D_lGb}{kT}\left(\frac{\sigma}{G}\right)^3 \tag{4.15}$$

The data for ceramics where $n \sim 5$ are assembled in Figs. 4.6 and 4.7, while for ceramics with $n \sim 3$ they appear in Figs. 4.8 and 4.9. The stress exponent, $n \sim 5$, suggests that the rate-controlling mechanism in creep is dislocation climb. In Figs. 4.8 and 4.10, the dashed line represents dislocation climb, according to the prediction in Eq. (4.15). Note that, in Fig. 4.9, the experimental agreement of BeO and SiC and, in Fig. 4.10, that of yttria-stabilized ZrO_2 fit well with the predicted theoretical lines.

It was observed in metals, deformed by power-law creep with $n \sim 5$, that grains become divided into subgrains with very small angles, typically >2. It has been established that the average grain size, λ, is inversely related to the applied stress. Measurements show that the normalized subgrain size may be expressed as

$$\frac{\lambda}{b} = \varsigma\left(\frac{\sigma}{G}\right)^{-1}. \tag{4.16}$$

λ has a value of 20 for metals. For the investigated ceramics, shown in Fig. 4.11, a value of 20–30 for λ is indicated higher than that of metals.

The dislocation density within the subgrains in metals, ρ, varies with stress squared. The normalized dislocation density, $b\rho^2$, is given as

$$b\rho^2 = \Psi\left(\frac{\sigma}{G}\right). \tag{4.17}$$

Fig. 4.6 Normalized creep
rate plotted against
normalized stress for
power-law creep in (△) NaCl
[67], (○) CiF [75], (▽) KCl
[76], (□) UC [77] and (◇)
reaction-sintered (RS) SiC
[78]. Cannon and Langdon
[7]. With kind permission of
Springer. References [67]
Burke; [75] Cropper and
Langdon; [76] Cannon and O.
D. Sherby; [77] Seltzer et al.;
[78] Carter et al.

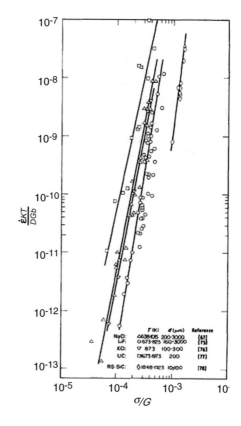

Here, Ψ is a constant having a value close to unity for all metals. The validity of this relation was tested for some ceramics and it was concluded that the agreement between metals and ceramics is very good. Figure 4.12 illustrates plots according to Eq. (4.17). Another point of importance is GBS in ceramics and its contribution to creep. This contribution is usually expressed as the ratio $\varepsilon_{gbs}/\varepsilon_t$, where ε_{gbs} is the strain due to GBS and ε_t is the total strain. An analysis of published data for metals shows that the magnitude of $\varepsilon_{gb}/\varepsilon_t$ tends to increase with decreasing stress and/or decreasing grain size. This concept has evaluated for only a few ceramics, namely in two sets of Al_2O_3 and MgO each. These results are summarized in Fig. 4.13. Although very few results are currently available for ceramics, the data confirm that, as in metals, sliding increases in importance at the lower stress levels and also with a decrease in grain size.

The experimental information indicates the similarities between the creep behaviors of ceramics and metals. Both metals and ceramics exhibit diffusion creep with $n = 1$ at low stresses and dislocation creep with $n \sim >3$ at high stresses. Creep, both in metals and ceramics in the steady state, is diffusion controlled and the homologue temperature of T/T_m, regarding the diffusion coefficient, also applies to both materials. Moreover, the creep rate in ceramics is smaller than in metals. One of

Fig. 4.7 Normalized creep rate plotted against normalized stress for power-law creep in UO₂ [58, 60, 61, 63] and ThO₂ [58]. Cannon and Langdon [7]. With kind permission of Springer. References: [58] Poteat and Yust; [60] Marples and Hough; [61] Seltzer et al.; [63] Burton and Reynolds

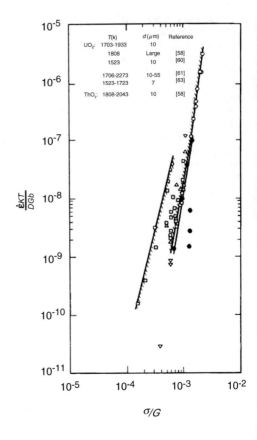

the reasons for this is related to the diffusion rate, which is smaller in ceramics than in metals at the same homologue temperature. Directly related to the diffusion-controlled creep rate is the higher activation energy, Q, required in ceramics (higher than in metals), causing a slower creep rate. Furthermore, in both types of materials, an inverse relation exists between the subgrain size and the applied stress (Eq. 4.16), and the dislocation density within the subgrains varies with σ^2. The major differences in creep between the two substances (metals and ceramics) are (a) the enhanced role of diffusion creep and (b) the division of creep behavior into two categories during power-law creep, with stress exponents of ~ 5 and ~ 3, for ceramics.

4.1   Creep in Single-Crystal Ceramics

The is much interest in single-crystal ceramics for creep-resistant applications, as for use in turbine blades, giving them the mechanical advantage of being able to operate at much higher temperatures than polycrystalline ones. Creep is a common cause of failure in turbine blades and is, in fact, their life-limiting factor. When the

Fig. 4.8 Normalized creep
rate plotted against
normalized stress for
power-law creep in Al_2O_3 [47,
52, 79, 80] and Fe_2O_3 [81]: the
short line is for Al_2O_3 single
crystal in $0°$ orientation [88].
Cannon and Langdon [7].
With kind permission of
Springer. References: [47]
Crosby and Evans; [52].
Cannon and Sherby [9]; [79]
Warshaw and Norton; [80]
Engelhardt and Thümmler;
[81] Crouch

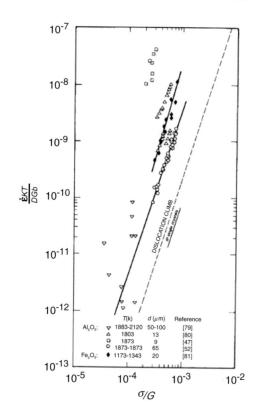

temperature of a material under high stress is raised to a critical point, its creep rate
quickly increases. A single-crystal structure is able to withstand creep at higher
temperatures than a polycrystalline structure can, due to the absence of grain
boundaries. Grain boundaries are areas in a microstructure where many defects and
failure mechanisms start, leading to creep. The lack of grain boundaries inhibits
creep and prevents GBS from occurring. Nonetheless, creep still occurs in
single-crystal turbine blades, caused by different high-temperature mechanisms.
Alas, pure ceramics, in general, are brittle, which hampers their use, although they
may become less brittle at high temperatures. Nevertheless, some ceramics, particularly Si_3N_4, is of interest for high-temperature use. This ceramic is one of the
many high-performance ceramic materials available to designers. It has a number of
desirable properties that make it attractive to the aerospace industry:

(a) low density: 3.2 g/cm^3, compared with ~ 8 g/cm^3 in superalloys;
(b) high hardness: 1800 kgf/mm, compared with 800 kgf/mm in steel;
(c) high strength at high temperatures: in some grades of Si_3N_4, strength decreases
    only at temperatures in excess of 1200 °C.

These properties make Si_3N_4 an ideal material for use at high temperatures,
especially when weight is critical and high resistance to wear and erosion is required.

Fig. 4.9 Normalized creep rate plotted against normalized stress for power-law creep in MgO [49, 82, 83], BeO [84] and chemically vapor-deposited (CVD) SiC [85]. Cannon and Langdon [7]. With kind permission of Springer. [49] Lessing and Gordon; [82] Hensler and Cullen; [83] Langdon and Pask; [84] Barmore and Vandervoort; [85] Carter et al.

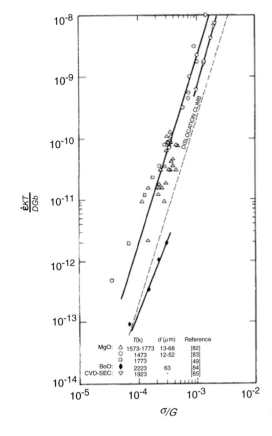

Since, in single crystal, no grain boundaries exist, creep may be described as the movement of dislocations through the crystal structure. Such movement is accommodated by slip along preferential planes. Thus, an additional factor (besides load, temperature and time) must be considered, namely the orientation of the crystal. Also, the possibility of twinning, resulting from homogeneous shear, may occur. In highly symmetric cubic structures, such as pure MgO, many planes are available for slip. At low temperatures, slip occurring along (100) planes in the [110] direction is usually indicated as (100) [110]. At high temperatures, where creep deformation is significant, slip occurs either on (001) [110] or (111) [110], resulting in five independent slip systems.

Generally, the slip systems that become operative and the stresses required at high temperatures depend on the bond strength of the particular ceramic material. Slip occurs at low temperatures and relatively low stresses in weakly bonded crystals, such as NaCl, while in strongly bonded materials, such as in the covalently bonded TiC, high temperatures and high stresses are required.

In single crystal with lower symmetry than cubic crystals, less slip systems are available for slip; for example, in the hexagonal Al_2O_3, only two independent slip systems exist. The slip systems are $\{0001\}\langle 1120\rangle$.

Fig. 4.10 Normalized creep rate plotted against normalized stress for power-law creep in ZrO_2 containing 10% Y_2O_3 [86]. Cannon and Langdon [7]. With kind permission of Springer. Reference: [86] Seltzer and P.K. Talty

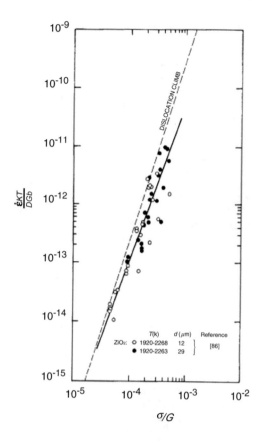

Fig. 4.11 Normalized subgrain size plotted against normalized stress for ceramics [67, 91 99]: each line is drawn through a single set of datum points with a slope of −1. Cannon and Langdon [7]. With kind permission of Springer. References: [67] Sherby and Burke; [91] Pontikis and Poirier; [99] Blum

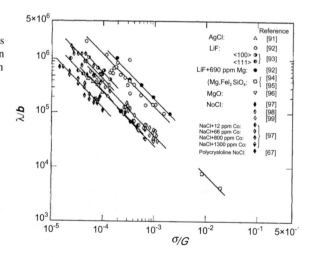

Fig. 4.12 Normalized dislocation density within the subgrains plotted against normalized stress for ceramics; [92, 96–98, 112, 113]: each line is drawn through a single set of datum points with a slope of 1. Cannon and Langdon [7]. With kind permission of Springer. References: [92] Streb and Reppich; [96] Huther and Reppich; [97] Schuh et al.; [98] Poirier; [112] Bilde-Sorensen; [113] Clauer and Wilcox

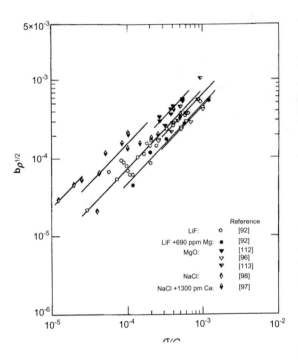

Fig. 4.13 The contribution of grain-boundary sliding to the total strain plotted against normalized stress for Al_2O_3 [52] and MgO [121]. Cannon and Langdon [7]. With kind permission of Springer. References: [52] Cannon and Sherby; [121] Langdon

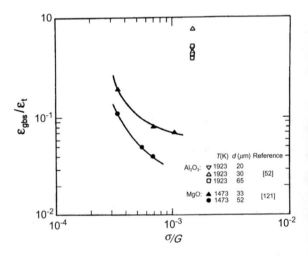

For measurable creep in a single crystal to occur, a dislocation must either be present or created. The energy required for forming, initiating, and moving dislocations by slip is supplied by stress and temperature. Clearly, the pinning of dislocations by obstacles blocking their motion decreases creep. A decrease in creep may also be achieved by the addition of certain additives, some forming solid solutions; the alloying and addition of various elements also reduces the creep by pinning dislocations.

Summing up this section on creep resistance, when choosing engineering materials (including ceramics), the following should be taken into account:

(i)   the potential use of single-crystal ceramics;
(ii) the use of materials with high melting points;
(iii) the use of strongly bonded ceramics; and
(iv) alloying for dislocation pinning.

The advantage of using single crystal is the absence of grain boundaries. In polycrystals, dislocation sliding is not a significant factor, because the random orientation of the individual grains making it difficult for dislocations to pass from one grain to an adjacent one. Thus, no GBS occurs, which enhances creep in polycrystalline materials. Recall that the diffusion rate of ions and vacancies through crystalline structures in grains and grain boundaries controls the creep rate together with GBS. In fact, GBS is often associated with porosity, which is a key factor in crack initiation (at triple points) leading to fracture. Cavities are shown in a micrograph obtained by TEM (Fig. 4.14) in a silicon nitride ceramic (Si_3N_4) crept at 1100 °C for 500 h under 80 MPa.

The cavities are at two triple-grain junctions. Strain contrast is visible in the micrograph along two grain boundaries attributed to GBS. Evidence of GBS between two grains is shown in Fig. 4.15 and, consequently, a gap develops between these grains.

Evidence also exists that the GBS and cavity formation, in addition to stress relaxation through nucleation of dislocations at the strain whorls, act together to produce a much shorter creep life to failure at high temperatures. The occurrence of dislocations in some silicon nitride grains is shown in Fig. 4.15. The silicon nitride was crept at 1295 °C for 3.95 h under 80 MPa. These dislocations are close to the grain boundaries and certain ones started from the grain boundaries and extended

Fig. 4.14 Cavities in GS44 crept at 1100 °C for 500 h under 80 MPa. Wei et al. [38]. With kind permission of Elsevier

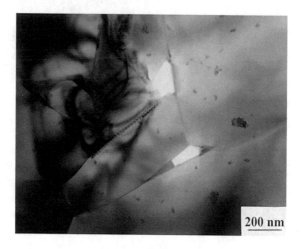

200 nm

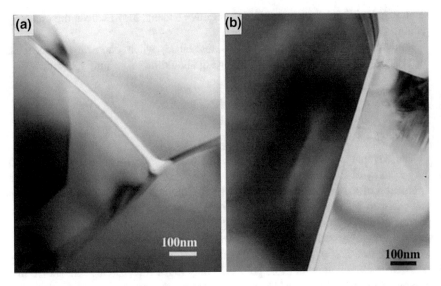

Fig. 4.15 Direct evidence of grain boundary sliding leaving gaps between two grains of silicon nitride. Wei et al. [38]. With kind permission of Elsevier

Fig. 4.16 Dislocation arrays in a silicon nitride grain of GS44 crept at 1275 °C under 80 MPa. Wei et al. [38]. With kind permission of Elsevier

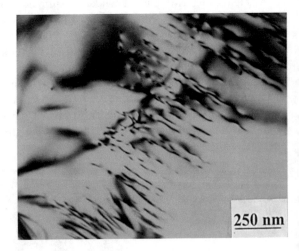

into the grains. Dislocation pile-ups are also observed. Figure 4.16 shows dislocation pile-ups and Fig. 4.17 illustrates arrays of dislocations in the silicon nitride grain, whereas Fig. 4.18 indicates those dislocations originating at the grain boundaries. The dislocation images shown in Fig. 4.18 extend into the silicon nitride grain. Additionally, in Fig. 4.19, dislocations emanate from the grain boundaries and a strain whorl is visible at the grain boundary, as well. The observations made from Figs. 4.14, 4.15, 4.16, 4.17, 4.18 and 4.19 indicate that, in addition to the diffusional processes, dislocation nucleation and dislocation motion

Fig. 4.17 Dislocation pile-ups in some silicon nitride grains of GS44 crept at 1275 °C under 80 MPa for 3.95 h. Wei et al. [38]. With kind permission of Elsevier

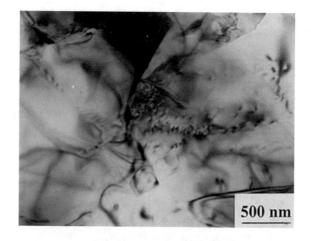

Fig. 4.18 Dislocations observed in GS44 crept at 1275 °C under 80 MPa showing a number of dislocations originated from the grain boundaries. Wei et al. [38]. With kind permission of Elsevier

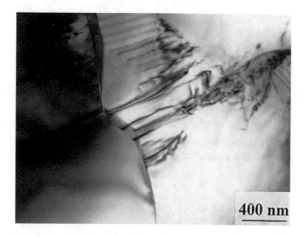

Fig. 4.19 Strain whorl at the grain boundary and dislocations starting from the grain boundary observed in GS44 crept at 1275 °C under 80 MPa. Wei et al. [38]. With kind permission of Elsevier

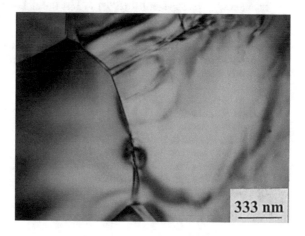

are alternate mechanisms for stress relaxation at grain boundaries during high-temperature creep. Nevertheless, the major creep strain in silicon nitride is not the dislocation mechanism, although it does relieve the interlocking stresses, enabling GBS and promoting more rapid stress rupture during high-temperature creep.

In regard to the contribution of dislocation motion, plastic deformation may be a plausible mechanism, but not as the major creep strain, because a large density of dislocations was observed only locally, in merely a few silicon nitride grains.

The strain, due to dislocation motion, may be written as

$$\dot{\varepsilon} = \rho b x, \tag{4.18}$$

where ρ is the density of the dislocations, b is the Burgers vector, and x is the average distance that a dislocation moved.

Recall, from Chap. 1, that

$$\dot{\varepsilon} = f(\sigma, t, T). \tag{1.2}$$

A more detailed expression for $\dot{\varepsilon}$ may be

$$\dot{\varepsilon} = A(\sigma, T, S, e)\sigma^n \exp\left[-\frac{\Delta H_c(\sigma, T, S, e)}{RT}\right]. \tag{4.19}$$

The preferential use of single-crystal materials is highlighted in all the above figures and texts, confirming that GBS and pore formation at triple points on grain boundaries cannot occur in single-crystal applications.

References

1. Barmore WL, Vandervoort RR (1965) J Amer Ceram Soc 48:499
2. Bohaboy PE, Asamoto RR, Conti AE (1969) Report No. GEAP-10054. General Electric Breeder Reactor Development Operation, Sunnyvale, CA
3. Bretheau T, Castaing J, Rabier J, Veyssiere P (1979) Adv Phys 28:835
4. Burke PM (1968) Ph. D. dissertation. Stanford University, Stanford
5. Burton B, Reynolds GL (1973) Acta Metall 21:1641
6. Cannon WR, Langdon TG (1983) J Mater Sci 18:1
7. Cannon WR, Langdon TG (1988) J Mater Sci 23:1
8. Cannon WR, Sherby OD (1970) J Amer Ceram Soc 53:346
9. Cannon WR, Sherby OD (1977) J Amer Ceram Soc 60:44
10. Carter CH, Davis RF, Bentley J (1984) J Amer Ceram Soc 67:409
11. Cropper DR, Langdon TG (1968) Phil Mag 18:1181
12. Crosby A, Evans PE (1973) J Mater Sci 8:1573
13. Davis RF (1985) Personal communication
14. Dixon-Stubbs PJ, Wilshire B (1982) Phil Mag A45:519
15. Engelhardt G, Thümmler F (1970) Ber Deut Keram Ges 47:571
16. Gordon RS (1973) J Amer Ceram Soc 56:147

17. Gordon RS (1975) Mass transport phenomena in ceramics. Plenum Press, New York, p 445
18. Herring C (1950) J Appl Phys 21:437
19. Marples JAC, Hough A (1970) Plutonium 1970 and other actinides. In: Miner WN (ed). Metallurgical Society of AIME, New York, p 479
20. Martinez-Fernandez J, Jimenez-Melendo M, Dominguez-Rodriguez A, Heuer AH (1990) J Amer Ceram Soc 73(8):2452
21. Mohamed FA, Langdon TG (1974) Acta Metall 22:779
22. Nabarro FRN (1967) Phil Mag 16:231
23. Nabarro FRN (1948) Report of a conference on strength of solids. Physical Society, London, p 75
24. Norton FH (1929) The creep of steels at high temperatures. McGraw Hill, New York
25. Oishi Y, Kingery WD (1960) J Chem Phys 33:905
26. Poteat LE, Yust CS (1968) Ceramic microstructures. In: Fulrath RM, Pask JA (eds). Wiley, New York, p 646
27. Rhodes WH, Carter RE (1966) J Amer Ceram Soc 49:244
28. Seltzer MS (1977) Bull Amer Ceram Soc 56:418
29. Seltzer MS, Clauer AH, Wilcox BA (1970) J Nucl Mater 34:351
30. Seltzer MS, Wright TR, Moak DP (1975) J Amer Ceram Soc 58:138
31. Tremper RT (1971) Ph.D. Dissertation. University of Utah, Salt Lake City
32. Vandervoort RR, Barmore WL (1963) J Amer Ceram Soc 43:180
33. Warshaw SI, Norton FH (1962) J Amer Ceram Soc 45:479
34. Weertman J (1957) J Appl Phys 28:362
35. Weertman J (1957) J Appl Phys 28:1185
36. Weertman J (1975) Rate processes in plastic deformation of materials. In: Li JCM, Mukherjee AK (eds). American Society for Metals, Metals Park, OH, p 315
37. Weertman J (1968) Trans ASM 61:681
38. Wei Q, Sankar J, Kelkar AD, Narayan J (1999) Mat Sci Eng A272:380
39. Yavari P, Miller DA, Langdon TG (1982) Acta Metall 30:871

Chapter 5
Testing Methods for Creep

Abstract The major techniques for collecting creep data are as follows:

(a) tensile creep testing;
(b) compressive creep testing;
(c) flexural (bend) testing for creep; and
(d) impression (hardness) creep testing.

The results of the creep test are plotted as strain versus time to obtain a curve characterizing it. Often a creep power law relates creep strain to the applied stress. In ceramics a creep test at constant load and temperature is generally performed at prolonged times, because of the bond character in ceramics. Creep tests are often performed by compressive loading to eliminate growth of cavities and their opening and thus creep rate is slower compared to tension. In ceramics which are inherently brittle flexural tests are preferable since machining of test specimens is difficult and also because the tendency to break in the grips when test is performed by tension.

With the growing use of ceramics in industry (for numerical and scientific applications) comes the increasing demand for the characterization and quantification of their properties. This may be achieved by improving the testing techniques, in order to yield more exact information, as required for design purposes, safety analyses, quality control, and basic scientific understanding. In this chapter, the various ceramic creep testing methods are discussed, with a focus on the relative advantages and disadvantages of each type of test. The major techniques for collecting creep data are as follows:

(a) tensile creep testing;
(b) compressive creep testing;
(c) flexural (bend) testing for creep; and
(d) impression (hardness) creep testing.

© Springer International Publishing AG 2017   63
J. Pelleg, *Creep in Ceramics*, Solid Mechanics and Its Applications 241,
DOI 10.1007/978-3-319-50826-9_5

5.1   Tensile Creep Testing

Nowadays, special machines may be purchased for conducting tensile creep experiments and tests. Generally, a creep test is performed using a tensile specimen, to which a constant stress is applied. Surrounding the specimen is a thermostatically controlled furnace. The temperature is controlled by a thermocouple attached to the specimen, usually in the gage length. The extension of the specimen under load is measured with a very sensitive extensometer. Modern tensile testing machinery is equipped with a high-temperature furnace and has an integrated slide table for use with a non-contacting laser extensometer. A creep tester of this kind is illustrated in Fig. 5.1.

The results of the creep test are plotted as strain versus time to obtain a curve, as schematically illustrated in Fig. 1.1a. An experimental curve, resembling Fig. 1.1a, is shown for Si_3N_3 in Fig. 5.2. The relation between the steady-state creep rate and the stress used in analyzing the creep data in Si_3N_4 is given in Eq. (4.19) and is reproduced here as

$$\dot{\varepsilon} = A(\sigma, T, S, e)\sigma^n \exp - \left[\frac{\Delta H_c(\sigma, T, S, e)}{RT}\right]. \tag{4.19}$$

Fig. 5.1 Special Series 2330 Lever Arm Creep Tester designed for tensile testing of ceramic specimens. Features include a high-temperature furnace and integrated slide table for use with non-contacting laser extensometer. With kind permission of R. Antolik of Applied Test Systems Inc.

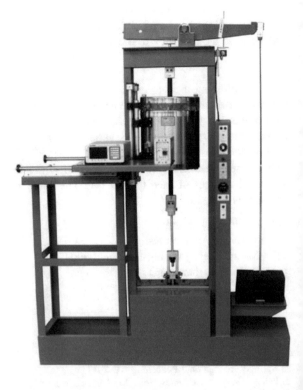

Fig. 5.2 Strain–time curves, hot-pressed Si_3N_a, creep tests in He. Kossowsky et al. [9]. With kind permission of Springer

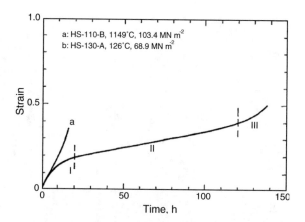

At elevated temperatures, $> \sim 1000$ deformation is controlled by GBS. No evidence of deformation within the grains was observed in Figs. 5.3 and 5.4. The micrographs indicate, however, that deformation and failure during creep are a direct result of the deformation and failure of the grain-boundary glassy phase. In these micrographs, grain separation and extensive cavitation are seen.

Fig. 5.3 Replica transmission micrographs of fracture surfaces. **a** Tensile specimen, room temperature test; **b** creep specimen, $1260 \sim C$ test, near center of specimen; **c** same specimen as in **b** in rough area of fracture. Bars $\equiv 1$ μm. Kossowsky et al. [9]. With kind permission of Springer

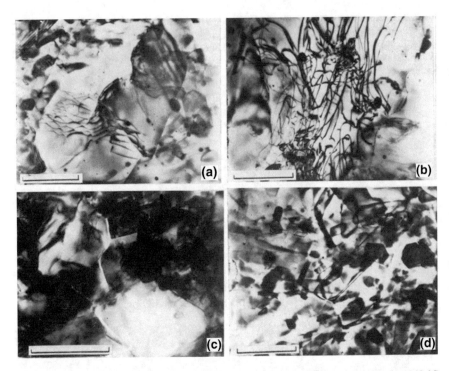

Fig. 5.4 Transmission electron micrographs of creep specimen. **a** Dislocation network, 1149 °C, 70.5 MN m^{-2}; **b** dislocation tangles, 1149 °C, 82.7 MN m^{-2}; **c** grain-boundary separations, 1260 °C, 70.5 MN m^{-2}; **d** extensive cavitation, 1260 °C, 82.7 MN m^{-2}. Bars ≡ 1 μm. Kossowsky et al. [9]. With kind permission of Springer

A number of specimen designs and gripping techniques for tensile testing have been used on ceramic materials. Various specimens may be used, such as simple, flat dog-bone-shaped, rectangular or round ones. Flat dog-bone specimens usually have holes drilled at each end of the specimen. The gripping of the specimens should have good alignment, avoiding the possibility of bending. Gripping techniques may be characterized by the temperature of the grips used to apply the load to the specimen. Thus, a hot-grip design for the loading fixture or a cold-grip design, where the tensile specimens extend outside the furnace and are gripped at or slightly above room temperature, are both in use in tensile creep tests. A major concern in tensile tests is the degree of alignment of the tensile specimen, to eliminate or reduce bending to ∼1% or less. A description of an experimental ceramic tensile creep testing technique, discussing tensile test specimens and the method of attaching them to the grips, is found in the work of Carrol et al. [2]. Bending is calculated by:

$$\%\text{bending} = \frac{\varepsilon_1 - \varepsilon_0}{\varepsilon_0} \times 100. \tag{5.1}$$

In Eq. (5.1), ε_1 is the elastic strain measured from one face of the gage section and ε_2 is the elastic strain measured from the opposite face. ε_0 is the average strain determined by $(\varepsilon_1 + \varepsilon_2)/2$.

Creep may occur at low temperatures, even at room temperatures, depending on the ceramic material. Thus for example, the most widely used piezoelectric ceramic, lead zirconate titanate (PZT), shows time-dependent deformation (creep) under both electric and mechanical loading. The primary interest here is in mechanical loading. The tensile load of a servohydraulic machine was transferred to a test piece (rectangular-shaped, with bore holes at both ends). The analysis was performed in accordance with

$$\varepsilon_c(t) = \varepsilon_{\text{total}}(t) - \left(\frac{\sigma}{E}\right) - \varepsilon_p = \varepsilon_{\text{total}}(t) - \varepsilon_{\text{total}}(0), \tag{5.2}$$

where $\varepsilon_{\text{total}}$ is the measured strain and ε_p is the plastic (time-independent) strain contribution. $E \cong 60\text{--}70$ GPa.

In Fig. 5.5, the spontaneous and time-dependent strains after loading appear in (a) and the reductions of the strains after unloading are found in (b).

Creep curves are illustrated logarithmically at various loads in Fig. 5.6.

Straight lines are observed; they are characteristic of primary creep behavior. Single creep tests are shown in Fig. 5.7 for $\sigma = 15.4$ MPa.

Stress influences creep deformation. A creep power law relates creep strain to stress, as indicated in Eq. (5.3):

$$\varepsilon_c = B\sigma^n t^m. \tag{5.3}$$

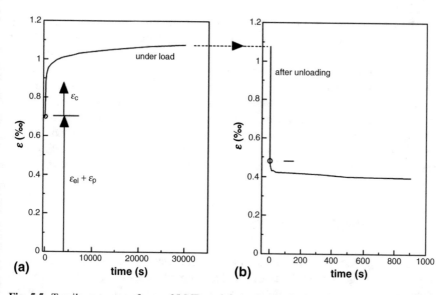

Fig. 5.5 Tensile creep curve for $\sigma = 25$ MPa: **a** deformations under load; **b** back deformations after unloading (unpoled material). The circles indicate the end of spontaneous deformations. (*Note* poled or unpoled material refers to electric loading.) Fett and Thun [4]. With kind permission of Springer

Fig. 5.6 Creep strains as a function of time (unpoled material). Fett and Thun [4]. With kind permission of Springer

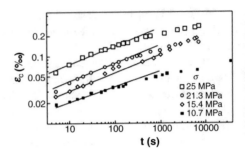

Fig. 5.7 Scatter in creep curves for a constant tensile stress. Unpoled material: *squares*; poled material: *circles*. Fett and Thun [4] With kind permission of Springer

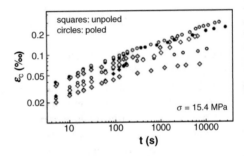

A plot, according to this relation of creep strain as a function of stress, is illustrated in Fig. 5.8, with $m = 0.27$; $n = 1.52$; $B = 2.7 \times 10^{-7}$ $(MPa)^{-n}s^{-m}$. The exponent, n, results from a plot of creep strain versus stress, as indicated for a certain time (here 120 s).

The method for determining creep properties (or stress relaxation behavior) in ceramics is to subject a ceramic specimen to prolonged, constant tension (or compression) at a constant temperature. The deformation is recorded at specified time intervals and a plot of the creep strain versus time is constructed. The slope of the curve at any point is the creep rate. If the test is carried out to failure, the time for rupture is recorded. If the specimen does not fracture within the test period, creep recovery may be measured. Stress relaxation may be measured as follows: a specimen is deformed a given amount and the decrease in stress over a long period of exposure at constant temperature is recorded. (For detailed creep testing procedure, the American Standard Test Method (ASTM) should be consulted.)

5.2 Compressive Creep Testing

Creep tests performed under uniaxial conditions, such as tension or compression, are advantageous, because the analyses of the uniform stress results may be simpler. However, in tensile stress applications, cavities have a considerable effect on creep. Materials (including ceramics) creep at much faster rates when exposed to tension, than under compression, as shown in Fig. 5.9.

Fig. 5.8 Creep strain for unpoled and poled PZT as a function of stress at $t = 120$ s. Fett and Thun [4]. With kind permission of Springer

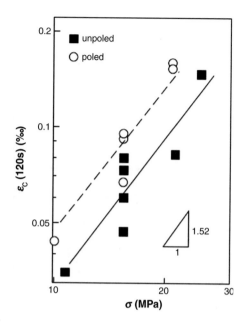

Ceramics are usually much stronger under compression than under tension. Cavitation contributes significantly to creep strain during tensile tests, while, under compressive loads, the cavitation effect is almost completely suppressed. With increased tensile creep, the volume fraction of the cavities increases linearly. Under low stresses, the tension–creep rate increases linearly, while under high stress it increases exponentially. Contrary to this observation, the creep rate increases linearly in compression creep tests. The tests at 1430 °C, shown in the above figure, were performed at stresses of 40–300 MPa. All brittle materials contain a certain population of small crack materials (or cavities) having different sizes, orientations, and geometries. The variation in these features affects the strength of the material, especially most ceramics, which are inherently brittle at low or room temperatures. Because of the variation in the features of brittle ceramics, ceramics with average strength are rarely used in design applications. The distribution of flaws is critical, since plastic deformation rarely occurs at room temperature. At high temperatures, deformation can occur. As such, to produce creep, ceramics are usually loaded under compression. The deformation of crystalline ceramics depends on the achievement of dislocation motion, which is more difficult than in metallic materials.

High-temperature creep curves tested under compression are illustrated in Figs. 5.10 and 5.11 in SiAlON. 'Syalons' are ceramics based on the elements silicon (Si), aluminum (Al), oxygen (O), and nitrogen (N).

The Syalons are the solid solution of silicon nitride (Si_3N_4) and exist in three basic forms. They are a special class of ceramics, high-temperature refractory ceramics, characterized by high strength, good thermal shock resistance, exceptional corrosion resistance, and resistance to wetting by molten ferrous metals.

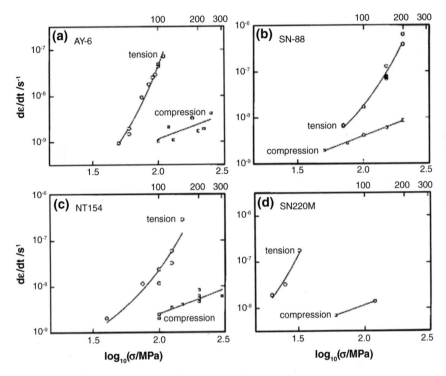

Fig. 5.9 Comparison of creep behavior of several silicon nitrides in tension and compression: **a** AY-6, a SiC-whisker-reinforced Si₃N₄ tested at 1200 °C; **b** SN-88, a gas-pressure-sintered Si₃N₄ tested at 1400 °C; **c** NT154, a HIPPed Si₃N₄ tested at 1430 °C; and **d** SN220 M, a sintered Si₃N₄ tested at 1200 °C. Luecke and Wiederhorn [10]. With kind permission of John Wiley and Sons

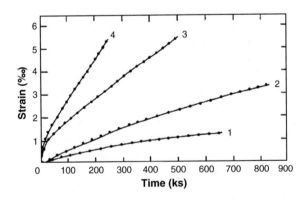

Fig. 5.10 Creep curves for Syalon 101-type material at 1573 K under compressive stresses of (*1*) 30, (*2*) 50, (*3*) 100, and (*4*) 150 MPa. The points show experimentally determined values, while the solid lines represent the fit achieved using Eq. (5.5). For clarity, less than one in ten of the measured creep strain/time reading recorded during each test are included in this figure. Wilshire [16]. With kind permission of Elsevier

Fig. 5.11 Creep curves for Syalon 201-type material at 1573 K under compressive stresses of (1) 60, (2) 80, (3) 100, (4) 140, and (5) 200 MPa. Wilshire [16]. With kind permission of Elsevier

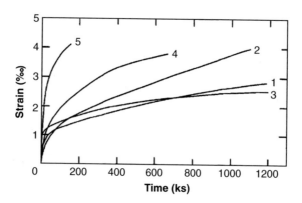

For the creep analysis of the above SiAlONs, the following equations are relevant:

$$\varepsilon = \theta_1\left(1 - e^{\beta_2 t}\right) + \theta_3\left(e^{\theta_2 t} - 1\right) \tag{5.4}$$

$$\varepsilon = \theta_1\left(1 - e^{\theta_2 t}\right) + \theta_3\theta_4 t, \tag{5.5}$$

where θ_1 and θ_3 scale the primary and tertiary stages with respect to strain, while θ_2 and θ_4 are rate parameters which quantify the curvatures of the primary and tertiary components, respectively. ε and t are strain and time, respectively. Note that there is a difference in the creep curves of the tested SiAlONs. In Syalon 101, the creep strain increased and the time to failure decreased systematically with increasing stress at the same temperature. Equation (5.4) describes the normal creep curves recorded at high temperatures with creep ductile materials, whereas Eq. (5.5) is used when only decaying primary curves are found in the experiments done on creep brittle ceramics.

When primary creep curves are analyzed according to Eq. (5.5), it is not possible to determine the tertiary parameters, θ_3 and θ_4, separately, but only their product, $\theta_3\theta_4$. Once θ_1, θ_2, and $\theta_3\theta_4$ have been obtained for each creep curve of Syalon 101, the results in Figs. 5.13 and 5.14 demonstrate that the θ parameters vary systematically with stress and temperature, i.e., the $\ln\theta_i$ values (where $i = 1, 2, 3, 4$) increase linearly with increasing stress at each temperature. Each curve, such as the one found in Fig. 5.11, is well represented by Eq. (5.5), and the θ values vary consistently with stress and temperature, as shown in Figs. 5.12 and 5.13. These θ relations provide a comprehensive description of the shapes of the individual creep curves and of the dependence of the creep curve shape on the test conditions.

The results in Fig. 5.13 show that the stress/$\ln\theta_2$ plots at different creep temperatures can be superimposed onto a single line by means of the temperature compensation of θ_2 using an Arrhenius term with an activation energy of approximately 430 kJ mol^{-1}.

Additional compressive creep stress examples of Si_3N_4 are illustrated below. Figure 5.14 compares creep in several materials.

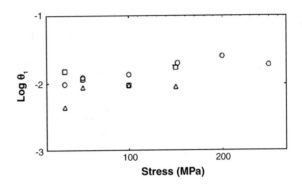

Fig. 5.12 The stress dependence of the values for $\ln\theta_1$ for the Syalon 101-type material at creep temperatures of 1598 ($\triangle$), 1573 ($\square$), and 1523 K ($\bigcirc$). Wilshire [16]. With kind permission of Elsevier

As indicated earlier, the interest in Si_3N_4-based ceramics was motivated by the expectation that they might be suitable for high-temperature applications in gas turbines, in transportation, and in power generation. Figures 5.10 and 5.14 indicate primary and secondary creep curves that are visible after initial, rapid creep, which then decreases continuously (primary creep) after loading, until steady-state creep is reached. The creep rate, as a function of stress, is shown in Fig. 5.15 for the same ceramics shown in Fig. 5.14. Some tests, those designated as 'reaction bonded silicon nitride' (RBSN) and 'hot-pressed silicon nitride (HPSN), were allowed to proceed at a constant strain rate to the tertiary (accelerated creep) stage, leading to fracture.

Equation (1.9), seen above and rewritten here, was used to analyze the stress dependence of the creep rate:

$$\dot{\varepsilon} = B\sigma^n \exp-\left(\frac{Q}{kT}\right). \tag{1.9}$$

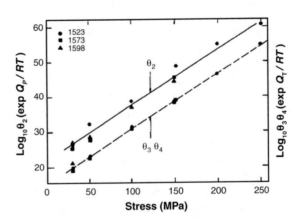

Fig. 5.13 Rationalization of the stress $\ln\theta_2$ and stress/$\ln\theta_3\theta_4$ relationships observed for Syalon 101-type material at 1598 ($\blacktriangle$), 1573 ($\blacksquare$) and 1523 K ($\bullet$). For both the θ_2 and $\theta_3\theta_4$ plots, temperature compensation was achieved through incorporation of an Arrhenius term with activation energies of about 433 kJ mol^{-1}. Wilshire [16]. With kind permission of Elsevier

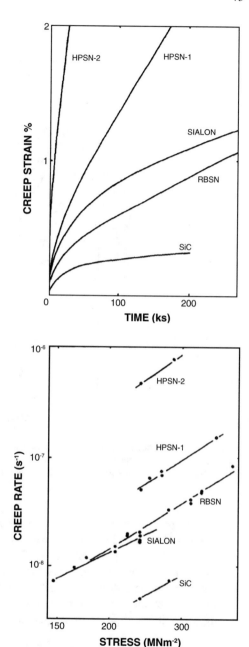

Fig. 5.14 Compression creep curves recorded at 238 MN m^{-2} and 1623 K for samples of reaction-bonded and hot-pressed silicon nitride, a SiAlON ($z = 1$) and sintered silicon carbide. The hot-pressed sample (designated HPSN-1) contained 2% MgO, compared with $\sim$5% MgO for the material labeled HPSN-2. Birch and Wilshire [1]. With kind permission of Springer

Fig. 5.15 Stress dependence of the secondary creep rate ($\dot{\varepsilon}_s$) for samples of reaction-bonded and hot-pressed silicon nitride, a SiAlON ($z = -1$) and sintered silicon carbide for compression creep tests carried out at 1623 K. (Fig. 5.1). Birch and Wilshire [1]. With kind permission of Springer

In Fig. 5.15, this relation is expressed graphically.

The temperature dependence of the creep rate is shown in Fig. 5.16. The stress exponent taken from Eq. (1.9) was 2.1, 2.2, 2.3 and 2.4 for the investigated ceramics. The activation energy for creep, Q, was $\sim$650 kJmol^{-1} for both the

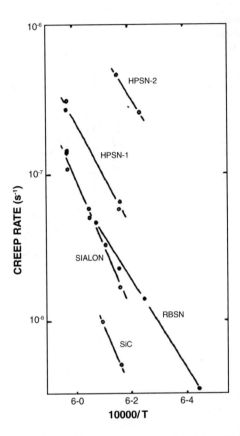

Fig. 5.16 Temperature-dependence of the secondary creep rate ($\dot{\varepsilon}_s$) for samples of reaction-bonded and hot-pressed silicon nitride, a SiAlON ($z = 1$) and sintered silicon carbide for compression creep tests carried out at 238 MN m^{-2}. Birch and Wilshire [1]. With kind permission of Springer

RBSN and the HPSN samples, and somewhat higher for the SiC and SiAlON specimens—730 and 850 kJmol^{-1}, respectively. As repeatedly indicated, several variables influence creep strength, even when tested under the same conditions. The most important of these variables are the impurity level, the amount of pores, pore size, pore shape, pore distribution, and the experimental atmosphere.

According to the Norton–Bailey equation for stain hardening/strain softening during creep, it is a function of temperature, stress, and creep strain, given as

$$\dot{\varepsilon}_{cr} = K(T) \cdot \sigma^n \cdot \varepsilon_{cr}^a. \tag{5.6}$$

In the above equation, K is a function of temperature, "n" is the stress, and "a" is a creep-strain exponent. In primary creep strain, hardening occurs and "a" is negative, while it is positive in tertiary creep and may be associated with strain softening. Steady-state creep (secondary creep) is represented by $a = 0$. The measured strain is elastic (creep deformation). In order to obtain the creep strain, the elastic (instantaneous) strain should be subtracted from the total strain measured, giving

$$\varepsilon_{cr} = \varepsilon_{tot} - \frac{\sigma}{E}. \tag{5.7}$$

The contribution of the elastic strain (the second term) is obtained from the ratio of the applied stress and Young's modulus. To learn the method for identifying the parameters of the Norton–Bailey creep law, the reader may consult the work of Jin et al. [8] The end result is given by

$$\varepsilon_{cr,i+1} = \left[\varepsilon_{cr,i}^{1-a} + \frac{(1-a) \cdot K \cdot \left(\sigma_{i+1}^n + \sigma_i^n \right) \cdot (t_{i+1} - t_i)}{2} \right]^{1/1-a}. \tag{5.8}$$

In the subscripts $i + 1$, the i of strain and stress refers to the respective time steps.

Magnesia–chromite bricks (56.6 wt%MgO, 25.5 wt%Cr$_2$O$_3$) were used for these creep measurements and analyzed in Eqs. (5.6)–(5.8). Figures 5.17, 5.18 and 5.19 illustrate creep curves at various temperatures and stresses.

The smooth lines in Figs. 5.17, 5.18 and 5.19 illustrate the inverse estimation of the results. The curves in Figs. 5.17, 5.18 and 5.19 indicate that they are primary-stage creep curves, since the values of "a" are negative at all the test temperatures. A typical three-stage curve is obtained at 1400 °C under a load of 9 MPa, as illustrated in Fig. 5.20. In Fig. 5.21, the normalized parameters are shown versus the number of iterations. Note the poor initial creep law parameters given (K is 0.1 MPa^{-n} s^{-1}; n is 20; and "a" is -20). A steep decrease of the residual is observed for the first 12 iterations and, after 20 iterations, the optimized creep law parameters are found with sufficient accuracy.

The Norton–Bailey creep parameters are listed in Table 5.1.

In the case of secondary creep, the exponent of creep strain, "a," was zero and only the value of $K_2\sigma^{n2}$ could be inversely calculated from this curve. At the third

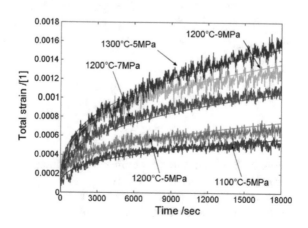

Fig. 5.17 Total strain/time curves of burnt magnesia–chromite bricks from experiments (fluctuating ones) and inverse estimations (smooth ones) at 1100–1300 °C. Jin et al. [8]. With kind permission of Elsevier

Fig. 5.18 Total strain/time curves of burnt magnesia–chromite bricks from experiments (fluctuating ones) and inverse estimations (smooth ones) at 1400–1550 °C. Jin et al. [8]. With kind permission of Elsevier

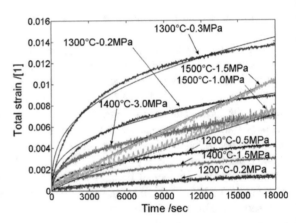

Fig. 5.19 Total strain/time curves of alumina castable from experiments (fluctuating ones) and inverse estimations (smooth ones) at 1200–1500 °C. Jin et al. [8]. With kind permission of Elsevier

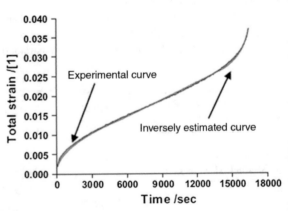

Fig. 5.20 A total strain/time curve of burnt magnesia–chromite bricks from the experiment and inverse estimation including three creep stages at 1400 °C under 9 MPa. Jin et al. [8]. With kind permission of Elsevier

stage, both a_3 and $K_3\sigma^{n_3}$ were inversely estimated. The corresponding creep law parameters are listed in Table 5.2.

In summary, it is possible to state that the measurements reveal that, at elevated loads, all three creep stages may be observed. Besides increasing the load, the

Fig. 5.21 The convergence behavior of the Levenberg–Marquardt method for inverse identification of the creep parameters. Jin et al. [8]. With kind permission of Elsevier

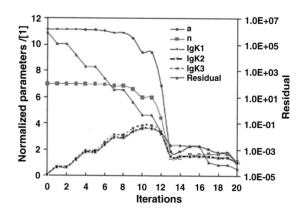

Table 5.1 Norton–Bailey creep law parameters of burnt magnesia chromite bricks corresponding to different temperatures. Jin et al. [8]. With kind permission of Elsevier

$T(C)$	L–M	GRG	L–M	GRG	L–M	GRG
	$K(MPa^{-n}S^{-1})$		n		a	
1100	1.18×10^{-16}	0.71×10^{-16}				
1200	2.77×10^{-16}	1.72×10^{-16}	2.86	2.91	-1.80	-1.85
1300	2.33×10^{-15}	1.47×10^{-15}				
1400	1.24×10^{-11}	–				
1500	2.20×10^{-10}	–	3.20	–	-1.04	–
1550	7.27×10^{-9}	–				

L–M Levenberg–Marquardt method
GRG Generalized Reduced Gradient method

Table 5.2 Norton–Bailey creep law parameters of three stages of burnt magnesia–chromite bricks. Jin et al. [8]. With kind permission of Elsevier

Creep law parameter	a_1	$K_1\sigma^{n1}(s^{-1})$	a_2	$K_2\sigma^{n2}(s^{-1})$	a_3	$K_3\sigma^{n3}(s^{-1})$
Value	-1.04	1.44×10^{-8}	0	1.47×10^{-6}	6.3	1.58×10^4

temperature and loading time may also produce the three stages of creep. In addition, the effect of certain impurities on the stages of creep is a significant factor for improving creep resistance.

5.3   Flexural (Bend) Tests

Tensile testing for creep in ceramics is problematic, especially in brittle ceramics (and most ceramics are brittle), because

(a) specimen preparation is difficult, due the absence of plasticity and, thus, unattainable by means of common machining procedures; and

(b) any small misalignment of the specimens in the grips may cause fracture in the vicinity of the grips, due to eccentricity.

After concerted attempts to solve the problem of ceramic tensile testing using pin-loaded specimens and a laser extensometer technique, a relatively uniform agreement about the reliability of tension tests was reached; nevertheless, the preferred and most reliable tests are by compression or bending (flexural) tests. Flexural-creep tests have several advantages:

(a) easy preparation of the specimens;
(b) relative ease of experimental design; and
(c) low costs, as a consequence of (a)–(b).

Therefore, many creep experiments employ flexure tests. There are two common testing methods, depending on the loading method. Three- and four-point bending tests are in general use, as indicated in Fig. 5.22.

The specimens are rectangular and without notches. The four-point bend setup is illustrated in Fig. 5.22 for two cases: for the loading spans of L/2 and L/3. The applied force (downward arrows) is compressive by nature, resisted by the tensional force (upward arrows). Thus, the longitudinal stresses at the lower surfaces (convex) in the specimens are tensile and compressive at their upper surfaces (concave). As a consequence, a calculable bending moment develops. A large variety of machines are available for flexural tests, such as MTS, Instron, Universal Testing Machine, etc. It is reasonable to believe that stress and strain are proportional to the distance from the neutral axis. The neutral axis is shown in the schematic specimen of Fig. 5.23 (at half of h).

The following relations apply to the L/2 span (see Figs. 5.23 or 5.22a). The load is clearly applied at L/2, expressed as

$$\sigma_f = \frac{3P(L - L_i)}{2tc^2}.$$
(5.9)

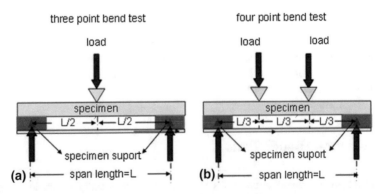

Fig. 5.22 Schematic bend test configurations: **a** three point, **b** four point. Pelleg [11]

Fig. 5.23 Schematic bend test, rectangular bar. Pelleg [11]

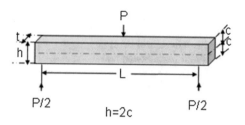

$P/2$ $P/2$

$h=2c$

Often in the literature (e.g., Ponraj and Iyer) [12], this relation is given along with Terwilliger et al. [13] as

$$\sigma = \frac{3}{2} \frac{(L-a)F}{bh^2}. \qquad (5.10)$$

Here, $\sigma(\equiv \sigma_f)$; L is the distance between the support points; $F (\equiv P)$ is the applied load; $a (\equiv L_i)$ is the distance between the load points; $b (\equiv t)$ is the width; and $h (\equiv c)$ is the height of the bar. Following the beam's elastic deformation, the strain, ε, in the outermost fiber from the measured deflection is

$$\varepsilon = \frac{6hx}{(L-a)(L+2a)}, \qquad (5.11)$$

where x is the deflection. Differentiating Eq. (5.11) with respect to time results in

$$\dot{\varepsilon} = \frac{6h\dot{x}}{(L-a)(L+2a)} = Kh\dot{x}. \qquad (5.12)$$

This equation is often used to calculate the strain rate, $\dot{\varepsilon}$, from the measured deflection rate at load points. A typical creep curve is shown in Fig. 5.24, and the four-point creep test specimen, with and without the deflection, appears in Fig. 5.25.

Equation (5.9) may be obtained by considering Fig. 5.23 as follows (Pelleg) [11]. The fracture stress, σ_f, is determined by

$$\sigma_f = \frac{Mc}{I} \qquad (5.13)$$

$$I = \frac{2tc^3}{3}. \qquad (5.14)$$

Replacing I in Eq. (5.13), it may be rewritten as

$$\sigma_f = \frac{3M}{2tc^2}, \qquad (5.15)$$

where M is the bending moment, c is half the specimen width, t is the thickness, and I is the moment of inertia of the cross-sectional area. Lists of the moments of inertia

Fig. 5.24 Typical creep curve of siliceous porcelain tested at 850° C at a stress level of 38 MPa. Ponraj and Iyer [12] With kind permission of Springer

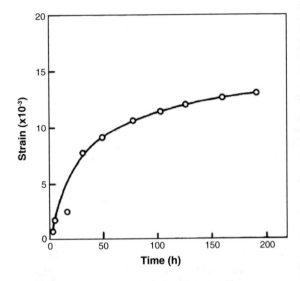

Fig. 5.25 Typical four-point bend-creep test specimens: (*bottom*) before testing and (*top*) after testing. Ponraj and Iyer [12] With kind permission of Springer

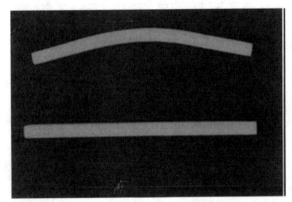

of plane figures and areas are found in the literature and also in the appendix of Timoshenko's [14] book. Basically, the plane under consideration is divided into small pieces and the contribution of each individual piece to the moment of inertia is evaluated by integration:

$$I = \frac{bh^3}{12} \equiv \frac{t(2c)^3}{12} = \frac{2tc^3}{3}. \tag{5.16}$$

Since I is the moment of inertia of the cross-sectional area, expressing the moment as the force times the lever allows Eq. (5.13) to be modified as

$$\sigma_f = \frac{2P\frac{L}{2}c}{\frac{2tc^3}{3}} = \frac{3PL}{2tc^2}. \tag{5.17}$$

Below, a method for evaluating the inertia is presented:

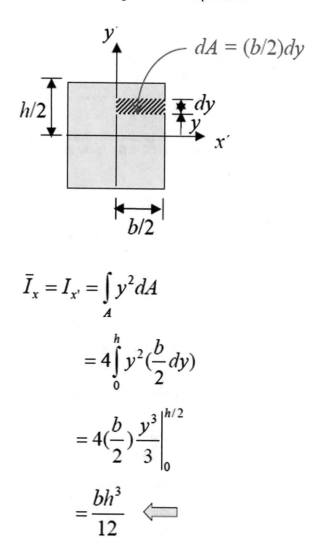

$$\bar{I}_x = I_{x'} = \int_A y^2 dA$$

$$= 4\int_0^h y^2 (\frac{b}{2} dy)$$

$$= 4(\frac{b}{2}) \frac{y^3}{3}\Big|_0^{h/2}$$

$$= \frac{bh^3}{12} \quad \Longleftarrow$$

Note the final relation for the inertia obtained above and shown in Eq. (5.16). When considering the notation in Fig. 5.23, Eq. (5.14) may be obtained. Both values for inertia, $\frac{bh^3}{12}$ and $I = \frac{2tc^3}{3}$ from Eq. (5.14) are indicated in Eq. (5.16).

Equation (5.17) gives the flexural strength for the three-point test of a rectangular bar. In the above relation, a force is acting on a lever of size $L/2$ ($1/2$ of the bar at the support) and this force is supported or balanced at the two supporting points marked by the arrows close to the ends of the rectangular bar (i.e., $M = PL$ force x arm), which yields the same answer as given in Eq. (5.17).

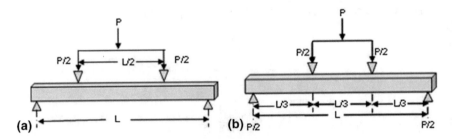

Fig. 5.26 Rectangular beams in a four-point bending test: **a** loading span $L/2$; **b** loading span $L/3$. *Note* that the loading span may be different from $L/2$ or $L/3$. In that case, it is customary to denote the load span as L_i

A four-point bend test setup is illustrated in Fig. 5.26 for two cases: for the loading spans at $L/2$ and at $L/3$.

The following relations apply to the $L/2$ span, using Eq. (5.15) with the appropriate substitution for M as

$$\sigma_f = \frac{3M}{2tc^2} = \frac{3P\frac{L}{2}}{2tc^2} = \frac{3PL}{4tc^2}. \tag{5.18}$$

Using Eq. (5.15) again, the L/3 span, shown in Fig. 5.26b, may be written as

$$\sigma_f = \frac{3M}{2tc^2} = \frac{3P\frac{L}{3}}{2tc^2} = \frac{PL}{2tc^2}. \tag{5.19}$$

In the general case, when the loading span is different from $L/2$ or $L/3$ in a four-point bend test, the stress is given as

$$\sigma_f = \frac{3P(L - L_i)}{2tc^2}. \tag{5.20}$$

Equation (5.20) is obtained in a manner similar to other bend test relations, namely

$$\sigma_f = \frac{3M}{2tc^2} = \frac{3P(L - L_i)}{2tc^2}. \tag{5.21}$$

The equations expressing the flexural strength, σ_f, actually represent the highest stress of the ceramics at the time of rupture. While tension or compression tests of metals are commonly used to characterize and development new materials for design purposes, bend tests of ceramics are the preferred test method. The flexural strength of a ceramic is dependent on its inherent properties, especially flaws and crack sizes (common features in ceramics). Variations in crack size, crack distribution, and the nature of such cracks cause a natural scatter in test-sample results, requiring the testing of several test specimens in order to get a statistical value for the inherent flexural strength.

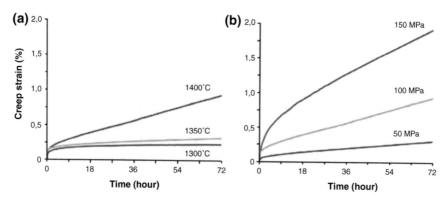

Fig. 5.27 Creep behavior of as-sintered α/β-SiAlON composite as a function of temperature **a** and stress **b**. Uludag and Turan [15] With kind permission of Elsevier

Note the relation between three and four-point bend tests and test specimen size. It is expected that the specimens having a larger volume will show a lower modulus of rupture than smaller sized specimens, since there is a higher probability that more defects (microcracks, for example) will exist in larger specimens. Therefore, test specimens must be standardized. The lower modulus of rupture in four-point bend tests, as opposed to three-point bend tests, is a consequence of the size effect.

It was indicated above that various creep testing facilities are available on the market. An Instron 5581 testing machine was used to obtain the creep curves shown in Fig. 5.27.

In Fig. 5.28, the additional flexural creep of Si_3N_4 at 1400 °C under 100 MPa is compared to a bend test of SiAlON under the same conditions. In Figs. 5.27 and 5.28, flexural-creep strain–time curves, obtained in air at a 1300–1400 °C temperature range and under stress levels ranging from 50 to 150 MPa, are illustrated.

Recall that the four-point bending-creep method involves supporting a test bar on two supports near its ends, heating it to the required, constant, elevated temperature, while applying a force to two symmetrically spaced loading points located between the support points, and then recording the deflection of the test bar over time (see Fig. 5.26a).

The steady-state creep rates were analyzed by Norton's equation, given as previously indicated and rewritten here as

$$\varepsilon_{ss} = A\sigma^n \exp\left(\frac{-Q}{RT}\right). \tag{5.22}$$

Equation (5.22) was used to obtain the stress and temperature dependences of the creep strain versus time curves (Figs. 5.27 and 5.28). Obviously, the aforementioned parameters in Eq. (5.22) are A the constant; σ the stress; Q the apparent activation energy; and n the stress exponent. By linear fitting of the strain rate versus stress, the stress exponent, n, is determined to be $n = (1.6 \pm 0.13)$ under the applied stress in the 50–150 MPa range (see Fig. 5.29a). From the plot of the strain

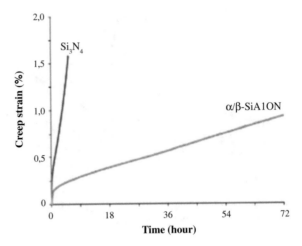

Fig. 5.28 Creep behavior of as-sintered α/β SiAlON composite, compared to Si_3N_4 at a temperature of 1400° C under 100 MPa. Uludag and Turan [15] With kind permission of Elsevier

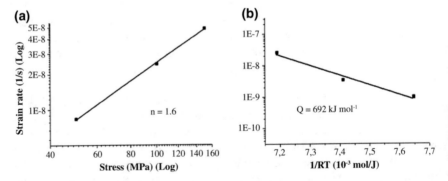

Fig. 5.29 a Stress and **b** temperature dependence of the steady-state creep rate of as-received α/β–SiAlON composite at a constant temperature of 1400 °C and under a constant stress of 100 MPa. Uludag and Turan [15]. With kind permission of Elsevier

rate versus $1/RT$, the activation energy of creep under a flexural load was determined to be in the 1300–1400 °C temperature range as (692 ± 37) kJmol^{-1}. This is shown in Fig. 5.29b.

5.4   Indentation (Hardness) Tests

As stated above, ceramics are increasingly applied where good wear, high strength, and creep resistance are prerequisite material properties. Long-term behavior over a relatively long operating lifetime is expected from these materials and, therefore, their evaluation by appropriate tests is mandatory. Indentation hardness technique, under sufficiently prolonged loading, is perhaps one of the simpler and less-expensive methods for collecting data to support estimations of the useful working lives of ceramics under loads (high stress) either at low or high temperatures. It is worth mentioning that various instrumented facilities are commercially available for the performance of indentation tests. Usually, microhardness tests or nanoindentations are performed on ceramics to eliminate crack development under the load.

The term 'microhardness testing' usually refers to static indentations made by loads of 1 kgf or less. The "Baby Brinell" hardness test uses a 1-mm carbide ball, while the Vickers hardness test employs a diamond with an apical angle of 136°, and the Knoop hardness test uses a narrow rhombus-shaped diamond indenter. In most cases, the test surface must be highly polished. The smaller the force applied, the higher the required metallographic finish. Microscopes with a magnification of around 500× are required to accurately measure the indentations produced. Furthermore, note that microhardness in ceramics is employed in cases where a "macro"-hardness test is not possible. Testing microhardness may also be useful when ceramic coatings must be evaluated for creep resistance. In fact, high-temperature tests are often referred to as 'hot-hardness tests.'

Microhardness is measured by taking the depth of the penetration of the indentor as a function of load. This enables the determination of the deformation under the load of indentations under increasing, fixed (stable), and decreasing loads over time. A schematic diagram of the variation of the load over time is shown in Fig. 5.30. This load increased monotonically up to 5 N, while the penetration depth of the indentor was measured as a function of that load, reaching a value of 3.19 mm. Then, at a fixed load of 5 N, the depth of penetration of the indentor (Δh) was measured as a function of time (see Fig. 5.30b). The load and the penetration depth of the indentor were registered on an X–Y recorder. Measurements were made at the start of the creep and the depth of penetration was measured after 10, 20, 30, and 60 s and, subsequently, every 60 s up to 600 s (see the enlargement taken from Fig. 5.30b). A Vickers pyramid indentor was used, and the unrelaxed microhardness was calculated by

$$H_V = \frac{0.3784P}{h^2},\tag{5.23}$$

where P is the load (N) and h is the measured depth of penetration (mm).

This relation is a consequence of the following (Pelleg) [11]. Hardness, H, is defined as the ratio of the applied load to the projected area of indentation and is generally expressed as

$$H = \alpha \frac{P}{d^2},\tag{5.24}$$

where d is the size of the measured impression with α, the indenter constant, taking the indenter geometry into consideration. The Vickers indenter has an angle of $\phi = 136°$ between the two opposite faces. The Vickers hardness is defined as the load divided by the surface area of the pyramid-shaped indentation (impression).

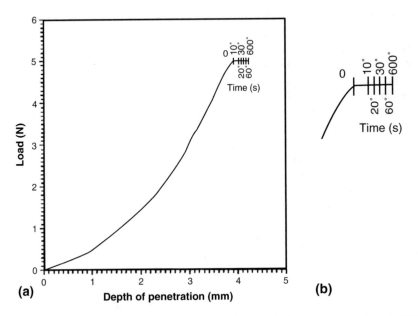

Fig. 5.30 Schematic diagram of the experiment: the load on the indentor increases up to 5 N; beginning from the point 0, the load is constant and the indentor penetrates into the material during the period of time **a**. Expanded portion at the 5 load, indicating the time in seconds. Yurkov [17]. With kind permission of Springer

This area is simple to evaluate from the geometry of the shape of the indentation, which requires measuring the diagonals and using the known angle between the two opposite faces. These two diagonals are measured on the screen of the Vickers tester and their average is used in the diamond pyramid hardness (DPH) formula. The area of the sloping surface of the indentation is calculated as shown in Fig. 5.31 and the steps are also shown for deriving the expression for the DPH measurements.

Thus, DPH is given (see Fig. 5.31) using the sine of the half angle of the Vickers indenter as

$$\text{DPH} = \frac{2P\,\sin(\phi/2)}{d^2} = \frac{1.854P}{d^2}. \tag{5.25}$$

The 0.02 mm increase in the load of the experimental tests and, recalling that $h \equiv d$, yields the following results:

$$H_V = \frac{0.3784P}{h^2} = \frac{0.3784P}{d^2}, \tag{5.26}$$

which is equivalent to Eq. (5.23).

The ceramics investigated were Si_3N_4 and AlN and the increase in the penetration depth of the indentor is illustrated for both in Fig. 5.32.

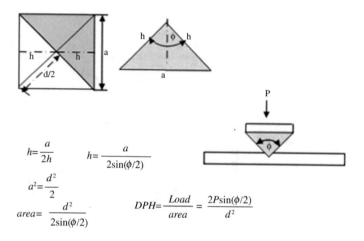

$$h = \frac{a}{2h}$$

$$h = \frac{a}{2\sin(\phi/2)}$$

$$a^2 = \frac{d^2}{2}$$

$$area = \frac{d^2}{2\sin(\phi/2)}$$

$$DPH = \frac{Load}{area} = \frac{2P\sin(\phi/2)}{d^2}$$

Fig. 5.31 Vicker's indentation. Reference [11].

Creep under nanoindentation is shown for SiCNs in Fig. 5.33. These plots represent the averaged data from at least five individual measurements, after discarding the data that showed large deviations from the average behavior. For convenient comparison of the displacements values, the initials h and t of all curves have been aligned to zero.

Here, creep is a consequence of plastic deformation and strain hardening. Details of the analysis in terms of the power law

$$\sigma = b\dot{\varepsilon}^m \tag{5.27}$$

Fig. 5.32 Plots of relative depth of penetration (h_t/h_o) versus time, where h_t is the depth of penetration at a time t and h_0 is the initial depth of penetration. (*) Si_3N_4-1 at a fixed load of 5 N, (O) Si_3N_4-2 at a fixed load of 5 N, (□) Al N at a fixed load of 5 N, (+) Si_3N_4-1 at a fixed load of 2 N and (x) Si_3N_4-2 at a fixed load of 2 N. Yurkov [17]. With kind permission of Springer

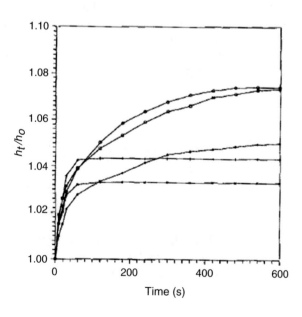

Fig. 5.33 Plots of hold-time creep displacement from Berkovich nanoindentation of Si–C–N ceramics. Closed and open symbols represent data from constant load hold segments at P_{hold} = 525 and 500 mN, respectively. Janakiraman and Aldinger [7]. With kind permission of John Wiley and Sons

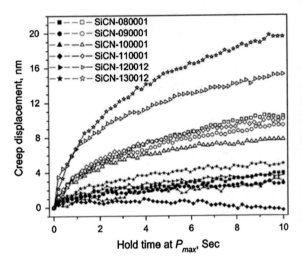

expressed in terms hardness (indentation) as

$$H = b\dot{\varepsilon}^m \tag{5.28}$$

may be found in the communications of Janakiraman; Han and Tomozawa [6]; and Grau et al. [5].

In Eqs. (5.27) and (5.28), σ is the flow stress, $\dot{\varepsilon}$ is the strain rate, and m is the strain rate sensitivity. In Eq. (5.28), H has been set equal to the flow stress, σ, and uses the strain rate relation.

Materials that are relatively strong at low (room) temperatures may fail at high temperatures over time, as is observed in the case of creep, for example. Failure due to the time-dependent deformation (creep) of a material, when subjected to a constant load or stress, is accelerated at high temperatures. There is another simple, low-cost method for measuring hardness (in order to evaluation strength) that may be used instead of tensile, compressive, or flexure tests. The Hall Petch equation, which relates strength to the grain size in a specimen, is rewritten here as

$$\sigma = \sigma_0 + \frac{k}{\sqrt{d}}, \tag{5.29}$$

where σ_0 and k are the material constants. This relation may also be adopted to assess hardness:

$$H = H_0 + \frac{k}{\sqrt{d}}. \tag{5.30}$$

The strength and hardness of materials decrease as the temperature increases. Therefore, in many applications involving high temperatures, it is critical to know the material properties at the service temperatures, particularly for predicting the service lifetime. 'Hot hardness' measurements are simpler, require smaller specimens, and are more cost effective than performing high-temperature tension, compression, or flexural tests.

Hot microhardness may be measured by a special tester—a Nikon *QM* Hot Microhardness Tester—as is used, for instance, at Arkansas University. This tester enables the observation and microhardness measurement of a variety of materials, such as metals, alloys, ceramics, composites, and even coatings at any temperature ranging from room temperature to about 1200 °C. It has been used for conducting tests on carbide cutting tool samples. Vickers hardness tests were conducted on the samples with a 500 g load at temperatures ranging from room temperature to 800 °C. These studies demonstrated that the hot microindentation technique is a very sensitive method for detecting and defining deformation mechanisms in structural materials at high temperatures. Figure 5.34 illustrates plots of hardness versus $1/T$.

The deformation of materials at elevated temperatures is often described by a phenomenological equation in the following form:

$$\dot{\varepsilon} = A\sigma^n \exp\left(-\frac{Q}{RT}\right),$$ (5.31)

where Q is the activation energy of deformation expressing the creep rate; A and n are the material constants. Since the rate of loading, $\dot{\varepsilon}$, of the indenter in a hardness test is preset to a constant value, and considering the proportionality between hardness and applied stress, Eq. (5.31) may be modified as (Fig. 5.34)

$$H_v^n \propto A \exp\left(\frac{Q_H}{RT}\right),$$ (5.32)

where H_V stands for Vickers hardness. Q_H is the activation energy needed for deformation by indentation (hardness), which is usually determined from the slope of the hardness versus $1/T$ plot.

Fig. 5.34 Plot of Vickers hardness against inverse of absolute temperature. The change in the slope of the lines suggests a change in the dominant deformation mechanism as temperature is increased. Reference [3].

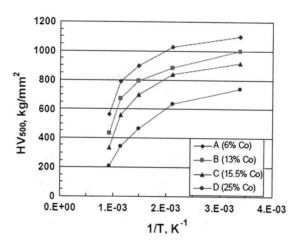

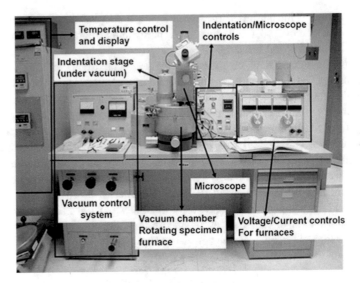

Fig. 5.35 Hot microhardness tester. Downloaded from the brochure of 2016 Case Western Reserve University, Advanced Manufacturing Center

Like other instrumented equipment for performing mechanical tests, hot indentation (hardness) instrumented facilities are also commercially available for application in creep studies. Figure 5.35 illustrates such a hot microhardness tester.

Such hot hardness units enable the observation and measurement of a variety of materials, such as metals, ceramics, and composites at temperatures ranging from room temperature to ~ 1200 °C. The equipment includes a power source and a vacuum control module. The facility allows for the setting of the indentation time, and control of the temperature, indenter furnace, and vacuum chamber. The various modules in the unit are detailed in Fig. 5.35

References

1. Birch JM, Wilshire B (1978) J Mater Sci 13:2627
2. Carrol DF, Wiederham SM, Roberts DE (1989) J Am Ceram Soc 72:1610
3. Chandelkar AA, Bhat DG (2005) In: Proceedings of the 2005 midwest section conference of the american society for engineering education, ASEE, Washington, DC, p 11
4. Fett T, Thun G (1998) J Mater Sci Let 17:1929
5. Grau P, Berg G, Meinhard H, Mosch S (1998) J Am Ceram Soc 81:1557
6. Han W-T, Tomozawa M (1990) J Am Ceram Soc 73:3626
7. Janakiraman N, Aldinger F (2010) J Am Ceram Soc 93:821
8. Jin S, Harmuth H, Gruber D (2014) J Eur Ceram Soc 34:4037
9. Kossowsky R, Miller DG, Diaz ES (1975) J Mater Sci 10:983
10. Luecke WE, Wiederhorn SM (1999) J Am Ceram Soc 82:2769
11. Pelleg J (2014) Mechanical Properties of Ceramics. Springer, Berlin, p 79

12. Ponraj R, Iyer SR (1992) J Mater Sci Let 11:1000
13. Terwilliger GR, Bowen HK, Gordon RS (1970) J Am Ceram Soc 53:271
14. Timoshenko S (1983) Strength of Materials Part 1, 3rd edn. Van Nostrand, New York
15. Uludag A, Turan D (2011) Ceram Int 37:921
16. Wilshire B (1993) Mater Design 14:39
17. Yurkov AL (1993) J Mater Sci Let 12:767

Chapter 6
Creep in Nanoceramics

Abstract Modern interest in nanosized specimens is a result of their possessing outstanding strength, superior hardness, and good fatigue resistance. This chapter is unique since no other book discusses the mechanical properties—including creep—of nanoceramics. All the tests performed in macro-sized material are used to evaluate the properties of nanosized ceramics. Thus results were collected by tension tests, compressive test, flexural, and hardness tests. The unique properties of nanoceramics are discussed in this chapter.

6.1 Introduction

Nanoceramics have been extensively studied over the past decades. The interest in this subject is associated with the many important properties of nanomaterials, in general, and of nanoceramics, in particular. One very important property exhibited by many nanomaterials is superplasticity (discussed later on in Chap. 8), which is probably due to the very small grain size (submicron, often in the 50 nm range). The consolidation techniques of the raw materials and their production methods generally determine the mechanical, physical, and other properties of the manufactured ceramics. This is especially true in the case of nanoceramics, since the distribution of the phases in composite ceramics, for example, and their location relative to the commonly found microcracks are decisive factors in obtaining strong, fracture-resistant substances. Nanoceramics show excellent physical, chemical and mechanical properties which generally desirable for various technological applications.

With regard to the study of creep and mechanical properties, nanoceramics are known to possess outstanding strength, superior hardness, and good fatigue resistance (Kuntz et al.; Mukhopadhyay and Basu; and Andrievski and Glezer). Since the properties of solids (and nanoceramics are no exemption) depend on chemical composition, atomic structure and microstructure, nanomaterials may exhibit properties (among them creep) that are very different from those of conventional polycrystalline materials, depending upon their method of production. As such, glassy phases are often observed in nanocrystalline structures resulting from their

© Springer International Publishing AG 2017
J. Pelleg, *Creep in Ceramics*, Solid Mechanics and Its Applications 241,
DOI 10.1007/978-3-319-50826-9_6

production method. Significantly improved creep resistance in nanoceramics may be obtained by eliminating the glassy phases at the grain boundaries via an appropriate production method. The formation of the glassy phase is associated with the presence of oxygen; the amount of distributed oxygen determines the amount of the glassy phase. Often, the oxygen is not distributed homogeneously in the grain boundary regions. Generally, creep behavior is tested by compression, rather than tensile testing, so as to avoid cavitation by crack formation. The next section discusses the creep testing of nanoceramics, providing some technologically important exemplars.

6.2 Testing of Nanoceramics

In the past decades, much effort has been devoted to the production of serviceable nanoceramics, expected to exhibit advanced mechanical properties useful for functional applications. The prerequisite for such nanoceramic structures is that they be produced without pores, cracks, and other flaws. To this end, modern production techniques have been developed in hopes of achieving such flawless structures. Another reason for striving to produce sound nanoceramics is the expectation that some of them may exhibit ductile properties, alongside high strength (which is absent in bulk ceramics). This anticipation is due to the fact that the fundamental behavior of nanostructures is quite different from that of bulk; in very small-sized structures, surface and atomistic properties dictate their performances. Thus, the study of nanoscale ceramics is an inevitable step toward understanding experimental observations. Note that, in order to facilitate the densification of nanoceramics and to solve other production problems, additives are usually incorporated into their nanostructures.

In general, all the test results are dependent on the grain size. The strength of materials, including ceramics, increases with the decrease in grain size and also when the structures are small, as in nanocrystals. This strengthening in small-sized structures is associated with the restriction of dislocation motion. (Note that when no dislocations are involved in deformation, high strength at a level approaching the theoretical strength is required to induce strain in the test specimen). Unlike conventionally sized test specimens, in which ductility usually decreases with increased strength, nanocrystalline-sized specimens show high strength combined with good elongation. Moreover, such nanospecimens may reach high values of plasticity, which, in some ceramics, may lead to superplastic behavior before fracture. Here, some of the testing methods discussed in earlier chapters with regard to macroscale materials will be reviewed in regard to nanomaterial behaviors. The nanoceramics considered are not necessarily monolithic, since various additives are incorporated into the specimens to improve and enhance their strength properties, as reflected by their improved creep resistance.

But before discussing the various nanoceramic creep tests, a brief review on the nanostructure is in order. A frequently presented illustration representing a typical nanostructure may be seen in Fig. 6.1. Since nanocrystalline materials are typically

Fig. 6.1 Reproduced from Fig. 1 of Jiang and Weng, according to Schiøtz et al.; **a** molecular dynamic simulation of grains and grain boundary in a nano-grained copper, showing the grain boundary has finite volume concentration. Jiang and Weng [6]. With kind permission of Elsevier

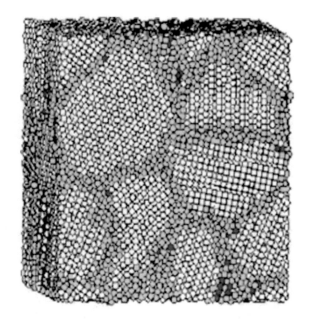

less than ~ 100 nm (1 nm = 10^{-9} m = 10 Å), their grain sizes are so small that a major part (or even all) of the microstructural volume consists of interfaces, mostly in the form of grain boundaries.

The next section on creep resistance in nanoceramics will start with the topic of compressive creep testing, which is the most common approach for avoiding premature fracture (failure) due to cavitation.

6.2.1 Compressive Evaluation of Creep

An example of the compressive testing of nanoceramics for high-temperature creep may be illustrated by yttria tetragonal zirconia (YTZ) nanocrystals (Lorenzo-Martín et al.). The usual motivation for such experiments is the desire to induce greater ductility in the ceramics and also to improve their sinterability. Another incentive is to test the possibility of obtaining superplasticity (discussed below in Chap. 8) by the reduction of grain size under the application of high strain rates. GBS is considered to be the principal mechanism during high-temperature deformation. As in the case of bulk ceramics, the results of the variation of strain over time are analyzed by the use of the high-temperature creep equation in ceramics, given as:

$$\dot{\varepsilon} = A \frac{Gb}{kT} \left(\frac{\sigma - \sigma_0}{G} \right)^n D_0 \left(-\frac{Q}{kT} \right), \tag{6.1}$$

where G is the shear modulus of the YTZ polycrystals, σ is the applied stress, and D_0 is the pre-exponential factor of the diffusion coefficient of the diffusing specimen controlling plasticity. A, n and Q are empirical parameters, where n is related to the stress exponent and Q is the activation energy. These three parameters are functions of grain size. In the above equation, n expresses the sensitivity to stress changes, while Q reflects the temperature changes. Equation (6.1) is an additional, similar, empirical equation to those indicated in earlier chapters (e.g., Chap. 4). The parameters, n and Q, define the mechanism of the creep rate, as seen in Eq. (6.1). The stress exponent may be evaluated from the stress changes at a fixed temperature according to:

$$n = \frac{\ln(\dot{\varepsilon}_2/\dot{\varepsilon}_1)}{\ln(\sigma_2/\sigma_1)} \tag{6.2}$$

$\dot{\varepsilon}_2$ and $\dot{\varepsilon}_1$ are the steady-state strain rates before and after the stress change, under identical conditions of strain and temperature. From the temperature changes under a constant load, the activation energy may be determined by:

$$\theta = \frac{kT_1T_2}{\Delta T} \ln\left(\frac{\dot{\varepsilon}_2T_2}{\dot{\varepsilon}_1T_1}\right) \tag{6.3}$$

Evaluations of n (Eq. 6.2) and Q (Eq. 6.3) were performed for nanocrystal YTZ with different contents of glassy phase. Figure 6.2 shows creep curves with 10% glassy phase in the YTZ.

Table 6.1 lists the values of n and Q with different amounts of glassy phase. The activation energies in all samples, including the pure one, are much higher than the $Q \sim 550$ kJ/mol obtained by the same authors in an earlier work in submicron-sized samples. This has been explained as follows: (a) the accommodation mechanism is different in the submicron specimens than in the nanosized ones. It involves Zr and Y cation bulk diffusion, which is no longer the case in nanoscale YTZ. Here, the accommodation mechanism is still unclear; and (b) the GBS is different in the two types of specimens; it is lower than the 5 MPa obtained from extrapolation from the measured values in the submicron specimen to the nanometric scale. The threshold stress has a1/d-dependence on the grain size in the submicrometric scale. This dependence predicts that the smaller the grain size, the higher the threshold stress will be. In YTZ, the threshold stress is as high as 160 MPa for d of ~ 50 nm.

The microstructures of YTZ, before and after deformation, with 5% glassy states are shown in Fig. 6.3.

Apparently, no changes in grain size or grain shape are observed between the as-received and the deformed specimens. Furthermore, neither cavitation, nor dislocation activity are detected; thus, the microstructure does not seem to have changed. Yttrium segregation to grain boundaries is observed in all the samples to a level of 0.7 ± 0.5 mol% in the pure YTZ polycrystalline samples and to 0.9 ± 0.5 in the impure ones.

The threshold stress affects the strain exponent, n, and if sufficient deviation from $n \sim 2$ exists it could be measured. The measured stress exponent is given by:

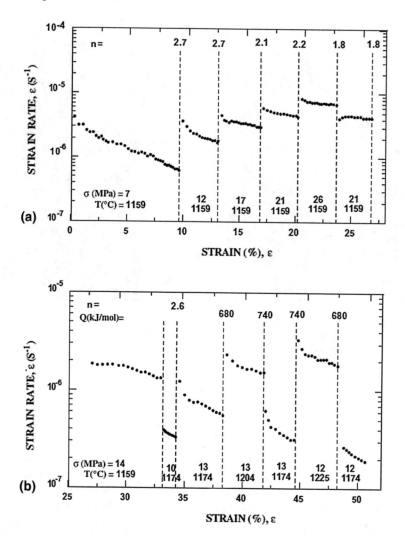

Fig. 6.2 Creep curve of nanocrystalline Y-ZTP with 10 wt% glassy phase. The curves illustrate calculation of (**a**) the stress exponent, *n*, and (**b**) the activation energy, *Q*. Lorenzo-Martín et al. [8]. With kind permission of Elsevier [The *P* in Y-TZP stands for polycrystalline specimens]

Table 6.1 *n* and *Q* for the different sets of samples. Lorenzo-Martín et al. [8]. With kind permission of Elsevier

Sample	*n*	*Q* (kJ/mol)
Pure	2.0 ± 0.4	680 ± 20
5% glassy phase	1.8 ± 0.5	690 ± 20
10% glassy phase	2.3 ± 0,5	710 ± 40
15% glassy phase	2.3 ± 0.4	740 ± 20

The impurity content does not change the mechanical properties, as displayed

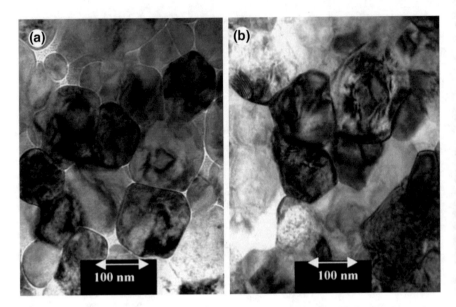

Fig. 6.3 TEM micrographs showing the microstructure of the Y-TZP with 5 wt% glassy phase: **a** as received and **b** after deformation. Lorenzo-Martín et al. [8]. With kind permission of Elsevier. The *P* in Y-TZP stands for polycrystalline specimens

$$n = 2\frac{\sigma}{\sigma - \sigma_0}, \tag{6.4}$$

which can be deduced from Eq. (6.1). For Y-TZP, the measured value of n is 2.3. Apparently, the cation segregation in grain boundaries is associated with the higher activation energies in the nanoceramic crystals. Moreover, this also predicts the small threshold stress for plastic deformation (creep in the present case) in nanoscale specimens. This may be explained by the effect of cation segregation at the grain boundaries, causing non-compensated charge density, with the consequence that local electric fields are created, capable of changing the mass transport of the charged species, and, in turn, leading to changes in the diffusion coefficient [see Eq. (6.1)].

For further details on the effect of yttria segregation in grain boundaries and its effect on the mechanical properties tested by compression in YZTP, the original research may be consulted. In addition, note that the amount of the glassy phase, which is dependent on the amount of oxygen and its distribution, had only a small effect on the activation energy in the ~ 50 to 60 kJ mol^{-1} range, as seen in Table 6.1.

For technological purposes, most of the nanoceramics contain various additives, often to such an extent that they become composite nanostructures, rather than monolithic ceramics. Therefore, it is of interest to consider one such important composite structure, such as silicon nitride/silicon carbide (S–C–N) nano-nanocomposite (studied by Wan et al.).

In this case, compressive stress was applied during the creep test. The test specimens were in the form of a 19 mm disk, 3–4 mm thick. Figure 6.4 shows the

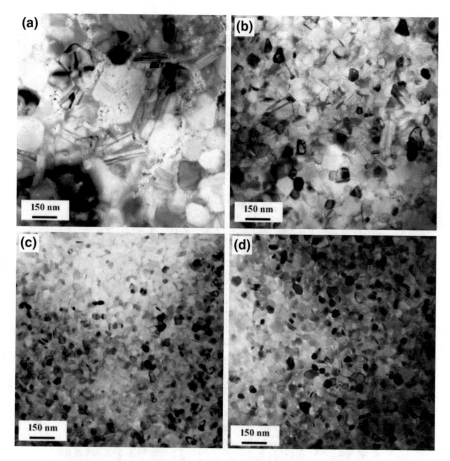

Fig. 6.4 Transmission electron microscopy (TEM) observations of Si_3N_4-SiC. **a** Sintered with 8 wt% Y_2O_3 at 1600 °C for 10 min, micronanostructure, **b** Sintered with 3 wt% Y_2O_3 at 1600 °C for 10 min, nano-nano structure, **c** Sintered with 1 wt% Y_2O_3 at 1600 °C for 10 min, nano-nano structure, **d** Sintered without additive at 1600 °C for 30 min, nano-nano structure. Wan et al. [18]. With kind permission of John Wiley and Sons

analysis of the initial S–C–N powder, having a nominal composition of $Si_{1.00}$ $C_{1.55}$ $N_{0.81}$ $O_{0.17}$. As mentioned earlier, a glassy phase can be induced by the presence of oxygen, which is a consequence of surface oxidation due to handling in air. The oxygen diffused into the specimens from their surfaces during the high-temperature processing. The amount of the glassy phase is oxygen dependent. The glassy phase may be observed in Fig. 6.5b. In this figure, a structure without a glassy phase is also shown for comparison. The mean particle size of the powder used for the sintering is ~1 μm, by scanning electron microscopy (SEM) evaluation. TEM reveals that the grain size of the composite decreases with the decrease in the amount of additive, which is associated with a transition from a micro–nano- to a nano–nano-type structure. The grain size varies when sintered without additives at

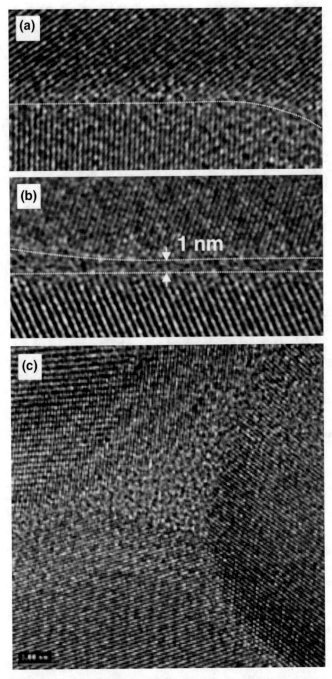

Fig. 6.5 High resolution transmission electron microscopy (HRTEM) analysis of the grain boundary of the nano–nano composite (no additive, 1600 °C/30 min sintered) **a** glass-free grain boundary, **b** grain boundary containing glassy layer, **c** triple junction. Wan et al. [18]. With kind permission of John Wiley and Sons

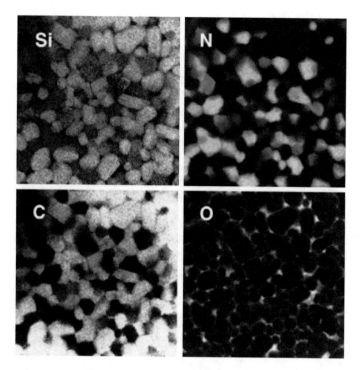

Fig. 6.6 Electron energy loss spectroscopy (EELS) analysis of the component elements in the Si₃N₄–SiC nanocomposite sintered at 1600 °C for 30 min without additive. Wan et al. [18]. With kind permission of John Wiley and Sons

1600 °C: for 10 min the grain size is 27 nm; for 30 min it is 40 nm (see Fig. 6.4d). The two phases comprising the composite, namely Si_3N_4 and SiC, have equal grain size and are randomly mixed, as indicated in Fig. 6.6 by electron energy loss spectroscopy (EELS). EELS reveals that oxygen is present in almost all the grain boundary regions. Only a small amount of glassy phase may be observed. In the ceramic under consideration, the oxygen was not homogeneously distributed in the grain boundary regions, some having more than others.

Most of the glassy grain boundary phase exists at multigrain junctions (e.g., see Fig. 6.5c). To avoid common complications, compression creep tests were conducted to examine the creep behavior of the nano–nano composites, rather than tensile creep tests, that are likely to induce cavitation. The steady-state creep of the nano–nano composites at various temperature and stress levels is shown in Fig. 6.7 by a plot of strain as a function of time. It is interesting to compare the results assembled in Fig. 6.6 for the nano–nano composites with those of microcrystalline Si_3N_4. Observe the very high creep resistance of the nano–nano ceramics and that the strain, as a function of time, is stress- and temperature-dependent.

The creep rate evaluated by the compression testing of the nano–nano ceramics is compared with other silicon nitride ceramics in Fig. 6.8. Specimens designated as C- and D-group nanoceramics, with various Y_2O_3 content, show higher creep

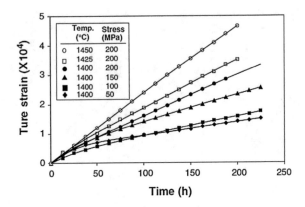

Fig. 6.7 Compression creep strain time curves for one of the nano-nano composites (1 wt% Y₂O₃, 1600 °C/10 min sintered). Wan et al. [18]. With kind permission of John Wiley and Sons

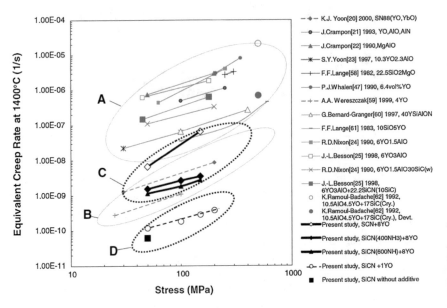

Fig. 6.8 Comparison of the compression creep property of nanocomposites with those of existing silicon nitride ceramics (additive in weight percentage unless specified, molecular formula simplified for clarity. For instance "6YO" in figure legend stands for "6 wt% Y₂O₃"). Wan et al. [18]. With kind permission of John Wiley and Sons

resistance than the A group. Note that a specimen without any yttria also appears among the D-group specimens evaluated by Wan et al. These D-group specimens, with and without yttria, show the highest creep resistance and represent the transition from micro-nano to nano-nano ceramics. This creep strength improvement is the consequence of the reduction of the oxygen-dependent glassy phase. Also note that the specimen in the D-group without yttria additive has a grain size in the 30–50 nm range. In order to obtain an improved nanoceramic, the oxygen should be distributed at the interfaces to prevent the formation of a glassy phase.

A phenomological equation equivalent to Eq. (1.9) was used to describe the steady-state creep deformation in the Si_3N_4–SiC composite, given as:

$$\dot{\varepsilon} = A \frac{\sigma^n}{d^p} \exp\left(-\frac{Q}{RT}\right) \tag{6.5}$$

The experimentally determined stress exponent, n, and the activation energy, Q, for various silicon nitride ceramics are in the range of $n = < 1 - > 3$ and $Q = 300$–1200 kJ/mol. The spread in these values is even higher when the creep test is carried out by tensile deformation, because the strain rate and the stress data are affected by cavitation. The grain size dependence, p, is in the range of 1–3, which is an indication of strong creep rate dependence on the grain size. As the grain size

Table 6.2 Activation energy for diffusion process in the Si_3N_4/SiC system. Wan et al. [18]. With kind permission of John Wiley and Sons

Medium	Diffusing particle	Temperature range (°C)	Activation energy (kJ/mol)	Note	References
α-Si$_3$N$_4$	Si	1400–1600	199	Self-diffusion	Kunz et al.[53]
	N	1200–1410	233	Lattice diffusion	Kijima and Shirasaki[54]
	Si	NA	NA	Grain boundary diffusion	NA
	N	NA	NA	Grain boundary diffusion	NA
β-S13N4	Si	1490–1750	390	Lattice diffusion (β with some α)	Batha and Whitney[35]
	N	1200–1410	777	Lattice diffusion	Kijima and Shirasaki[54]
	Si	NA	NA	Grain boundary diffusion	NA
	N	NA	NA	Grain boundary diffusion	NA
β-SiC	Si	1960–2260	911	Lattice diffusion	Ziegler et al.[2]
	Si	2010–2270	612	Grain boundary diffusion	Hon et al.[56]
	C	1860–2230	841	Lattice diffusion	Ziegler et al.[2]
	c	1855–2100	564	Grain boundary diffusion	Hon and Davis[57]
In GB of HPSN	Si/N	1450–1550	448	Grain boundary diffusion	Ziegler et al.[2]
(10 wt% Y$_2$O$_3$)	Si/N	1550–1760	695	Grain boundary diffusion	Ziegler et al.[2]

decreases, the creep rate is expected to increase by about 1–3 orders of magnitude, from micron size (50–1000 nm) to nanosize. Table 6.2 assembles the activation energies of the diffusion processes in various specimens, since they are relevant to diffusion-controlled creep. However, all these values are higher than the activation energy of ~205 kJ/mol found in nanoceramics (for example with 1 wt% yttria additive), which may indicate the action of a different creep mechanism in the nanocomposite than in microcrystalline silicon nitride. This exceptionally high creep resistance is believed to be a consequence of the prevention of the intergranular glassy phase formation by the dispersive distribution of the oxygen.

6.2.2 Tensile Testing of Creep

It has been observed that the creep resistance of nanoceramics may be greatly improved by additives. To achieve such a substantial improvement, those additives must be nanosized particles and dispersed in the grain boundaries of the ceramic matrix. For example, the dispersal of 5 vol% SiC nanoparticles into an alumina ceramic matrix, increases the room temperature fracture strength by more than three times (Ohji et al.). Furthermore, the achieved high strength was maintained up to 1200 °C, suggesting that this ceramic has good creep resistance. Monolithic alumina may be compared with alumina/17% silicon nanocomposite powder at 1200 and 1300 °C, tested at 50 and 150 MPa. Figure 6.9 compares both at 1200 °C under a stress of 50 MPa. The observed creep life in the nanocomposite was 10 times longer (150 h and 4% creep strain at fracture) than in the monolith (see Figs. 6.9 and 6.10). A large number of microcracks were also observed. The monolithic ceramic consists of primary, steady state and a small (barely visible) tertiary creep.

Fig. 6.9 Tensile creep curves of the monolith and nanocomposite at 1220 °C and 50 MPa. Slight accelerated creep and steady-state creep were present in the monolith, while they were little observed in the nanocomposite. Ohji et al. [11]. With kind permission of John Wiley and Sons

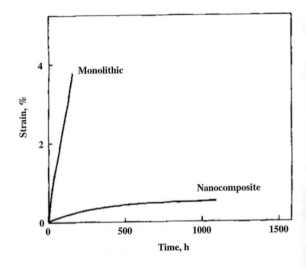

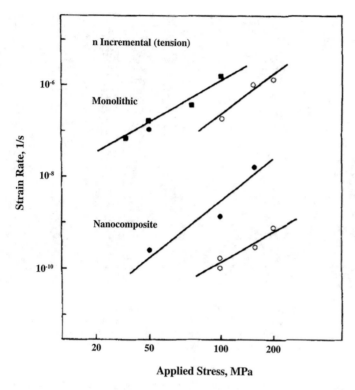

Fig. 6.10 Stress-dependence of steady-state or minimum creep rates in the tension (*closed symbol*) and the flexure (*open symbol*) for the monolith and the nanocomposite. The temperature is 1200 °C. The stress exponent for creep rate is 2.2 for the monolith and 3.1 for the nanocomposite in tension, and 2.9 for the monolith and 2.2 for the nanocomposite in flexure. Ohji et al. [11]. With kind permission of John Wiley and Sons

Compared with the monolith, the nanocomposite has a lifetime of 1120 h under the same test conditions (1200 °C and 50 MPa), while the creep strain was only 0.5%, which is eight times smaller. Furthermore, the nanocomposite showed little steady-state creep and no microcracks were observed via optical microscopy. The superior strength of the nanocomposite was also observed during flexural creep testing.

The steady-state creep rates of the monolith and of the nanocomposite ceramics are compared in Fig. 6.10. The steady-state creep rate of the nanocomposite is about three orders of magnitude lower under tension than the monolith (the flexural creep rate is 3–4 orders of magnitude lower). The stress exponent of the monolith is ~ 2.2 under tension (2.9 in the flexural test). In the case of the nanocomposite, the data points under tension were widely scattered and, therefore, not identified.

In order to inhibit creep, it is important to carry out observations on fractured surfaces. One of the most characteristic changes of the microstructure during creep

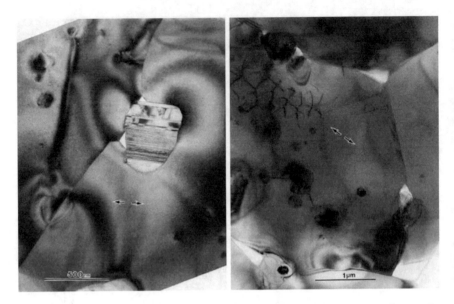

Fig. 6.11 Transmission electron micrographs of microstructures of the nanocomposite tested at 1300 °C and 50 MPa in tension, showing examples of rotating and plunging of intergranular silicon carbide particles and associated cavity formation. The stress direction is indicated by *arrows*. Ohji et al. [11]. With kind permission of John Wiley and Sons

is a rotation of the intergranular silicon carbide particles, which accompanies GBS and small cavity formation around the particles (Fig. 6.11a). Figure 6.12, shows the occurrence of intergranular crack propagation the alumina–alumina grain boundary, where small cavities formed around the SiC particles as a result of the GBS. The nanoparticle was transgranually fractured when the crack propagated through the boundary.

The intergranular small cavities, which are generated by the plunging of the interfacial particles, weaken the interfacial bonding and induce crack formation at the grain boundaries.

In an additional work on Al_2O_3 composite, the tensile creep results of Ohji et al. were substantiated by Thompson et al. It was found that, in nanocomposites of alumina containing 5 vol% SiC tested in air at 1200–1300 °C at a stress level of 100 MPa, the creep rate of Al_2O_3 is reduced by 2–3 orders of magnitude by SiC. The test was performed at a constant load. It was observed that no primary or secondary stages are present and only tertiary creep exists. The excessive cavitation observed is associated with SiC particles that are located at the Al_2O_3 grain boundaries, which ultimately lead to failure by creep rupture. The reduction of creep rate is associated with the inhibition of GBS by the SiC particles located at the grain boundaries. For the reduction of the tensile creep of fully dense Al_2O_3, 5 vol% SiC is sufficient. In Fig. 6.13a, the well-developed tertiary creep and the stress rupture of the Al_2O_3–SiC nanocomposite are illustrated, while in Fig. 6.13b, the strain rate versus the total strain is shown.

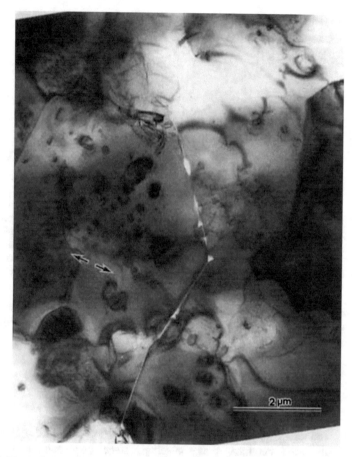

Fig. 6.12 Transmission electron micrograph of a trace of intergranular crack propagation. The sample was tested at 1220 °C and 100 MPa in tension. Note the transgranular fractured nanoparticle. The stress direction is indicated by an *arrow*. Ohji et al. [11]. With kind permission of John Wiley and Sons

The temperature step tests at 50 °C intervals, using single test specimens, are illustrated in Fig. 6.14 as strain rates versus strain in the temperature range of 1200–1300 °C.

Comparing the strain rates at constant strain in the temperature interval indicated in the form of an Arrhenius plot provides the activation energy for creep at various strains from the slopes in Fig. 6.15.

The Arrhemius relation is expressed, for the present case, as:

$$\dot{\varepsilon} = f(\varepsilon)\exp\left(-\frac{Q}{RT}\right) \tag{6.6}$$

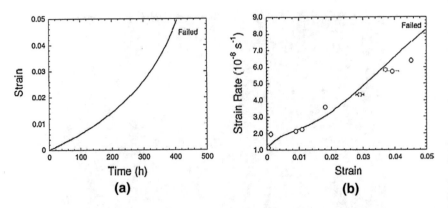

Fig. 6.13 Creep rupture of Al_2O_3–SiC nanocomposite at 1250 °C (100 MPa tensile stress). **a** Plot of strain versus time, displaying a well-developed tertiary stage; and **b** strain rate plotted as a function of total strain. *Solid line* represents the strain rate derived from the data shown in Fig. 6.11a; the *open circles* are strain rate data obtained at 1250 °C during two separate temperature step tests. Thompson et al. [16]. With kind permission of John Wiley and Sons

where $f(\varepsilon)$ represents the strain-dependence of the strain rate and the others are the usual symbols. In Fig. 6.15, the lines are parallel, indicating that the creep mechanism did not change within the investigated temperature range.

6.2.3 Flexural (Bending) Testing of Creep

As an example of flexural testing, the technologically important silicon-nitride-based ceramics may be cited, since they exhibit outstanding mechanical and thermomechanical properties at high temperatures and are used in various structural applications, particularly in energy conversion systems, engines, turbines, etc. As mentioned above, SiC additives are an important compositional part in silicon nitride, dispersed as nanoparticles in the matrix. Recent experiments reveal that the microstructure and the high-temperature strength of Si_3N_4+SiC nanocomposites are strongly influenced by the nucleation step of the nanoparticles before full densification and that the high-temperature strength is improved only when the SiC nanoparticles are located intergranulary. The characteristic microstructures of the C-derived nanocomposites prepared by the optimized processing route (Dusza et al.) appear in Fig. 6.16.

The composite nano- and submicron-sized SiC particles, distributed intragranularly between the Si_3N_4 grains, are in the 40–150 nm size range. All the materials contained the same volume fraction of Y_2O_3 additives and were prepared by the same processing steps (Dusza et al.) to enable the study of the influence of SiC nanoparticles on the creep behavior of Si_3N_4. In the four-point bending tests, the inner and outer span lengths were 20 and 40 mm, respectively. A silicon carbide fixture was used, and the loading was performed in air in the 1200–1450 °C

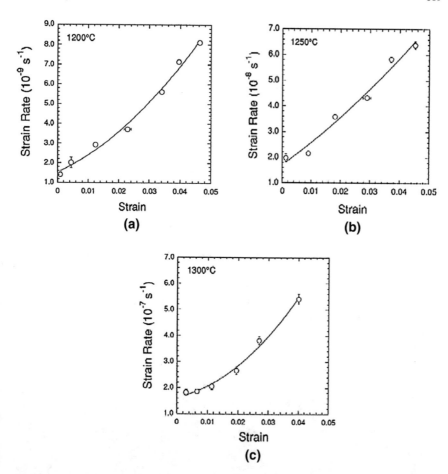

Fig. 6.14 Series of strain-rate-strain plots from a single specimen that was step-tested at 100 MPa tensile stress, shown at 50 °C intervals; the *solid lines* are polynomial fits to the data. Thompson et al. [16]. With kind permission of John Wiley and Sons

temperature range, with the outer fiber stress of 50–150 MPa. The deflection of the specimens was recorded continuously during the creep tests, from which the strain (taken to be the creep strain, ε) was calculated as a function of time. From the creep, ε, versus t, the creep rate was evaluated from the slope of the plots. Norton's equation, shown as Eq. (6.7), was used to describe the steady-state creep rate as follows:

$$\dot{\varepsilon} = A\sigma^n \frac{1}{d^m} \exp\left(-\frac{Q_c}{RT}\right) \qquad (6.7)$$

Here, A is a constant that depends on the specific material and the microstructure; n and m are the stress and grain size exponents, respectively;

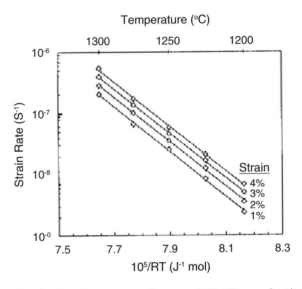

Fig. 6.15 Arrhenius plot of strain rates at a tensile stress of 100 MPa, as a function of total strain. Strain rates were extracted from the polynomial fits shown in Fig. 6.12. The apparent activation energy is calculated to be 840 ± 5 kJ/mol. Thompson et al. [16]. With kind permission of John Wiley and Sons

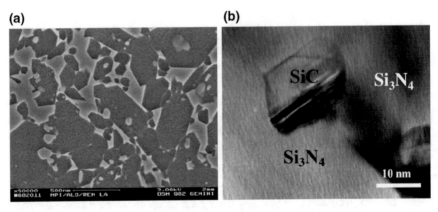

Fig. 6.16 Characteristic microstructure of the carbon-derived nanocopmposite, **a** scanning electron microscopy—plasma etched, **b** transmission electron microscopy. The intragranularly located SiC particles are not visible in Fig. 2a and appear as an intergranular phase. Dusza et al. [3]. With kind permission of John Wiley and Sons

d is the grain size; and σ represents the stress. The creep deformations of both materials were characterized by both primary and steady-state (secondary) creep, without tertiary creep. Up to 1300 °C, only minimum creep deformation was observed in the composite ceramics. However, pronounced creep was observed at 1350 °C and above it. In spite of the fact that the grain size was larger in the nanocomposites, compared to the monolithic ceramics, their creep resistance was significantly higher. The monolithic and the nanocomposite are compared in Fig. 6.17 at 1350 °C. The stress exponents of the substances tested for creep were 0.8–1.25 and 1.0–2.0 in the monolithic and nanocomposite ceramics, respectively.

The respective activation energies for creep are 480 and 372 kJ/mol for the nanocomposite and the monolithic specimens, respectively.

TEM observations indicate that the main phase in both materials is β-Si_3N_4 and a small amount of α-Si_3N_4 is also present. In addition to the Si_3N_4 (and SiC in the composite) in both the monolithic and the composite materials, some additional crystalline phases were detected, which were mainly $YSiO_2N$ and $Y_2Si_3O_3N_4$. The volume fraction of these phases increases during the creep test, mostly in the nanocomposites. The glassy phase rapidly undergoes complete crystallization during creep within the multigrain junction, but complete devitrification of the intergranular phase was not observed. TEM and high-resolution TEM (HREM) observations revealed that, due to the intergranularly located SiC particles and the crystallized glassy phase, there are no glassy-phase triple points present in the microstructure of the composite (Fig. 6.18) and there is no equilibrium thickness of the intergranular phase between the Si_3N_4/Si_3N_4 grains and Si_3N_4/SiC grains. The average thickness of the intergranular glassy phase between the Si_3N_4/SiC is approximately 15 A, as shown in Fig. 6.19, but with a favorable orientation; no boundary phase is present.

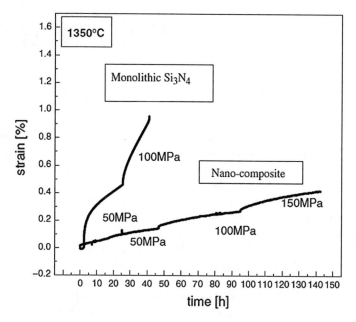

Fig. 6.17 Comparison of the creep deformation of monolithic silicon nitride and the C-derived nanocomposite. Dusza et al. [3]. With kind permission of John Wiley and Sons

Fig. 6.18 Triple point and Si₃N₄/SiC grain boundaries in the carbon-derived composite with different thickness of the intergranular phase between Si₃N₄ and SiC. Dusza et al. [2]. With kind permission of John Wiley and Sons

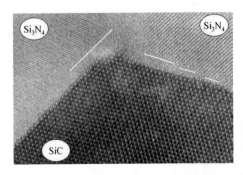

Fig. 6.19 Si₃N₄/SiC grain boundaries in the carbon-derived nanocomposite. Dusza et al. [2]. With kind permission of John Wiley and Sons

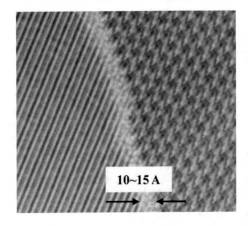

In addition to Fig. 6.17, characteristic creep curves of the nanocomposite, obtained by stepwise loading at various temperatures and at stresses in the 50–150 MPa range, are shown in Fig. 6.20.

As previously mentioned, the creep deformations of both materials were characterized by primary and steady-state (secondary) creep, without tertiary creep.

The influence of the applied stress on the creep rate is illustrated in Fig. 6.21, which was used to evaluate the stress exponent.

TEM observation of the crept nanocomposite specimen did not reveal cavitation over the temperature range investigated, but above 1400 °C, limited cavitation was observed in the monolithic ceramics. The role of the SiC particles is significant in hindering grain growth in the silicon nitride, but what seems to be more crucial is the interlocking of the Si₃N₄ grains, preventing GBS. This may be seen in the schematic Fig. 6.22.

In closing, it may be stated that the creep resistance of Si₃N₄ (in the experiments referred to as monolithic) is improved by the addition of intergranular SiC particles that modifying the creep mechanism by: (a) limiting the grain growth in the Si₃N₄ and (b) hindering its GBS.

Fig. 6.20 Creep deformation at different temperatures and stresses. Dusza et al. [2]. With kind permission of John Wiley and Sons

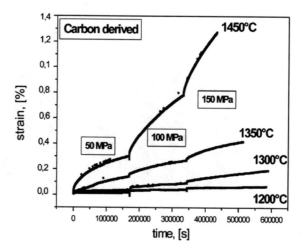

Fig. 6.21 Figure 14 determination of the stress exponent of the C-derived nanocomposite. Dusza et al. [2]. With kind permission of John Wiley and Sons

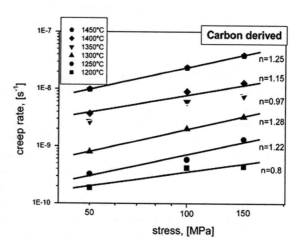

6.2.4 Indentation (Hardness) Testing of Creep

Other than in polymers, not much information is available on creep and hardness testing in general, and there is little specifically on nanoceramics. Nevertheless, the characterization of creep deformation by indentation is of interest, as from theoretical works, such as the article: "Analysis of indentation creep" by Stone et al. and the report: *Computer-Aided Multi-Scale Design of SiC–Si₃N₄ Nanoceramic Composites for High-Temperature Structural Applications* by Tomar and Renaud. The characterization of creep is very important for high-temperature applications, regardless if the material is macrodimensioned or nanoscaled. As a matter of fact, in miniature devices, creep is even more critical, due to the very small dimensions and the generally lower temperature of the applications. Again, creep failure occurs in

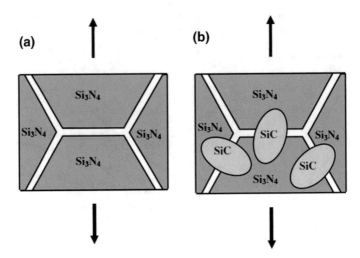

Fig. 6.22 Schematic of the monolithic Si_3N_4 (**a**) and nanocomposite (**b**) microstructure with the intergranularly located SiC particles, which interlock the Si_3N_4 grains and prevent the grain boundary sliding. Dusza et al. [3]. With kind permission of John Wiley and Sons

materials exposed, for prolonged time periods, to loads below the elastic limit, basically involving an increase in length in the direction of the applied stress. Creep deformation can occur at ambient temperature in low melting materials, which is quite unlikely in ceramics, where the deformation is so slow that it is barely detected (if at all) and, therefore, insignificant. However, in high-temperature applications, deformation at a given load, even in ceramics, is likely unless the component has been safely designed to avoid the potential, premature failure.

Testing is generally carried out in air at atmospheric pressure. But if it is necessary to produce creep data for materials that react with air, they may be tested in a chamber containing an inert atmosphere, such as argon, or in a vacuum. If a material is meant to serve in an aggressive environment, then the testing should be carried out in a controlled environment simulating the intended service conditions.

Coming back to creep indentation tests, one good example is the case of creep in Si–C–O ceramics (Gan and Tomar). In their experiments, both the sample and the indenter tip were heated to the required testing temperature. Parts of the setup, which must be kept at room temperature, are isolated by heat shields. The tests were performed at 6 different temperatures: room temperature, 100, 200, 300, 400, and 500 °C.

The tip radius of the nanoindenter is approximately 20 nm and that of microindenter is approximately 200 nm. Both indenters are of the Berkovich type. During these tests, the samples were mounted on the indentation stage using glue. Indentation locations were selected randomly on the sample surfaces. In each chosen location, the tests were performed in either a 3 by 3 matrix or in a 4 by 4 matrix pattern, with equal longitudinal and transverse spacing between each indentation spots of approximately 5 μm in the nanoindentation and 20 μm in the microindentation. The nanoindentation depths were in the range of 200–500 nm and the microindentation depths were in the range of 1–3 μm. At such depths, the effect of the tip radius, as well as of the substrate, is negligible. The projected area of the nanoindenter is approximately 0.5 μm^2 and of the microindenter is approximately 5 μm^2. Figure 6.23 shows an example of the procedure used to extract the indentation creep data. The creep deformation profile is extracted from the dwell period at the peak load (see Fig. 6.23b). The dwell period may affect the creep data. A number of dwell periods were tried. A dwell period of 500 s was chosen, based on convergence in the measurements. Indentation profiles were imaged before and after the tests to ensure that similar surface conditions exist before and after the tests. At each load, more than nine independent indentations were performed.

During the experiments, care was taken to minimize thermal drift by allowing thermal equilibrium to be reached. Such drift might occur if the stability of the indentation equipment is not maintained during the measurements, resulting in a shift in the measured indentation depth away from the actual value as a function of time. For the minimization of the thermal drift and its measurement, one is referred to the work of Gan and Tomar. Note that, during the calculation of the creep data, the thermal drift rate is multiplied by the dwell time and, then, the result is subtracted from the measurement. The following steps, involved in the evaluation of creep, have been borrowed from the measurement of plastic indentation in metals and alloys. The modulus and hardness were calculated using the Oliver–Pharr method. A typical indentation unloading curve (seen in Fig. 6.23a), may be described by the Oliver–Pharr method as:

$$P = A\left(h - h^{\text{final}}\right)^m \tag{6.8}$$

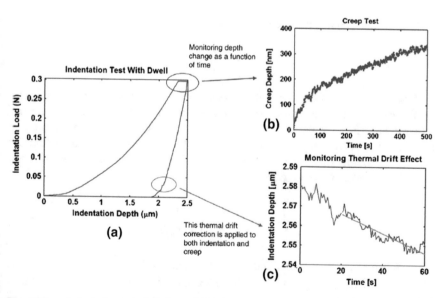

Fig. 6.23 **a** A typical nanoindentation profile showing **b** how the creep data is extracted and **c** how the thermal drift rate is extracted. Gan and Tomar [5]. With kind permission of Elsevier

Here, A and m are material constants; h^{final} is the indentation depth after complete unloading; h_c is the contact depth at the maximum load, P_{max}, which corresponds to the total indentation depth, h_{max}. These are written as:

$$h_c = h_{max} - \varepsilon \frac{P_{max}}{S} \tag{6.9}$$

where S is the stiffness. In this case, stiffness is the initial unloading stiffness (SP). Parameter ε is a correction factor; $\varepsilon = 1$ for a spherical tip and $\varepsilon = 0.75$ for a Berkovich tip. The area of contact is calculated as a function of contact depth for the Berkovich indenter as:

$$A_c = 3\sqrt{3}h_c^2 \tan^2 65.3 \approx 24.5h_c^2 \tag{6.10}$$

The reduced modulus is given by:

$$\frac{1}{E_r} = \frac{1 - v_i^2}{E_i} + \frac{1 - v_s^2}{E_s} \tag{6.11}$$

v is Poisson's ratio and E is the elastic modulus. The subscripts i and s represent the indenter tip and specimen, respectively. For the diamond tip, $v_i = 0.07$ and $E_i = 1141$ GPa. E_r may be obtained by Sneddon's solution as:

$$E_r = \frac{\sqrt{\pi}}{2} \frac{S}{\sqrt{A_c}} \tag{6.12}$$

The stiffness, S, is calculated as follows:

$$S = \frac{dP}{dh} \tag{6.13}$$

P is the applied load and h is the displacement of the indenter. S is represented by the slope of the indentation curve upon unloading. A correction is required (Ngan et al.), since the elastic modulus calculated in this way can be greatly affected by creep in particular under the following conditions: when the tested material is soft, or the holding time is short, or the peak load is large, or if the unloading rate is slow. The corrected elastic stiffness is:

$$\frac{1}{S_e} = \left(\frac{1}{S} - \frac{\dot{h}_h}{\dot{P}_u} \right) \frac{1}{1 - \dot{P}_h/\dot{P}_u} \tag{6.14}$$

$\dot{h}_h$ is the creep rate at the end of load holding; $\dot{P}_h$ is the load-decaying rate at the end of load holding; and $\dot{P}_u$ is the unloading rate. In the case of constant load holding, $\dot{P}_h$ is zero. The effect of creep may be corrected along with Feng and Ngan as:

$$C = \frac{\dot{h}_h S_e}{|\dot{P}_u|} \tag{6.15}$$

Replacing S by S_e in Eqs. (6.9) and (6.12), the elastic modulus may be calculated by Eq. (6.11), thus taking the creep effect into account. Hardness is given as:

$$H = \frac{P_{max}}{24.5 h_c^2} \tag{6.16}$$

The strain rate is given by a power law as:

$$\dot{\varepsilon} = A\sigma^n \exp\left(-\frac{Q}{RT}\right) \tag{6.17}$$

The stress exponent, n, is an indication of the creep mechanism, which may be obtained from a plot of $\dot{\varepsilon}$ versus σ. The power law relation is valid when the homologous temperature is $<0.57 T_m$. The strain rate, stress, and indentation depth are related by:

$$\dot{\varepsilon} \sim \frac{\dot{h}}{h} \text{ and } \sigma \propto \frac{P}{h^2} \tag{6.18}$$

The derivation of the strain rate requires differentiation of the $h(t)$ curve, which may be seen in Fig. 6.24.

The data are scattered around the fitted line. The fitting function used for $h(t)$ in the experiments is:

$$h(t) = h_i + \alpha t^b + kt \tag{6.19}$$

Figure 6.25 plots the hardness data corresponding to the moduli plots. As seen on the nanoscale, the trend for the elastic moduli (shown in the work of Tomar) is repeated for the hardness values. A clear trend emerges from the nanoindentation data regarding the increase in the Meyer's exponent with the increase in temperature. The microindentation hardness trend is opposite to the trend observed in the case of microindentation elastic moduli–hardness decreases as the temperature rises. In addition, hardness diminishes with the increase in the peak indentation load, signifying the strain softening of the material. The reduction of hardness with increased temperature is attributable to stronger $TiSi_2$ particles pressing into the relatively softer SiCO matrix, getting progressively softer as the temperature increases at the microscale. At the nanoscale, not enough contact area is available to have such an effect.

Apart from the influence of strain hardening or softening, material pileup around the indent and indentation creep may also contribute to observed indentation hardness behavior. The creep depth, as a function of dwell time, at three different temperatures and two different nanoindentation and microindentation peak loads, is illustrated in Fig. 6.26.

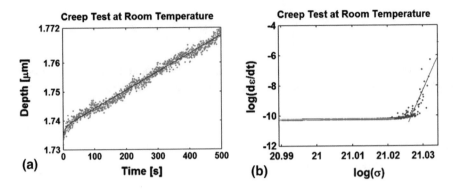

Fig. 6.24 Illustration of stress exponent calculation. **a** The creep raw data file was plotted and fitted using Eq. (6.17); and **b** the equivalent strain rate and stress, based on the fitted data. The *horizontal line* is used to calculate the steady-state stress exponent. Gan and Tomar [5]. With kind permission of Elsevier

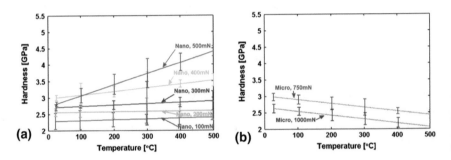

Fig. 6.25 **a** Nanoindentation hardness and **b** microindentation hardness of Si–C–O coatings as a function of temperature and maximum indentation load. Gan and Tomar [5]. With kind permission of Elsevier

Creep data is normalized by subtracting the initial depth of each creep test. For comparison, the plots also show the uncorrected thermal drift, as well as the corrected data. The effect of thermal drift correction is the highest at the highest temperature and at the nanoscale. The creep depth vs. time plots reached a steady state within first 100 s of the plotting in all the cases. The effect of temperature is to increase the creep rate. The temperature effect is more pronounced on the microscale, than on the nanoscale. This trend is particularly strong when transitioning from 250 to 500 °C.

The stress exponent and creep strain rate, as functions of temperature and peak-indentation load on both length scales, are shown in Fig. 6.27.

The nanoscale stress exponent is in the 4–5 range, indicating a dislocation-climb-related creep deformation mechanism occurring by means of the bulk or pipeline diffusion of dislocations. This exponent decreases with the increase in temperature, marking the transition of the mechanism from that of dislocation

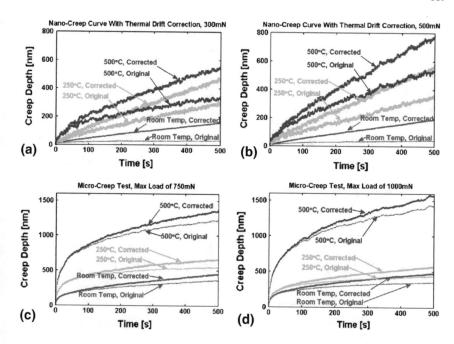

Fig. 6.26 Indentation depth (original and thermal drift corrected) as a function of time at different temperatures in the case of **a** nanoindentation test at the peak load of 300 mN, **b** nanoindentation test at the peak load of 500 mN, **c** microindentation test at the peak load of 750 mN, and **d** microindentation test at the peak load of 1000 mN. Gan and Tomar [5]. With kind permission of Elsevier

climb to that of diffusion. On the microscale, the stress exponent is considerably higher, indicating that the primary mechanism of deformation is volumetric densification. Such a transition suggests a rapid change of mechanism–from linear diffusion flow to the power law mechanism (e.g., climb) and eventually to rate-insensitive plastic flow (dislocation glide), as the indentation size moves toward the microscale. The creep strain rate generally increases as a function of the increase in temperature, as well as with increasing peak indentation load.

In the case of microscale measurements, the indenter covers an area that is approximately 10 times larger than the nanoindenter area. At the microscale, since hardness diminishes and the creep strain rate increases with the increase in temperature, the strain rate sensitivity index (the constant, k, in $\sigma = b\dot{\varepsilon}^k$) indicates strain softening. These same analyses reveal strain hardening also at the nanoscale.

The low-temperature creep deformation strain rate, $\dot{\varepsilon}$, of materials may be expressed by an Arrhenius-type flow function, given as:

$$\dot{\varepsilon} = \dot{\varepsilon}_0 \exp\left[-\frac{\Delta G(\sigma)}{k_B T}\right] \tag{6.20}$$

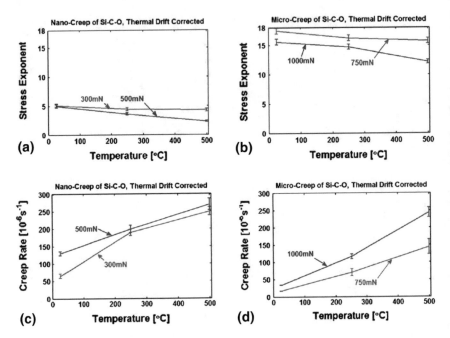

Fig. 6.27 **a** Stress exponent as a function of load during nanoindentation creep measurements, **b** stress exponent as a function of load during microindentation creep measurements, **c** creep strain rate as a function of load during nanoindentation creep measurements, and **d** creep strain rate as a function of load during microindentation creep measurements. Gan and Tomar [5]. With kind permission of Elsevier

Here, $\dot{\varepsilon}_0$ is the reference strain rate, which is a function of the type of material, stress level, and microstructure; ΔG is the activation energy for creep or other rate-dependent processes; and k_B and T have the usual meaning. The parameter, $\dot{\varepsilon}_0$, is proportional to the concentration of elementary defects that cause plastic strain. The sensitivity of the strain rate to stress is mainly determined by the ΔG term. The thermal activation volume (V^*) is expressed by the partial derivative of ΔG with respect to stress, σ, approximated by:

$$V^* \equiv k_B T \frac{\partial \ln \dot{\varepsilon}}{\partial \sigma}\Big|_{\varepsilon,T} \qquad (6.21)$$

In Fig. 6.28, plots of the thermal activation volume, calculated as a function of temperature, peak indentation load and length scale, are shown as an example of indentation creep.

The creep deformation mechanism relates to the microstructure of the discussed ceramic material, which has $TiSi_2$ particles, with higher melting points, embedded in a Si–C–O matrix, with a lower melting point. As the temperature increases, the matrix softens, particles slide, and particle rearrangement contributes to the deformation mechanism. The effect of these factors is more pronounced at the microscale, due to the higher surface area sampled.

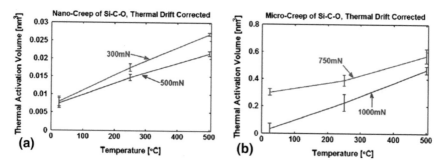

Fig. 6.28 Thermal activation volume as a function of peak indentation load and measurement temperature at **a** nanoscale and **b** microscale. Gan and Tomar [5]. With kind permission of Elsevier

In addition to the creep exponent and the creep strain rate measured by indentation, the properties investigated include the elastic modulus and hardness. At the nanoscale, the deformation mechanism is dominated by dislocation climb and diffusion. Overall, analyses reveal that both indentation creep and the capability of strain hardening determine the length-scale-dependent indentation behavior in the material system. The effect of temperature is to introduce strain hardening at the nanoscale level and strain softening at the microscale level.

References

1. Andrievski RA, Glezer AM (2009) Phys Usp 52:315
2. Dusza J, Kovalčík J, Hvizdoš P, Šajgalík P, Hnatko M, Reece MJ (2004) J Eur Ceram Soc 24:3307
3. Dusza J, Kovalčík J, Hvizdoš P, Šajgalík P, Hnatko M, Reece MJ (2005) J Eur Ceram Soc 88:1500
4. Feng G, Ngan AHW (2002) J Mater Res 17:660
5. Gan M, Tomar V (2010) Mater Sci Eng A 527:7615
6. Jiang B, Weng GJ (2004) Int J Plasticity 20:2007
7. Kuntz JD, Zhan G-D, Mukherjee AK (2004) MRS Bull 29:22
8. Lorenzo-Martín C, Gómez-García D, Gallardo-López A, Domínguez-Rodríguez A, Chaim R (2004) Scripta Mater 50:1151
9. Mukhopadhyay A, Basu B (2007) Int Mater Rev 52:257
10. Ngan AHW, Wang HT, Tang B, Sze KY (2005) Int J Solids Struct 42:1831
11. Ohji T, Nakahira A, Niihara K (1994) J Am Ceram Soc 77:3259
12. Oliver WC, Pharr GM (2004) J Mater Res 19:3
13. Poirier JP (1985) Creep in crystals. Cambridge University Press, Cambridge, p 195
14. Sneddon IN (1965) Int J Eng Sci 3:47
15. Stone DS, Jakes JE, Puthoff J, Elmustafa AA (2010) J Mater Res 24(3):1279
16. Thompson AM, Chan HM, Harmer MP (1997) J Am Ceram Soc 80:2221
17. Tomar V, Renaud JE (2010) Computer-aided multi-scale design of SiC-Si$_3$N$_4$ nanoceramic composites for high-temperature structural applications, 2010, unpublished
18. Wan J, Duan RG, Gasch MJ, Mukherjee AK (2006) J Am Ceram Soc 89:274

Chapter 7
Creep Rupture

Abstract Creep—a time-dependent deformation—occurs below its yield strength usually at elevated temperatures. Creep terminates in rupture if steps are not taken to bring it to a halt. The purpose of creep rupture tests is to determine the time-to-failure. For such tests higher stresses are applied until the specimen fractures. The objectives of the tests are to determine the minimum creep rate at stage II creep and to evaluate the time at which failure sets in. Such information is essential so that the proper ceramics will be selected to eliminate failure during service and to assess the time period of safe use during high-temperature applications. Creep rupture (failure) is the objective of this chapter.

7.1 Introduction

Creep is a time-dependent deformation of a material under an applied load that is below its yield strength. It most often occurs at elevated temperatures, but some materials creep at room temperature. In ceramics intended for high-temperature applications, room temperature creep seldom, if ever, occurs. Therefore, there is concern regarding creep in ceramics at elevated temperatures. Creep terminates in rupture, if steps are not taken to bring it to a halt. Basically, creep rupture tests are used to determine the elapsed time-to-failure. Generally, higher stresses are used for creep rupture testing than in conventional creep tests, which are carried out until the specimen fractures.

The objectives of the respective tests are to determine the minimum creep rate at stage II creep, on the one hand, and to evaluate the time at which failure sets in, on the other. Such information is essential so that the proper ceramics will be selected to eliminate failure during service and to assess the time period of safe use during high-temperature applications, when structural stability is essential. Various ceramic components operate at high temperatures and may experience creep. As in creep testing, stress rupture testing involves the same testing elements (for example, a tensile specimen) and is performed under a constant load (or stress) at a constant temperature. Not surprisingly, creep failures may appear ductile or brittle, due to the

© Springer International Publishing AG 2017
J. Pelleg, *Creep in Ceramics*, Solid Mechanics and Its Applications 241,
DOI 10.1007/978-3-319-50826-9_7

Fig. 7.1 A schematic creep
curve. ε_f and t_f are the strain
and time-to-creep failure.
Reference [1]

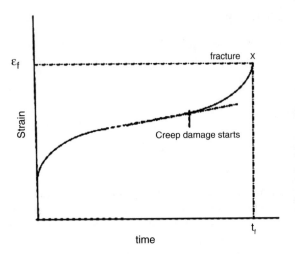

nature of ceramics, but they are all temperature-dependent. Cavities, believed to be responsible for cracking by cavity coalescence, may be either r-type or w-type and either transgranular or intragranular (see Chap. 2). Figure 7.1 shows a schematic illustration (based on the earlier illustrated schematic creep curve in Chap. 1), showing where creep damage starts.

The location of the creep damage coincides with the place of origin of the tertiary creep and represents the minimum creep rate. This is one of the thoughts regarding the time at which cavitation develops (either as microcavities or voids). However, there are experimental indications (density measurements) that intergranular cavities may be observed before tertiary creep sets in and they may be well developed at the end of second-stage creep. Creep data for general design use are usually obtained under conditions of constant uniaxial loading and constant temperature. Test results are usually plotted as strain versus time-to-rupture. The experimental data indicate that, of the great variety of creep curves described by various laws, their shape is near-linear when the data are presented on a log-strain vs. log-time basis. Time is expressed in hours. Very often, instead of using the rupture time, the time until reaching a steady state or minimum creep is preferred, because then a much shorter period of testing time is needed to collect the creep data. Creep rupture in technologically important ceramics will be considered in Part B of this book.

7.2 Elements of Creep Rupture

Stress rupture is the sudden and complete failure of a material held under a definite constant load (or stress) for a given period of time at a specific temperature. In stress rupture testing, loads may be applied by tensile bending, flexural, biaxial, or hydrostatic methods. It is essential to predict the lifetime to stress rupture in order to know the applicability of each ceramic. An accepted method for predicting longevity until stress rupture is the Monkman–Grant (MG) relationship, which has been applied to various materials, among them the ceramic, silicon nitride. The MG relationship states that the rupture life, t_f, is uniquely related to the minimum creep rate (secondary creep), $\dot{\varepsilon}_s$. The basic element of the MG relationship may be described by the following equation:

$$t_f = K(\dot{\varepsilon}_s)^{b_i}, \qquad (7.1)$$

where K and b are constants. This advantageous equation has been applied to both metals and ceramics, because of its relatively short test duration. Another advantage is that it has often been observed to be temperature-independent. All that is needed in order to estimate failure-by-rupture for a new set of conditions is the value of the secondary creep rate. This is a relatively brief procedure, since the time required to reach the secondary creep rate is only a small portion of the failure time (namely of the overall testing time). A modified MG equation has been suggested, given as

$$\ln(t_f) = b_0 + b_1 \ln(\dot{\varepsilon}_s) + \frac{b_2}{T}, \qquad (7.2)$$

where b_0 and b_2 are constants. A comparison of Eqs. (7.1) and (7.2) shows that the first and last terms in Eq. (7.2) are equal to $\ln(K)$ in Eq. (7.1). Equation (7.1) is plotted on a log–log scale as shown in Fig. 7.2. Using this modified equation, one can obtain the time-of-failure of the secondary creep, as shown in Fig. 7.3. Equation (7.2) was modified due to the fact that when the stress rupture life was plotted against the minimum creep rate on a log–log scale, the data were found to be stratified with respect to temperature in such a way as to suggest that K in Eq. (7.1) is an increasing function of temperature.

Fig. 7.2 Temperature dependence of the Monkman–Grant lines correlated with an additional temperature term (Eq. 7.2). Menon et al. [2]. With kind permission of John Wiley and Sons

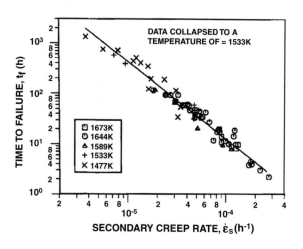

Fig. 7.3 Comparison of the
average lines predicted by
Eq. (7.3) with the unmodified
data. Menon et al. [2]. With
kind permission of John
Wiley and Sons

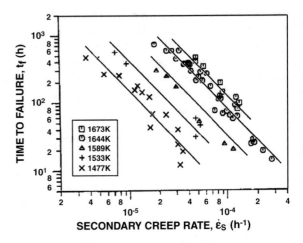

Fig. 7.3 Comparison of the average lines predicted by Eq. (7.3) with the unmodified data. Menon et al. [2]. With kind permission of John Wiley and Sons

The data may be correlated with an additional temperature term, specifically with an inverse function of temperature, which may be used to correlate the data plotted in Fig. 7.2. The data collapsed to a temperature of 1533 K. Figure 7.3 shows the prediction from Eq. (7.3) for the individual temperatures, along with the unmodified data. The values for b_0, b_1, and b_2 in Eq. (7.2) for NT154 silicon nitride are 15.87, -1.53, and -4.2×10^4, respectively. Note that b_2 is negative, which means that rupture life increases with an increase in temperature for the same creep rate value.

The stress, σ, is uniquely related to $1/T$ at a constant value for creep rate, $\dot{\varepsilon}_s$, by the following relation:

$$\ln(\dot{\varepsilon}_s) = \ln(A) + n \ln\left(\frac{\sigma}{E}\right) - \left(\frac{Q}{R}\right)\left(\frac{1}{T}\right), \tag{7.3}$$

where A and n are, respectively, a constant and a stress exponent of the applied stress, σ, with E being the Young's modulus. The secondary creep rate is usually modeled through temperature- and stress-dependent terms, using an activation-energy approach:

$$\dot{\varepsilon}_s = A\left(\frac{\sigma}{E}\right)^n \exp\left(-\frac{Q}{RT}\right). \tag{7.4}$$

Equation (7.3) is the logarithmic expression of Eq. (7.4). This is basically Eq. (1.9) shown again, except that instead of having the stress ratio with the Young's modulus, only the stress is indicated as

$$\dot{\varepsilon} = B\sigma^n \exp -\left(\frac{Q}{kT}\right) \tag{1.9}$$

This is because the relation between σ and T, through the strain rate, $\dot{\varepsilon}_s$, means that the stress rupture may also be expressed by the stress term as

$$\ln\left(t_f\right) = c_0 + c_1 \ln(\dot{\varepsilon}_s) + c_2 \ln\left(\frac{\sigma}{E}\right) \tag{7.5}$$

This relationship of the rupture life (as time-to-failure) to the secondary creep rate is plotted in Fig. 7.4.

The values of c_0, c_1, and c_2 are, respectively, -27.79, -1.27, and -2.71. Note that c_2 is negative, which indicates that rupture life decreases with an increase in stress for the same value of creep rate (Eq. 7.5). Figure 7.5(a–d) show SEM fractographs of NT154 showing crack initiation and growth region in internal initiations in specimens tested at 1477, 1533, 1644, and 1673 K. The creep failure is localized in a well-demarcated region, with subcritical crack growth (SCG) observable on the fracture surface.

The arrows indicate the periphery of the SCG region, which is circular in shape. This region is known as the 'slow crack growth region.' Beyond this region, a mirror-like region is seen, which is indicative of fast fracture. The SCG-zone measurement resulted in a quantity, $\sigma\sqrt{a}$, where a is the radius of the zone, which is approximately constant for a variety of stresses and temperatures, indicating that the failure occurs when the SCG grows to a size corresponding to a constant stress intensity. In Fig. 7.6, a typical failure, initiated at the surface, appears in a SEM fractograph of a specimen tested at 1644 K. The fast fracture emanates from the region (mirror-like in appearance) in which the SCG meets the surfaces.

The rupture time, as a function of creep rate, is shown in Fig. 7.7. The failure process associated with crack advancement occurs by diffusion of the material away from the crack tip. The crack-growth rate, da/dt, is a function of temperature and the local stress, σ_{loc}, at the crack tip, given as

Fig. 7.4 Stratification of the Monkman–Grant lines can be correlated with an additional stress term (Eq. 7.5). Menon et al. [2]. With kind permission of John Wiley and Sons

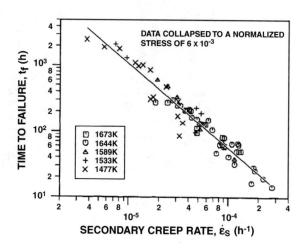

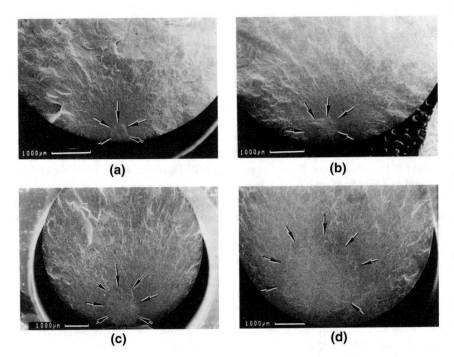

Fig. 7.5 SEM of subcritical crack growth zones in specimens tested at **a** 1477 K/325 MPa/68 h; **b** 1533 K/295 MPa/560 h; **c** 1644 K/180 MPa/19 h; **d** 1673 K/150 MPa/80 h. Note the circular shape of the SCG regions, all internally initiated. Menon et al. [2]. With kind permission of John Wiley and Sons

Fig. 7.6 SEM of the fracture surface in a surface-initiated failure. The radiating features, indicative of fast fracture, start at the junction where the periphery of the SCG region meets the surface of the specimen (where the *larger arrow points*). Menon et al. [2]. With kind permission of John Wiley and Sons

Fig. 7.7 Monkman–Grant plot for GN-10 silicon nitride. The plot shows no temperature stratification for the relation between rupture life and creep rate. Menon et al. [2]. With kind permission of John Wiley and Sons

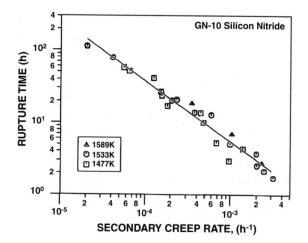

$$\frac{da}{dt} = A_1 \exp\left(-\frac{Q'}{RT}\right) f(\sigma_{\text{loc}}),\qquad(7.6)$$

where A_1 is a constant and, here, Q' is the activation energy for matter transport away from the crack tip. Assume that Q' equals Q, the activation energy for matter transport involved in creep. Let the stress-dependent function, $f(\sigma_{\text{loc}})$, be expressed by a power law of the type $(\sigma)^z$. When the stress exponent n equals z, a unique MG relationship is obtained. Tip blunting and general material degradation ahead of the crack tip does not affect these calculations.

An integral,

$$\int \left[\frac{da}{(da/dt)}\right],\qquad(7.7)$$

integrated from an initial flaw size at a_0 to the final crack length, a_f, provides a value for the time spent in crack propagation, t_p, given as

$$t_p = A_2 \exp\left(\frac{Q}{RT}\right)(\sigma)^{-z} f'(a_0),\qquad(7.8)$$

where A_2 is a constant and $f'(a_0)$ is a function of the initial flaw size. Equation (7.8) may be rearranged to provide an expression similar to Eq. (7.5), as follows:

$$t_p = A_3 (\dot{\varepsilon}_s)^{-1} \left(\frac{\sigma}{E}\right)^{(n-z)},\qquad(7.9)$$

where a_3 is a constant, as seen from a comparison with Eq. (7.5), where $z > n$. When $n = z$, i.e., when the stress dependence of the failure mechanism is the same as that of creep, Eq. (7.5) becomes similar to the original MG relationship and, if $z > n$, stratification of the MG lines may be expected. The variation of stress

over time-of-failure at various temperatures is shown in Fig. 7.8. Thus, lower
stresses at higher temperatures should result in longer lives, or alternately, higher
stresses at lower temperatures should result in shorter lives than a correlation with
creep rate alone would predict.

In addition, the prediction of failure strength, as a function of temperature, is
shown in Fig. 7.9, based on the modified MG relationship (Eq. 7.2). The data show
a change in slope between the two regimes, i.e., at <1.589 K and at >1589 K.
A change in slope in a plot of stress versus failure time is often taken as an
indication of a possible change in the failure mode (Fig. 7.8).

In the above discussion, it was indicated that the MG relationship is applicable to
the prediction of the stress rupture life of NT154 silicon nitride, as a function of
temperature or stress and that the original MG equation may be modified for
application to rupture-life prediction (e.g., of NT154). The MG lines may be

Fig. 7.8 Rupture data plotted
with prediction from the
modified Monkman–Grant
relation (Eq. 7.2). Note the
change in slope in the data, as
well as in the predicted line
at >1589 K. Menon et al. [2].
With kind permission of John
Wiley and Sons

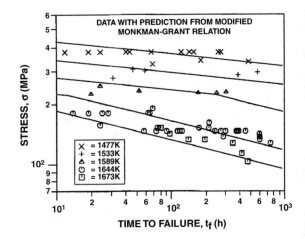

Fig. 7.9 Predicted failure
strength from the modified
Monkman–Grant relationship
is plotted against temperature
for specific failure times.
Menon et al. [2]. With kind
permission of John Wiley
and Sons

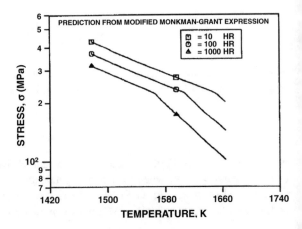

explained based on crack growth. The results point to a failure process in which the crack dependence on stress is greater than that required for creep.

As indicated earlier, creep is a time-dependent deformation and, therefore, failure in ceramics may also show time-dependent. As such, this failure may be considered as a delayed failure, because, in the accepted application, the fracture is not expected to occur suddenly. Notably, two mechanisms lead to failure: (a) slow crack growth and (b) creep rupture. In (a), slow crack growth, initiated by a pre-existing flaw, continues until a critical crack size is reached, causing catastrophic failure. The failure is 'catastrophic,' since a manufactured part is designed for long-term use; (b) creep rupture is a bulk damage that occurs in the material. Void nucleation and their coalescence lead to eventual macrocracks, which then propagate to failure by fast fracture. The successful application of advanced ceramics depends on the proper characterization of material behavior and the use of an appropriate design and manufacturing techniques. It is essential for the prediction of reliable ceramic service lifetimes under creep conditions to have knowledge of fast fracture and SCG. Modern techniques have developed ceramics for high-temperature applications that are highly resistant to creep deformation. For instance, special ceramics have been produced for turbine engine components with a service lifetime on the order of 10,000–30,000 h (usually at low stresses). It is general practice to add additives to the monolithic ceramics in order to increase the stress level. Techniques have been developed for predicting the lifetimes of ceramic structural components operating under creep conditions, where creep rupture might set in.

Again, a MG creep rupture criterion or its modification may be used to predict the cumulative damage. Si_3N_4 continues to provide a good example of creep rupture. Table 7.1 summarizes tests performed on hot-pressed Si_3N_4. Whisker-strengthened composite Si_3N_4 is included. Some of the illustrations compare these systems. The Si_3N_4undeformed and superplastically deformed specimens were doped with 3% alumina and 5% yttria. The resulting composite was reinforced with 20 wt% SiC whiskers. The powder from these components was hot-pressed at 1750 °C in a nitrogen atmosphere. The whiskers were 10–100 μm long, with a diameter of 0.1–1.0 μm. A glassy phase was present continuously around the silicon nitride grains. The thickness of the glassy phase between two grains typically ranged from 1 to 3 nm. The glassy phase in the silicon nitride is shown in Fig. 7.10. Tensile creep testing was applied to the cylindrical specimen (25 mm long and 2.8 mm in diameter). Tensile creep is the most frequently used stress, but loading by other stresses are also used in creep tests (compressive, torsional). The elongation (change in the distance) was measured by an optical extensometer. These creep experiments were conducted in air and performed at 1200, 1250, 1300, and 1350 °C, with applied stresses ranging from 70 to 250 MPa, as indicated in Table 7.1. The creep curves at 1200 °C are shown in Fig. 7.11.

Failure is seen at this temperature at 250 MPa, in the transient creep range; at 200 MPa and at lower stresses, the results showed the steady state, but at 1300 °C, signs of accelerated creep are observed, as shown in Fig. 7.12. Substantial

Table 7.1 Summary of creep tests for hot-pressed silicon nitride. Ohji and Yamauchi [3]. With kind permission of John Wiley and Sons

Test temperature (C)	Stress (MPa)	Minimum strain rate (l/s)	Life (h)	Total strain (%)	Note
Monolith					
1200	250	5.0×10^{-8}	45	1.0	1
1200	250	4.4×10^{-8}	40	0.9	1
1200	250	4.1×10^{8}	51	1.0	1
1200	200	1.8×10^{-8}	148	1.4	2
1200	150	8.2×10^{-9}	202	1.0	2
1200	150	7.5×10^{-9}	350	1.6	2
1200	100	1.4×10^{-9}			4
1200	70	5.0×10^{-10}	3280	0.8	2
1250	100	5.8×10^{-9}	360	1.3	2
1300	150	1.6×10^{-6}	2	1.1	3
1300	100	4.1×10^{-8}	12	1.4	3
1300	70	8.7×10^{-8}	25	1.4	3
1300	70	7.2×10^{-8}	19	1.2	3
1350	70	8.1×10^{7}	4	2.7	3
1350	40	9.0×10^{-8}	55	3.4	3
Composite					
1200	250	4.0×10^{8}	56	1.0	1
1200	200	1.6×10^{-8}	95	0.9	2
1200	200	1.5×10^{-8}	132	1.1	2
1200	200	1.4×10^{-8}	195	1.5	2
1200	150	4.8×10^{-9}	254	0.8	2
1200	100	8.0×10^{10}			4
1200	100	6.4×10^{10}			4
1250	100	3.1×10^{-9}	605	1.1	2
1300	100	2.5×10^{-7}	14	1.4	3
1350	70	8.2×10^{-7}	6	2.9	3
1350	40	1.2×10^{-7}	48	3.2	3

Notes 1 Fracture in a transient creep regime. *2* Fracture in a steady-state creep regime. *3* Fracture in an accelerated creep regime. *4* Interrupted before fracture

accelerated creep exists at 1350 °C, as shown in Fig. 7.13. The accelerated creep rate at 1350 °C causes creep rupture under 70 MPa in the specimen, associated with microcracks distributed throughout the entire specimen. In Fig. 7.14, the spread of a crack is illustrated. These microcracks reduce the overall cross section of the specimen and, thus, the load-bearing capacity is also reduced with the consequent accelerated failure (creep rupture).

Cracks are usually associated with accelerated creep (namely tertiary creep). The creep curves of SiC-whisker-reinforced silicon nitride are similar to those of the monolithic curves (Fig. 7.15). Note that the transient part of the curves in Fig. 7.15

Fig. 7.10 Typical example of glassy phase between two silicon nitride grains. Lattice fringe spacing is 0.66 nm. Ohji and Yamauchi [3]. With kind permission of John Wiley and Sons

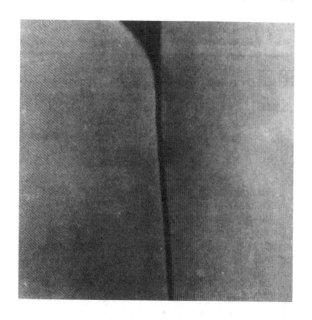

Fig. 7.11 Creep curves of monolithic hot-pressed silicon nitride at 1200 °C. While the failure occurred in the transient creep regime at 250 MPa, steady-state creep was apparently observed at 200 MPa and lower. The test with 100 MPa was interrupted at 1000 h. Ohji and Yamauchi [3]. With kind permission of John Wiley and Sons

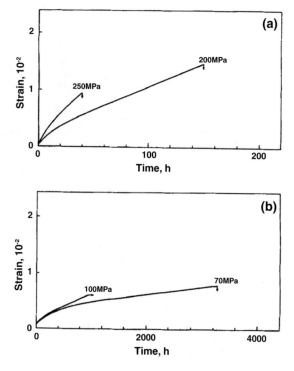

Fig. 7.12 Creep curves of monolithic hot-pressed silicon nitride at 1300 °C. Signs of accelerated creep appear at the ends of the creep curves. Ohji and Yamauchi [3]. With kind permission of John Wiley and Sons

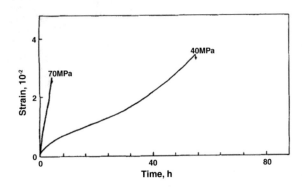

Fig. 7.13 Creep curves of monolithic hot-pressed silicon nitride at 1350 °C. Both creep curves showed substantial accelerated creep regimes. Ohji and Yamauchi [3]. With kind permission of John Wiley and Sons

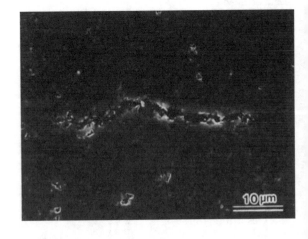

Fig. 7.14 Crack in polished gauge section of a specimen tested at 1350 °C, 70 MPa. This kind of crack was observed only in specimens which exhibited accelerated creep regimes. Ohji and Yamauchi [3]. With kind permission of John Wiley and Sons

is the same as that in the monolithic. The departure of the reinforced composite from the monolithic occurs in the steady state.

The strengthening of the whiskers is apparent in the reduced creep strain of the composite. However, the observed effects of the whisker reinforcement on creep

Fig. 7.15 Creep curves of whisker-free (*monolithic*) and whisker-reinforced (*composite*) hot-pressed silicon nitride at 1200 °C and 100 MPa. The tests were interrupted at l000 h. Ohji and Yamauchi [3]. With kind permission of John Wiley and Sons

Fig. 7.16 Stress dependence of steady-state creep rate. Creep exponent is 3.2 for the monolithic material (■) and 4.3 for the composite material (●) at 1200 °C, 4.5 for the monolithic material at 1300 °C, and 4.0 for both materials at 1350 °C. Ohji and Yamauchi [3]. With kind permission of John Wiley and Sons

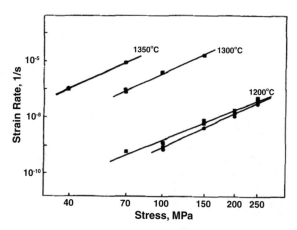

resistance are not so large, when compared with those reported for SiC-whisker-reinforced silicon nitride and alumina (Ferber, Chokshi, Lin).

As on several occasions in earlier chapters, the steady-state creep rate is generally given by Eq. (7.4):

$$\frac{d\varepsilon}{dt} = A\sigma^n \exp\left(-\frac{Q}{RT}\right) \tag{7.4}$$

For the sake of convenience, the known parameters are re-indicated: ε is the creep strain, t is the time, A is a constant, σ is the stress, n is the creep exponent, Q is the apparent activation energy, R is the gas constant, and T is the absolute temperature. The stress–strain relation is plotted in Fig. 7.16. The creep exponents obtained from the best-fit straight lines are in the range of 3–5. The creep rate (log scale), as a function of inverse temperature, is shown in Fig. 7.17. The activation energies evaluated for the monolithic and whisker-reinforced specimens are 1065 and 1190 kJ mol^{-1}, respectively.

The MG relationship was also used for this composite to evaluate the time-to-failure [given above in Eq. (7.1)], but rewritten somewhat differently here, yet with the same meaning, as

Fig. 7.17 Temperature
dependence of the
steady-state creep rate. The
activation energy is
1065 kJ/mol for the
monolithic material (■) and
1190 kJ/mol for the
composite material (●) at
100 MPa, and 1032 kJ/mol
for the monolithic material at
70 MPa. Ohji and Yamauchi
[3]. With kind permission of
John Wiley and Sons

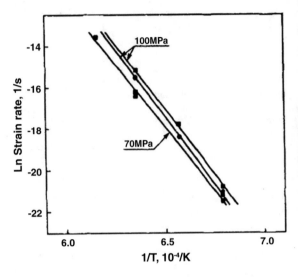

Fig. 7.18 Relationship
between time-to-failure and
steady-state creep rate for the
monolithic (■) and composite
(●) materials. The *downward
arrows* indicate tests with
large accelerated creep (tests
at 1350 °C). The *upward
arrows* indicate interrupted
tests. Ohji and Yamauchi [3].
With kind permission of John
Wiley and Sons

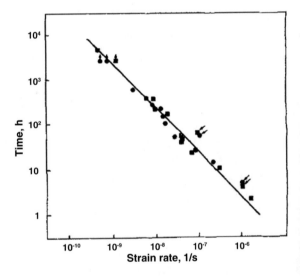

$$C = t_f \left(\frac{d\varepsilon}{dt}\right)^m, \tag{7.1}$$

where C is a constant, t_f represents time-to-failure, and m is a strain-rate exponent.
The collected data forms an almost straight line, with an exponent $m = 1$, regardless
of stress and temperature, as shown in Fig. 7.18.

The fractured surface after creep failure is illustrated in Fig. 7.19. The SEM
shows a fractured surface at 1200 °C and 250 MPa exhibiting slow crack growth,
originating from a preexisting flow (Fig. 7.19a).

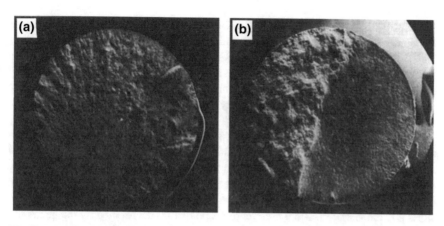

Fig. 7.19 Fracture surfaces of tensile creep specimens. **a** 1200 °C, 250 MPa and **b** 1300 °C, 100 MPa. Ohji and Yamauchi [3]. With kind permission of John Wiley and Sons

Creep damage zones on the fractured surface are shown in Fig. 7.19b, but, in both figures, a flat, mirror-like region is also present. In Fig. 7.19a, the fracture surface indicates a slow crack region. The creep-damage zone tends to appear in the steady- and (accelerated) tertiary-creep stages, whereas the slow crack region (originating from a preexisting flow) is associated with transient creep. The creep-damage zone seems to be produced by the linkage of a number of grain-boundary microcracks of facet-sized cavities. Thus, it can be suggested that steady-state creep stage is associated with cavity formation.

The time-to-failure (fracture), t_f, in a creep test (Eq. 7.1, rewritten) may be related to

$$t_f = A\sigma^{-N}, \tag{7.10}$$

where A and N are constants. Here, the exponent N determines the stress-dependent lifetime of the specimen. As such, creep fracture consists of three aspects: (a) the formation of a facet-sized cavity, (b) the coalescence of a facet-sized microcrack, and (c) the coalescence (or growth) of the microcracks into a critical-sized macrocrack (being the creep-damage zone). Based on these three aspects, the total time-to-failure is

$$t_f = t_p + t_{ml} + t_{m2}. \tag{7.11}$$

In this equation, t_p is the time of facet-sized cavity formation; t_{ml} is the time of the facet-sized cavity coalescence into a microcrack; and t_{m2} is the time it takes the microcrack to grow into a macrocrack. As previously stated, the total failure time is dependent on both t_{ml} and t_{m2}. It was observed that, at 1200 °C, the specimens ruptured in the steady-state stage of creep and only facet-sized cavities occasionally or rarely appeared, while microcracks could not be detected at 1350 °C. It seems that a critical-sized crack forms in such a short time that microcracks and macro-cracks are rarely produced. Thus, t_{m1} and t_{m2} are negligibly small compared to t_p, which is almost equal to t_f. At 1350 °C, however, the cavity coalesces up to a microcrack, thought further crack growth is decreased by crack blunting or by grain (or whisker) pull-out. As such, the contribution of t_{m1} and t_{m2} to t_f is significant. The time required to create a facet-sized cavity may be considered under two conditions: (a) a constrained condition, where cavity formation is isolated and constrained by the surrounding grains, or (b) an unconstrained one, where cavitation occurs uni-formly throughout the microstructure. Under the constrained condition, a product of t_p and $d\varepsilon/dt$ depends primarily on the geometric parameters and is relatively independent of the material transport mechanism. In contrast, t_p, under the unconstrained condition, is mechanism-sensitive and is a function of the applied stress and the viscosity. If the product of t_p and $d\varepsilon/dt$ is determined by the geometric parameters (the initial thickness of the glassy phase, the grain-boundary length, the hole radius at nucleation, and the half spacing between the holes), the validity of the MG relationship, irrespective of temperature and stress, explains creep rupture without accelerated creep. Thus, specimen lifetimes follow the MG relationship, except for fractures with large, accelerated creep regimes. Fractures with large, accelerated creep deviate from the predicted line, due to the appreciable time for crack coalescence, i.e., the sum of t_{m1} and t_{m2}.

References

1. Pelleg J (2014) Mechanical properties of ceramics. Springer, Berlin, p 79
2. Menon MN et al (1994) J Am Ceram Soc 77:1235
3. Ohji T, Yamauchi Y (1995) J Am Ceram Soc 76:3105
4. Monkman FC, Grant NJ (1956) Proc Am Soc Test Mater 56:593

Chapter 8
Superplasticity

Abstract Some materials are capable of undergoing large tensile extension of the order of hundreds and even thousands of percent. These superplastic materials are strong and ductile at low temperatures and exhibit high plasticity at high temperatures. They deform without showing necking. The large plastic extension enables fabrication into intricate shapes by simple forming process which is quite important in the case of ceramics where usually machining is a difficult process. A much investigated ceramics is the Y-TZP which can show an extension over ~100%. Structural considerations are grain size, grain boundaries and cavitation. Fine-grained ceramics can be superplastic at elevated temperatures.

8.1 Introduction

Some metals and alloys are capable of enduring large, tensile extension on the order of hundreds and even thousands of percent. Such materials, called 'superplastic', are generally strong and ductile at low temperatures and exhibit high plasticity (and low strength) at high temperatures, which enables their fabrication into intricate and complex shapes by means of simple forming processes. The creep resistance of these materials may be introduced after some post-forming treatment. Considerable attention has been devoted to the study of superplasticity, not only because of the forming capability of such materials, but also due to the effects of such phenomena on their mechanical properties. Lately, there has also been a rise of interest in exploring the possibility of superplasticity specifically in ceramic materials. Y-TZP, among the first ceramics studied for superplasticity, showed an extension over 100% [9, 10]. This observation initiated a great interest in the possibility of obtaining superplastic characteristics in other ceramics and regarding the potential application of ceramic components in high-temperature applications.

© Springer International Publishing AG 2017

J. Pelleg, *Creep in Ceramics*, Solid Mechanics and Its Applications 241,
DOI 10.1007/978-3-319-50826-9_8

8.2   The Principles of Superplasticity

The atoms in ceramic materials are held together by chemical bonds. The two most common chemical bonds in ceramic materials are covalent and ionic, which are responsible for their brittleness (unlike metallic bonding in metals). Despite the common knowledge that ceramics are mostly brittle materials, almost without plastic deformation at ambient temperatures, some fine-grained ceramics can be superplastic at elevated temperatures. An illustration of superplasticity in ceramics appears in Fig. 8.1, showing an elongation of over 470%.

Some of the requirements for superplasticity may be classified as—(a) mechanical and (b) structural. The mechanical requirement pivots around plastic stability and the structural one is considered with the grain size, grain boundaries and cavitation. When (a) and (b) are satisfied during deformation, superplastic behavior can be anticipated in some ceramics, just as in metals and alloys, where superplasticity was first observed and studied. One of the first ceramics exhibiting superplasticity was Y-TZP. Tensile-test curves of Y-TZP, with two-yttria content, is illustrated in Fig. 8.2, and the properties and crystalline phases of the Y-TZP materials are summarized in Table 8.1. Ce additives are included both in the curve and in Table 8.1.

In the last column of Table 8.1, t and c stand for tetragonal and cubic, respectively. As known, polycrystalline materials generally break under tension after a modest (usually less than 50%) elongation. Under superplastic conditions of tensile testing, however, very high elongations, of more than 500% or, in specific cases, even higher, may be obtained.

Usually superplasticity is attained in the low-strain range, from 10^{-5} to 10^{-3} s^{-1}. In commercial applications, there is interest in forming a component at high-strain rates, on the order of 10^{-2}–10^{-1} s^{-1}. However, in the case of ceramic materials, superplastic deformation has been restricted to low-strain rates on the

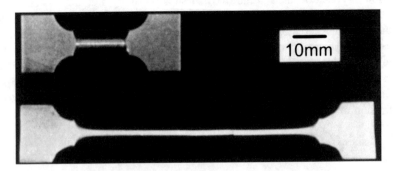

Fig. 8.1 Undeformed and superplastically-deformed Si$_3$N$_4$ specimens. An elongation of over 470% is noted (Reprinted with permission from *J. Ceram. Soc. Japan*). Wakai et al. [9]. With kind permission of Elsevier

Fig. 8.2 True stress-true
strain curves for 2Y, 4Y and
6Ymaterials at 1450 °C in
tension test. Wakai [11]. With
kind permission of Elsevier

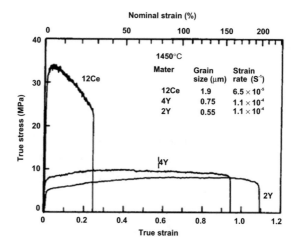

Table 8.1 Properties of ZrO_2 polycrystals adopted for experiments. Wakai [11]. With kind permission of Elsevier

Material composition		Density (g cm^{-3})	Grain size (μm)	Phase
2Y	2 mol% Y_2O_3	6.04	0.55	t
3Y1	3 mol% Y_2O_3	6.07	0.51	t + c
3Y2	3 mol% Y_2O_3	6.07	0.50	t + c
4Y	4 mol% Y_2O_3	6.03	0.75	t + c
6Y	6 mol% Y_2O_3	5.95	5.5	c
12Ce	12 mol% CeO_2	6.22	1.9	t

order of 10^{-5}–10^{-4} s^{-1} for most oxides and nitrides [6], with the presence of intergranular cavities leading to premature failure. In ceramics, high-strain-rate superplasticity (HSRS) requires a small initial grain size, enhanced diffusivity, the suppression of dynamic grain growth, a low defect content and an homogeneous microstructure. Furthermore, HSRS may be induced by doping Y-TZP, with the aim of suppressing cavitation damage during deformation, as realized by post-deformation microstructure. Thus, it was observed that a composite ceramic material consisting of tetragonal zirconium oxide, magnesium aluminate spinel and alumina phases exhibits superplasticity at strain rates up to 1 s^{-1} and this composite also attains a large tensile elongation, exceeding 1050%, with a strain rate of 0.4 s^{-1} [6]. The deformed microstructure of the material indicates that this superplasticity is due to a combination of limited grain growth in the constitutive phases and the intervention of dislocation-induced plasticity in the zirconium oxide phase.

Plastic stability was mentioned above as one of the prerequisites for superplasticity. Deformation under tension may be expressed as:

$$\sigma = K\dot{\varepsilon}^m \exp{(\gamma\varepsilon)} \qquad (8.1)$$

γ is the strain-hardening exponent, m is the strain-rate sensitivity index and K is a coefficient. Hart has indicated that the following relation (8.2) should be satisfied for tensile instability and uniform elongation by:

$$m + \gamma \geq 1 \qquad (8.2)$$

To have superplasticity (large elongation), high values of m, $m > 0.3$, are required when $\gamma = 0$. This means that the necking rate is slow. The relation for creep during a diffusion-controlled process (see Chap. 4, Eq. (4.2), for example) may be given as:

$$\dot{\varepsilon} = \frac{AGb}{kT}\left(\frac{b}{d}\right)^p \left(\frac{\sigma}{G}\right)^n D_0 \exp\left(-\frac{Q}{RT}\right) \qquad (8.3)$$

Rewriting the familiar parameters from Eq. (8.3): G is the shear modulus, b is the Burger's vector, d is the grain size, n is the stress exponent ($n = 1/m$), p is the exponent of the inverse grain size, D_0 is the frequency factor and A is a dimensionless coefficient. Clearly, k and Q are the Boltzmann's constant and the activation energy, respectively. The stress exponent for creep for fine-grained ceramics was found to be 1–3, when 3 is for dislocation glide and climb, while 1 is related to diffusional creep, corresponding to $0.3 < m \leq 1$ and, thus, satisfying the requirements for uniform elongation. Usually, fine-grained ceramics fracture under tension and are not superplastic. One of the models developed for superplasticity involves GBS accommodated by diffusion. Ashby and Verral developed such a model and gave the following:

$$\dot{\varepsilon}_D 100 \frac{\Omega}{kTd^2}\left[\sigma - \frac{0.72\Gamma}{d}\right]D_L\left(1 + \frac{3.3\delta D_{gb}}{dD_L}\right) \qquad (8.4)$$

Γ is the grain-boundary free energy. The assumption in Eq. (8.4) is that the grain-boundary surface is a perfect source and sink for vacancies. If this assumption is not satisfied, the diffusional creep is controlled by the interface reaction:

$$\dot{\varepsilon}_i = B\frac{\sigma^2}{d} \qquad (8.5)$$

The altered stress exponent may be explained by the assumption that the total-strain rate, $\dot{\varepsilon}_t$, is expressed as:

$$\frac{1}{\dot{\varepsilon}_t} = \frac{1}{\dot{\varepsilon}_D} + \frac{1}{\dot{\varepsilon}_i} \qquad (8.6)$$

According to Eq. (8.6), the mechanism of superplasticity is GBS, accomodated by interface-reaction-controlled diffusion.

In two-phase composites, flow behavior is different than that in single-phase materials, particularly when the creep properties of the second-phase grains differ markedly. A much-investigated composite is Y-TZP with Al_2O_3 additives. The deformation of the composite, in which the second-phase grains are uniformly distributed among the matrix grains, is explained by means of a rheological model. The flow (non-Newtonian) of the matrix is expressed as:

$$\dot{\varepsilon} = A\sigma^n \tag{8.7}$$

Chen's rheological prediction was that the constitutive equation for the composite containing spherical inclusions of volume fraction, p, should become:

$$\dot{\varepsilon} = (1-p)^q A\sigma^n \tag{8.8}$$

q being an exponent. Since the superplasticity of Y-TZP and the interface-reaction-controlled diffusion creep of fine-grained Al_2O_3 were expressed by a stress exponent of 2, the stress exponents of about 2 (for the composites) were consistent with the rheological model. From a rheological point of view, super-plastic flow may be expected in a wide spectrum of ceramic composites with various second-phase grains.

Coming back to Y-TZP, Fig. 8.3 shows the effect of alumina additives to TZP. The composite ZrO_2/Al_2O_3 has been considered as a model material for studying superplasticity in two-phase composites. The addition of Al_2O_3 reduces elongation (see Fig. 8.3) and, as the amount increases, a further decrease in elongation is observed. Superplastic elongation (more than 120%) may be achieved by composites containing 86 vol.% of Al_2O_3 at 1550 °C, as shown in Fig. 8.3. The stress exponent was around the value of 2, being high at an intermediate volume fraction of Al_2O_3, but decreasing to slightly lower values at both high and low volume fractions. The activation energy of the composites changed

Fig. 8.3 Effect of Al_2O_3 content on stress–strain curves of composites at 1550 °C. Wakai [11]. With kind permission of Elsevier

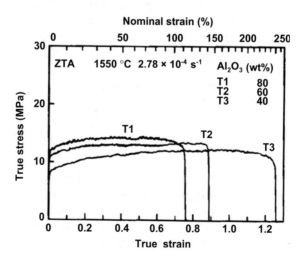

Fig. 8.4 Effect of Al_2O_3 content on activation energy at various stresses. Wakai [11]. With kind permission of Elsevier

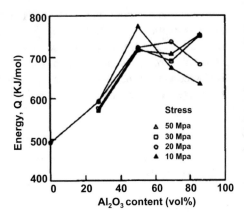

from 550 to 700 kJ/mol when the Al_2O_3 content increased to more than 50 vol.%, as seen in Fig. 8.4.

In summary, the strain rate of a composite was given in Eq. (8.7), while the strain rate for the rheological model with spherical grains is found in Eq. (8.8). The interface-reaction-controlled diffusional creep in fine-grained Al_2O_3 polycrystals was affected by the dispersion state of the ZrO_2 particles. A small amount of ZrO_2 particles at the grain boundaries reduced the creep rate and increased the activation energy for creep. Small-grained ceramics are essential for superplasticity, and further reduction of the grain size to a nanometer level can enhance superplasticity, with the prospect of forming ceramics more effectively.

HSRS is essential for industrial applications, such that further studies are desirable for the enhancement of knowledge on advanced ceramic engineering materials, especially in regard to the factors limiting the strain rate of superplastic deformation. As mentioned above, intergranular cavitation and dynamic grain growth are among those factors of considerable interest. Furthermore, in the case of ceramic materials, reduction of the initial grain size, enhanced diffusivity, suppressed dynamic grain growth, an homogeneous microstructure and a reduced number of residual defects are essential for HSRS. Some of this will be presented below.

Ideally, superplastic deformation may be expressed by the stress–strain rate equation given as:

$$\dot{\varepsilon} = A \exp\left(-\frac{Q}{RT}\right)\sigma^n d^{-p} \qquad (8.9)$$

Equation (8.9), for a certain stress and temperature, means that the strain rate can be increased by decreasing the grain size. For example, decreasing the grain size to $0.32d$–$0.46d$ for a value of p of 2–3, increases the strain rate by a factor of 10. Similarly, an increase in the term of $A \exp(-Q/RT)$ enhances diffusion along the grain boundaries and within the grains and, thus, enhances the strain rate. More additives (doping) with different valences, enhance diffusion and also have an effect on grain growth, stress concentration and relaxation during deformation. In Eq. (8.9), no consideration is given to the microstructural changes; however, such changes, which are observed during superplastic deformation (like cavitation), are of great importance. Note the cavitation in Fig. 8.5 during the Y-TZP superplastic deformation. Cavitation leads to premature failure and degrades the post-deformation strength. It is believed that intergranular cavitation occurs due to the stress concentration, as a consequence of microstructural inhomogeneity and of the chemical composition in composites. However, also in monolithic ceramics, grain size and shape are not always homogeneously distributed. Thus, stress distribution, chemical potential and, consequently, GBS and grain accommodation, associated with diffusion, also become inhomogeneous, leading to breakdown in local regions. As such, the stress concentrations cannot be sufficiently relaxed. Those regions are multiple grain junctions and phase boundaries, in which cavity nuclei are frequently observed after superplastic deformation (Fig. 8.5b). It is possible to estimate [3] the relaxation length, Λ, over which grain-boundary diffusion can relax the stress concentrations caused by the deformation, as follows:

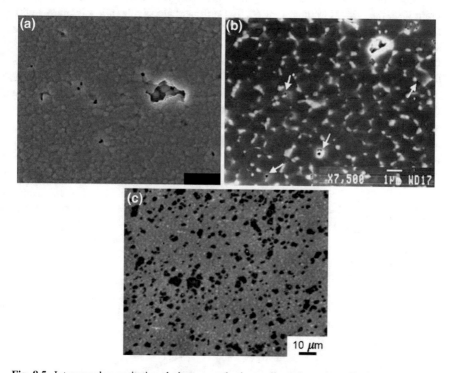

Fig. 8.5 Intergranular cavitation during superplastic tensile deformation. Cavity nucleation at multiple junctions in **a** 3Y-TZP; **b** 10-vol.% ZrO_2 (3Y)-dispersed Al_2O_3 and; **c** micrometer-sized cavities grown from cavity nuclei and pre-existing defects in 3Y-dispersed Al_2O_3. The tensile axis is horizontal. Hiraga et al. [4]. With kind permission of Elsevier

$$\frac{\Lambda}{d} = \left(\frac{L}{d}\right)^{1/[1-s(n-1)/3]} \tag{8.10}$$

L is the characteristic diffusion length and s is the extent of the stress singularity at the triple junction in the absence of diffusion. L is given as:

$$L = \left(\frac{\delta D_b \sigma \Omega}{k_B T \dot{\varepsilon}}\right)^{\frac{1}{3}} \tag{8.11}$$

Clearly, δD_b is the product of the grain-boundary coefficient and its width, which is known to be proportional to $\exp(-Q/RT)$. Ω is the atomic volume. Figure 8.6 indicates the critical grain size over which stress concentration may be relaxed by grain-boundary diffusion. For a detailed evaluation see [3]. Figure 8.6 shows that the critical grain size for a given temperature decreases rapidly with an increase in the strain rate.

The estimation indicates that the critical grain size at $T = 1400$ °C is 0.26 μm for ordinary superplasticity at a strain rate of 10^{-4} s^{-1}, whereas it is reduced to 0.06 μm for HSRS at 10^{-2} s^{-1}. Furthermore, this figure suggests that, for a certain constant grain size, the critical temperature, T_c, for sufficient stress relaxation increases with strain rate. For instance, for $d_c = 0.2$ μm, T_c increases from 1400 to 1600 °C as the strain rate increases from about 2×10^{-4} to 10^{-2} s^{-1}. In addition, enhanced diffusion is also indicated in the figure below. Note that enhanced diffusion increases the critical grain size for a given strain rate. Specifically, when grain-boundary diffusivity increases by a factor of 50 at 1400 °C, the critical grain size increases from 0.06 to 0.2 μm at 10^{-2} s^{-1}. This estimation also shows that enhanced diffusion, by factors of 10 and 50, correspond to increases in the critical temperature by about 100 and 200 °C, respectively. In fact, when the width of grain-boundary product with the boundary diffusion is $50\delta D_b$, stress relaxation at 10^2 s^{-1} is

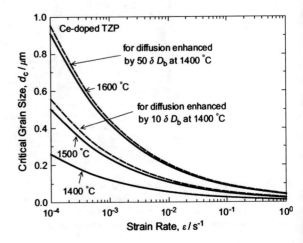

Fig. 8.6 Estimation of the critical grain size over which grain-boundary diffusion can relax stress concentrations caused by deformation. Hiraga et al. [4]. With kind permission of Elsevier

expected to become sufficient at a grain size of $d = 0.2$ μm, even at 1400 °C. Thus, grain-size reduction, suppressed grain growth and enhanced diffusion are essential for achieving sufficient stress relaxation and suppressed cavity nucleation during high-strain-rate deformation. Summarizing this aspect of cavity formation, one can state that the probability of cavity nucleation depends on grain-boundary diffusivity, surface energy, γ_s, grain-boundary energy, γ_b, and geometrical factors, such as the dihedral angle between the grain facets. For a fixed grain size, for example, the probability of cavity nucleation decreases with increasing δD_b and γ_s, and with decreasing γ_b.

As inferred from the above, to reduce cavitation, it is essential to suppress dynamic grain growth, in order to maintain a lower level of flow stress (Eq. 8.9) and a shorter length of grain facet to be relaxed by diffusion (Eq. 8.10). A model for dynamic grain growth, which agrees with experimental data, gives the differential of grain size during deformation as the sum of the dynamic and static components, which are expressed by the first and second terms of the right-hand side, respectively, of Eq. (8.12), given below as:

$$d(d) = \alpha d(d\varepsilon) + \frac{k}{m} d^{1-m}(dt) \tag{8.12}$$

where a is a constant dependent on the grain-shape and grain-size distribution, and m and k are the grain-growth exponent and kinetic constant, respectively, of the static grain-growth law, $d^m - d_0^m = kt$, where t is the heating time. Under an initial condition of $d = d_0$ at $\varepsilon = 0$, Eq. (8.12) yields the following equation for the constant displacement-rate loading:

$$d = \left[d_0^m \exp\left(\alpha m\varepsilon\right) + \frac{k\{\exp\left(\varepsilon\right) - \exp\left(\alpha m\varepsilon\right)\}}{\dot{\varepsilon}_0 \left(1 - \alpha m\right)} \right]^{1/m} \tag{8.13}$$

Equation (8.12) implies that the grain size for a given strain becomes smaller with a reduction in the initial size and with an increase in the deformation rate. In addition, the experimental data indicate that the value of α for superplastic deformation is insensitive to chemical compositions. α is about 0.5–0.6 for various oxide materials, such as ZrO_2–Al_2O_3, ZrO_2-spinel-Al_2O_3, $ZrO2(3Y)$ and superplastic metals, Zn–Al. Such data, used with Eq. (8.13), suggest that the mechanism of dynamic-growth rate may be estimated from k and m. Consequently, it is essential to suppress the static grain growth by grain-boundary pinning and/or dragging. Highly-limited grain growth may be expected in a microstructure consisting of three or more phases, where the amount of each phase is similar. This situation is beneficial for achieving suppressed grain growth, since grain growth occurs by the migration of such grain boundaries and/or by the growth of grains through interphase boundary diffusion.

For a given combination of chemical composition, grain size and relative density, superplastic properties may vary with differences in the homogeneity of the microstructure or the number of fine residual defects, which have negligibly small volume and, hence, have little effect on the relative density. A typical example is shown in Fig. 8.7 for 10 vol.% ZrO_2 (3Y)-dispersed Al_2O_3 [5]. For a grain size of 0.45 mm and a relative density of 99.5%, the tensile ductility of a material prepared by conventional dry processing (D in the figure) was about half that of a material prepared by colloidal processing (designated in the figure as C). This difference is the result of enhanced cavity damage, due to accelerated grain growth. The former is a consequence of the higher density of fine residual defects, while the latter is the result of less effective grain-boundary pinning by ZrO_2. Both of these arise from the agglomeration of ZrO_2 and Al_2O_2, which is process-dependent.

Table 8.2 lists ceramics in which HSRS may be attained.

Factors necessary or desirable for attaining HSRS are shown in the first column of Table 8.3. The second column summarizes the relationship between these factors and superplastic deformation or cavitation. The third column indicates the dependence of these factors on the process, P, chemical constituents, C, and phases, Ph. For a given combination of chemical constituents and phases, factors (a), (b), (d) and (e) are strongly process-dependent, particularly in composites. Note that some of the factors listed in Table 8.3 appear to conflict with each other. The additive should simultaneously enhance grain refinement and second-phase pinning. By the use of additives and the doping of Y-TZP and composites synthesized from ZrO_2, Al_2O_3 and MgO_2, HSRS may be attained with a tensile ductility of 300–2500% at a strain rate of $0.01–1.0 \text{ s}^{-1}$. The deformed microstructure in the ZrO_2 revealed the spinel grains' dense intragranular dislocations and the dislocation-related substructures, such as sub-boundaries (Fig. 8.8). Such dislocation substructures and grain elongations were not found in undeformed materials.

Fig. 8.7 Process-dependence of **a** tensile ductility and **b** accumulated cavity damage in 10-vol.%-ZrO_2 (3Y)-dispersed Al_2O_3 [5]. Colloidal processing and conventional dry processing are designated as C and D, respectively. Hiraga et al. [4]. With kind permission of Elsevier

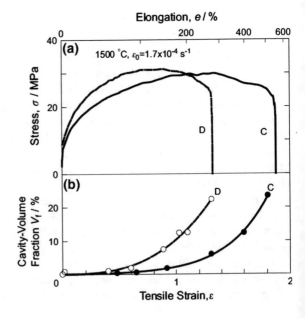

Table 8.2 Oxide ceramics exhibiting high-strain-rate superplasticity. Hiraga et al. [4]. With kind permission of Elsevier

Material	Prerequisites	$\dot{\varepsilon}$ (s^{-1})	e_f (%)	T (°C)	Ref.
20Al$_2$O$_3$–ZrO$_2$(3Y)^a	(b), (c)	0.04	300	1650	[19]
5SiO$_2$–ZrO$_2$(2.5Y)^b	(b), (e)–(g)	0.01	360	1400	[48]
2CaO–2TiO$_2$–ZrO$_2$(3Y)^c	(b), (c)	0.01	400	1400	[49]
0.2Al$_2$O$_3$–ZrO$_2$(3Y)^a	(a), (c)–(e)	0.03	370	1450	[50]
3(Y$_2$O$_3$,MgO)–97 (Zr$_{0.95}$Ti$_{0.05}$)O$_2^b$	(a), (c)–(e)	0.01, 0.01	220, 300	1350, 1450	[57]
0.2Mn$_3$O$_4$–0.3Al$_2$O$_3$–ZrO$_2$(3Y)^b	(a), (c)–(e)	0.01	600	1450	[58]
40ZrO$_2$(3Y)–30spinel-Al$_2$O$_3^c$	(a)–(d), (h)^d	0.01, 0.08, 1.0	500, 2500, 390	1500, 1650, 1650	[59–61]
30MgAl$_2$O$_4$–ZrO$_2$(3Y)^c	(a)–(d), (h)^d	0,02, 0.7	660, 250	1450, 1550	[62–65]

^{a}mass%
^{b}mol%
^{c}vol.%
^{d}See Table 8.1

Table 8.3 Microstructural and compositional conditions for attaining high-strain rate superplasticity. Hiraga et al. [4]. With kind permission of Elsevier

Prerequisites	Relationship with superplastic deformation or cavitation	Note^a
(a) Reduced initial grain size	Strain rate, stress relaxation, cavity nucleation (processing-dependent)	P, C, Ph
(b) Suppressed dynamic grain growth	Stress concentrations, cavity nucleation (second-phase pinning and dragging: processing-dependent)	P, C, Ph
(c) Enhanced diffusivity	Strain rate, stress relaxation, cavity nucleation (doping of aliovalent cations)	C, Ph
(d) Homogeneous microstructure	Dynamic grain growth, cavity nucleation (processing-dependent)	P, Ph
(e) Reduced residual defects	Damage due to micrometer-sized cavities (processing-dependent)	P, C, Ph
(f) Low γ_b and high γ_a	Cavity nucleation (grain-boundary segregation)	C, Ph
(g) Enhanced accommodation (by a viscous phase)	Cavity nucleation (glass-phase dispersion, intergranular Si-segregation)	C, Ph
(h) Additional accommodation (by limited intragranular plasticity^b)	Cavity nucleation [relates to (c) and (d)]	C, Ph

^{a}Dependence on processing (P), chemical compositions (C) and phases (Ph)

Elongation along the stress axis was also observed in the spinel grains of the MgAl$_2$O$_4$-dispersed 3Y-ZrO$_2$ in Fig. 8.9. These observations indicate that the grains in these phases may not be perfectly rigid, namely that these grains may deform to some extent by dislocation mechanisms during high-strain-rate loading.

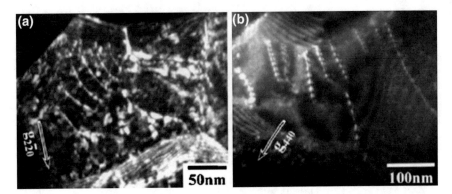

Fig. 8.8 Dislocation substructures observed in **a** ZrO$_2$ and **b** spinel grains of ZrO$_2$-spinel-Al$_2$O$_3$ superplastically deformed at 1500 °C and at ∼0.2 s^{-1}. Hiraga et al. [4]. With kind permission of Elsevier

Fig. 8.9 Comparison of grain shape in MgAl$_2$O$_4$-dispersed ZrO$_2$ **a** before and **b** after superplastic deformation. While the white zirconia grains are equiaxial after deformation, the *dark spinel grains* are noticeably elongated by deformation. The tensile axis is horizontal. Hiraga et al. [4]. With kind permission of Elsevier

Another observation shows the cavitation behavior in Fig. 8.10: (a) indicates elongated cavities along the stress axis, with some small amount of micrometer sized cavities remaining in the matrix. Such a microstructure is obtained when cavity nucleation is suppressed during the growth of residual defects during deformation. If cavity nucleation is active (observed in conventional materials), a large number of micrometer-sized voids grow from pre-existing defects (Fig. 8.5c) during superplastic deformation. The void growth, normal to the stress axis, is followed by coalescence and microcracking, as seen in Fig. 8.10b. Stress concentration and cavitation may be suppressed by adding Si^{4+} and, thereby, enhancing HSRS.

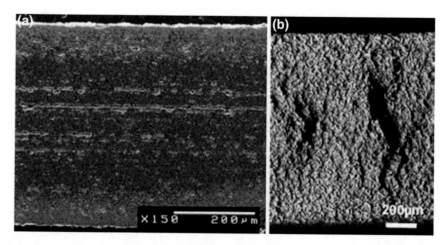

Fig. 8.10 Comparison of cavitation damage between **a** $40ZrO_2$-30spinel-Al_2O_3 deformed to 2500% at 1650 °C and at 10^{-1} s^{-1} and **b** ZrO_2-$10Al_2O_3$ deformed to 550% at 1500 °C and at 10^{-4} s^{-1}. Hiraga et al. [4]. With kind permission of Elsevier

References

1. Ashby MF, Verral RA (1973) Acta Metall 21:149
2. Hart EW (1967) Acta Metall 15:351
3. Hiraga K, Kim B-N, Morita K, Suzuki TS, Sakka Y (2005) J Ceram Soc Jpn 113:191
4. Hiraga K, Kim B-N, Morita K, Yoshida H, Suzuki TS, Sakka Y (2007) Sci Tech Adv Mater 8:578
5. Hiraga K, Nakano K, Suzuki TS, Sakka Y (2002) J Am Ceram Soc 85:2763
6. Kim B-N, Hiraga K, Morita K, Sakka Y (2001) Nature 413:288
7. Nieh TG, McNally CM, Wadsworth J (1988) Scripta Metall 1297:22
8. Nieh TG, Wadsworth J (1990) Ann Rev Mater Sci 20:117
9. Wakai F, Kondo N, Shinoda Y (1999) Curr Opin Solid St M Sci 4:461
10. Wakai F, Sakaguchi S, Matsuno Y (1986) Adv Ceram Mater 1:259
11. Wakai F (1991) Ceram Int 17:153

Chapter 9
Creep and Recovery

Abstract The rate of decrease in deformation after load removal, following a long application during a creep experiment is discussed in this chapter. Constant temperature must be maintained in such experiments in order to eliminate the contributions of thermal expansion (contraction). Monolithic ceramics typically do not possess the elevated-temperature toughness required for safety-critical designs. For this reason, a considerable amount of effort has been devoted to the development of ceramics reinforced with continuous fibers. Most of the ceramics for various applications are usually in composite forms (and contain various additives in specified amounts), depending on their use for a definite purpose. A knowledge of creep-strain recovery behavior may be used to increase the lifetimes of components subjected to sustained- and cyclic-creep load.

The recovery of mechanical properties is the 'work-hardening' (or 'cold-work') rate, which is temperature-dependent. Work hardening involves a thermal-recovery process, superimposed on an thermal deformation process. The above generality applies to creep and its recovery, since creep is a specific sort of a deformation process. Thus, interest is focused on the rate of decrease in deformation after load removal, following a long application during a creep experiment. Constant temperature must be maintained in such experiments in order to eliminate the contributions of thermal expansion (contraction). Monolithic ceramics typically do not possess the elevated-temperature toughness required for safety-critical designs. For this reason, a considerable amount of effort has been devoted to the development of ceramics reinforced with continuous fibers. Most of the ceramics for various applications are usually in composite forms (and contain various additives in specified amounts), depending on their use for a definite purpose. As such, one has to consider the possibility that the various components and the matrix do not necessarily deform or recover simultaneously during a test period. Hence, in this chapter, consideration is given preferentially to composite materials. One exemplary composite ceramic is the silicon carbide Nicalon fiber/calcium aluminosilicate matrix. The Nicalon fiber serves as a reinforcing agent in this ceramic. Nicalon is a ceramic multi-filament silicon carbide fiber (Nippon Carbon Co. Ltd., Japan) of

© Springer International Publishing AG 2017
J. Pelleg, *Creep in Ceramics*, Solid Mechanics and Its Applications 241,
DOI 10.1007/978-3-319-50826-9_9

homogeneously composed ultra-fine beta-SiC crystallites and an amorphous mixture of silicon, carbon, and oxygen. This fiber has excellent strength, retains its properties at high temperatures, and is highly resistant to oxidation and chemical attack.

The fiber content of this ceramic matrix was 40% Nicalon. An important aspect of the development of matrix cracking in continuous-fiber-matrix composites is that creep stress rupture will not necessarily occur, since the possible fiber bridging often has sufficient creep strength to transfer loads across the crack faces. (Nicalon was also used as whiskers in other experiments). This additive improves ceramic toughness and reliability. The redistribution of the stress, between the matrix and the fibers, has a significant effect on the overall creep behavior and on microstructural damage occurring during creep. The damage modes in fiber-reinforced ceramics have been classified by Holmes and Chennant, according to the use of a creep rate mismatch ratio (CMR), defined as the ratio of the creep rate of the fibers to that of the matrix:

$$\text{CMR} = \frac{\dot{\varepsilon}_f}{\dot{\varepsilon}_m}. \tag{9.1}$$

Clearly, $\dot{\varepsilon}_f$ and $\dot{\varepsilon}_m$ are the strain rates of the fiber and the matrix, respectively. For composites with CMR <1, periodic fiber rupture can occur during long-duration, tensile- or flexural-creep loading. This happens as a result of the redistribution of the stress from the matrix to the more creep-resistant fibers. For composites with CMR >1, matrix fracture has been identified as a characteristic creep-damage mode for tensile and flexural loading. The redistribution under stress, from the fibers to the matrix, occurs during creep loading. The composite was hot-pressed at 0″ panels, with either 16 or 32 plies and as 2D panels with a [0/90]$_{4s}$ ply layup (16 plies). The specimens, shown in Fig. 9.1, had a gage length of 33 mm and the thickness was about 3.0 mm for the 16-ply panels and 6.0 mm for the 32-ply panels.

Fig. 9.1 Monotonic tensile behavior of [O]$_{16}$ and [0/90]$_{4s}$ Nicalon SIC/CAS-II composites at 20 and 1200 °C. The experiments were conducted in high-purity argon (≤10 ppm O$_2$) at a loading rate of 100 MPa/s. Wu and Holmes [1]. With kind permission of John Wiley and Sons

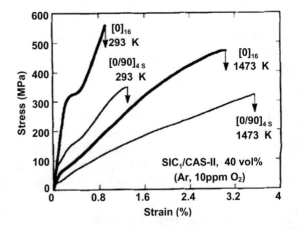

All creep experiments were conducted at 1200 °C. At this temperature, the matrix will contribute little to the overall creep resistance of the composite and most of the load is concentrated in the fiber. The creep experiments of the $[0]_{32}$ specimens were conducted at nominal stresses of 60, 120, 200, and 250 MPa (these stresses ranged from 13 to 52% of the monotonic strength at 1200 °C). The strain recoveries of the unloaded composites that survived 100 h at a stress of 2 MPa were evaluated (temperature maintained at 1200 °C). The experimental recovery test conditions appear in Table 9.1, indicating the effects of the fiber its layup, the strain and strain rates and the respective recoveries. In addition, the effect of cycle duration on strain recovery was examined.

Curves of strain versus time and strain rate versus stress are shown in Fig. 9.2 for the 16- and 32-ply unidirectional specimens. There is no apparent, significant influence of the number of plies on tensile creep for 200 MPa shown in Fig. 9.2a. Furthermore, no stress rupture was observed in the 32-ply specimens between 60

Table 9.1 Summary of loading histories and experimental results. Wu and Holmes [1]. With kind permission of John Wiley and Sons

Fiber layup	Loading history	ε_{100h} (%)	$\dot{\varepsilon}_{100h}$ (s^{-1})	$R_{cr,100h}$ (%)	$R_{t,100h}$ (%)
$[0]_{16}$	200 MPa/100 h	3.38	2.2×10^{-8}		
$[0]_{32}$	120 MPa/100 h	1.36	1.1×10^{-8}	23	32
$[0]_{32}$	60 MPa/100 h + 2 MPa/100 h	0.58	4.6×10^{-9}	27	33
$[0/90]_{4S}$	60 MPa/100 h + 2 MPa/100 h	0.59	4.0×10^{-9}	49	56
$[0/90]_{4S}$	60 MPa/100 h + 2 MPa/100 h	0.55	2.7×10^{-9}	45	52
$[0/90]_{4S}$	60 MPa/100 h + 2 MPa/100 h	0.62	3.9×10^{-9}	51	56
$[0/90]_{4S}$	60 MPa/40 min + 2 MPa/40 min			57/80^a	73/70^a

^{a}Two 40-min cycles (first cycle/second cycle)

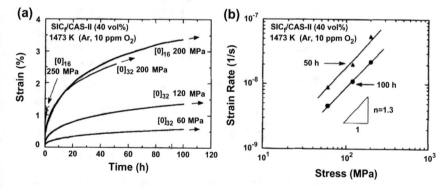

Fig. 9.2 a Total strain versus time for $[0_{16}-$ and $[0]_{32}-$ Nicalon SiC/CAS-II composites crept at 1200 °C in high-purity argon. **b** Stress dependence of 50- and 100-h creep rate for the unidirectional specimens. The experiment with the 32-ply specimen was stopped at 50 h to investigate creep-damage accumulation. Wu and Holmes [1]. With kind permission of John Wiley and Sons

and 200 MPa after exposure to 100 h experiments. Increasing the creep stress to 250 MPa resulted in stress rupture at about 70 min (Fig. 9.2). The strain rate dependence on stress is shown in Fig. 9.2b for the indicated times for the 32-ply unidirectional specimens.

The strain recovery that occurs after cyclic creep is shown in Fig. 9.3. Two methods can be used to indicate recovery, R_{cr} and R_t, which stand for the creep recovery ratio and the total strain recovery ratio, respectively. The strain recovery behavior of 32-ply specimens, which were crept at stresses of 60 and 120 MPa for 100 h and then unloaded to a stress of 2 MPa and held for 100 h, is shown in Fig. 9.3a.

R_{cr} is defined as the creep strain recovered during a particular unloading cycle, $\varepsilon_{cr,R}$, divided by the creep strain for the cycle, ε_{cr} : $R_{cr} = \varepsilon_{cr,R}/\varepsilon_{cr}$. R_t is defined as the total strain recovered within a particular cycle $\varepsilon_{el,R} + \varepsilon_{cr,R}$ divided by the total accumulated strain, ε_t : $R_t = (\varepsilon_{el,R} + \varepsilon_{cr,R})/\varepsilon_t$. As may be seen in Table 9.1, the creep strain recovery ratio, R_{cr}, for specimens crept at 60 MPa is 27%, while for the percentage for those at 120 MPa is 23%. Note that the 0°/90° composite has about the same amount of strain accumulation as the 0° composite, but much more recovery (Fig. 9.3a). In Fig. 9.3b, the recovery of 0°/90° composite crept at 60 MPa is shown. The creep recovery values are also indicated in this figure. Furthermore, it may be seen, by comparing the 100-h-creep/100-h-recovery experiments, that the strain recovery is much higher for shorter durations of the creep cycle (Fig. 9.3b). In the first cycle, $R_{cr} = 57\%$ and $R_t = 73\%$ and in the second cycle, these ratios were 80 and 70%, respectively (note that R_t decreases as the accumulated strain increases).

Figure 9.4 shows a schematic representation of the load transfer in a composite system, where the matrix has a lower creep resistance than the fibers, indicating a redistribution of the axial stress. On application of a creep load, the stress increases

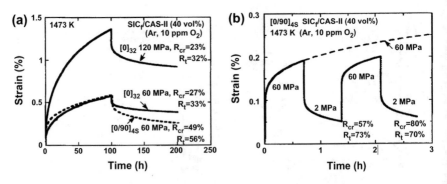

Fig. 9.3 Isothermal tensile-creep and creep strain recovery behavior of Nicalon SiC/CAS-II composites at 1200 °C (the total strain is shown): **a** cyclic-creep behavior of [0]$_{32}$ and [0/90]$_{4S}$ composites. The creep rate of the [0/90]$_{4S}$ composite was similar to that of the unidirectional composite; **b** short-duration cyclic-creep behavior of [0/90]$_{4S}$ composites. The recovery-creep strain is similar for both cycles; however, because of a reduction in transient creep, R_{cr} increased significantly for the second cycle. Wu and Holmes [1]. With kind permission of John Wiley and Sons

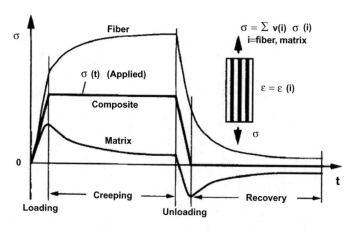

Fig. 9.4 Schematic representation of the redistribution in axial stress between the fibers and matrix during tensile creep and creep recovery. The *curves* assume that the fibers have a higher creep resistance and elastic modulus than the matrix. During creep, the fiber stress progressively increases, while the matrix stress relaxes. Upon unloading, the stress in the fibers and matrix decreases. Note that the initial loading and unloading transients have been expanded for clarity. The actual stresses in the fibers and matrix during loading and unloading will depend upon the constitutive law and volume fraction of each constituent. Wu and Holmes [1]. With kind permission of John Wiley and Sons

in both the fiber and the matrix. At a later stage (following full load application), the stress in the matrix relaxes, while it continues to increase in the fibers, since the load is transferred to the more creep-resistant fibers. Upon unloading, elastic contraction causes residual compression the matrix and residual tension in the fibers, until, at a later time, creep strain recovery results in the relaxation of both these residual stresses: compressive in the matrix and tensile in the fibers. This is shown schematically in Fig. 9.4. After concluding this experiment, it may be stated that the application of creep loading to a composite ceramic tends to emphasize the difference in the stress distribution between its various constituents, while the recovery of the system tends to diminish those differences.

Parallel changes occur in the structure: (a) grain growth in the fibers and (b) phase changes in the matrix. Moreover, no fiber and matrix fractures occurred in the specimens subjected to 100 h creep at 60 MPa and 100 h recovery, but cavity formation was observed in the matrix, generally located in the fiber-rich regions of the microstructure (Fig. 9.5).

Increasing creep stress to 120 MPa resulted in limited matrix microcracking and in fiber fracture, attributed to the inhomogeneous distribution of the fibers. This is based on the observation that microcracking occurred in those regions of specimens that were matrix-rich. After 100 h creep at 200 MPa, periodic fiber fracture and void formation in the matrix occurred with only random matrix microcracking. The periodic fiber fracture, in the absence of matrix fracture, was attributed to the redistribution of stress from the matrix to the more creep-resistant fibers. This creep-damage mode is considered to be a fundamental damage mechanism in

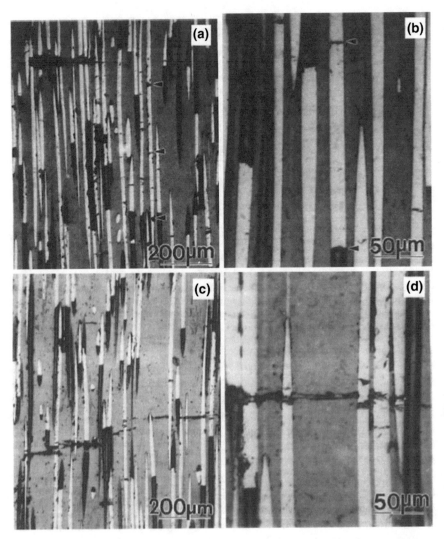

Fig. 9.5 Microstructural damage found after creep at 1200 °C: **a, b** Periodic fiber fracture observed after 100 h at 200 MPa (the specimen did not fail). The *arrows* show the locations of periodic fiber fracture along one of the fibers. **c, d** Matrix fracture and rupture of bridging fibers observed after 70 min at 250 MPa (the micrographs were taken approximately 5 mm from the failure location of the specimen). Wu and Holmes [1]. With kind permission of John Wiley and Sons

composites, where the creep rate of the matrix significantly exceeds that of the fibers.

The creep recovery in a SiC-fiber-Si_3N_4-matrix composite (a technologically important ceramic) is considered in the following. Tensile-creep, cyclic-creep and creep recovery experiments performed on this composite are of great interest. These experiments consider the SiC matrix, as in the previous case, but here the fiber is different. As such, the following experiments complement the earlier ones. Again, the test temperature is 1200 °C in air. The cyclic-creep strain recovery is of considerable interest, because the prior creep strain can be recovered to a high value. Furthermore, matrix cracking in continuous-fiber composites does not necessarily result in rupture, since bridging fibers have sufficient creep strength to transfer loads across the crack faces in whisker-reinforced ceramics.

The applied stresses by tensile loading of the composites are 0, 60, 75, 90, 150, 200, and 250 MPa. A zero-load experiment was performed to ascertain whether a change in gage-section length, caused by relaxation of processing-related residual stresses or the closure of matrix porosity, would influence the strains measured during the creep and strain recovery experiments conducted at higher stresses (for example, a volume contraction caused by additional sintering might lower the apparent creep rate or be confused with creep strain recovery). The loading and unloading used in the sustained-load (tensile) and cyclic-creep experiments were performed at the rate of 100 MPa/s. More stress-increment experiments (during which the creep stress was increased in small steps) were also performed to determine the stress dependence of the creep rate. The creep stress was increased in 10-MPa increments from 60 to 80 MPa, followed by 20-MPa increments to 100 and 120 MPa. The cyclic-creep experiments and creep recovery were performed at 1220 °C, with various hold times at the creep stress of 200 MPa and recovery stress of two MPa. Specimens were loaded and unloaded at a rate of 100 MP/s between these two stresses. These cyclic-creep/creep recovery experiments are summarized in Table 9.2. The loading histories of the composites and their corresponding, idealized strain responses are schematically shown in Fig. 9.6a, b.

One may define the amount of strain recovered after unloading the specimen either as (a) the total strain recovery ratio or (b) the creep strain recovery ratio. As shown in Fig. 9.7, the total strain recovery ratio is the sum of the elastic and creep strains recovered in a given cycle, i.e., $(\varepsilon_{el.R} + \varepsilon_{cr.R})$ divided by the total accumulated strain, ε_t, including the elastic strain that exists before unloading or

$$R_t = (\varepsilon_{el.R} + \varepsilon_{cr.R})/\varepsilon_t. \qquad (9.2)$$

Table 9.2 Loading histories used in the cyclic-creep/creep recovery experiments. Holmes et al. [2]. With kind permission of John Wiley and Sons

Creep per cycle (s)	Recovery per cycle (s)	Total cycles[a]	Total test duration (h)
300	0	2384	200
300	300	1192	200
180,000 (50 h)	0	4	200
180,000 (50 h)	180,000 (50 h)	4	400
720,000 (200 h)	90,000 (25 h)	1	125

[*]In all cases, the creep stress was 200 MPa and the recovery stress was 2 MPa
[a]Note that the loading and unloading transients were approximately 2 s each (100 MPa/s)

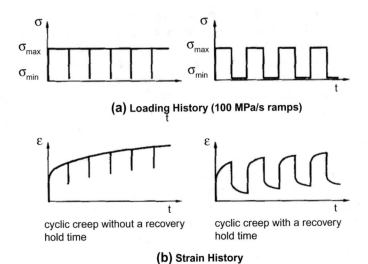

(a) Loading History (100 MPa/s ramps)

cyclic creep without a recovery   cyclic creep with a recovery
hold time   hold time

(b) Strain History

Fig. 9.6 Cyclic-creep experiments. **a** Schematic representation of the cyclic-loading histories examined and **b** idealized strain response. For all cyclic-creep experiments, the creep stress (σ_{max}) was fixed at 200 MPa and the recovery stress (σ_{min}) at 2 MPa. The loading and unloading ramps were prefixed at 100 MPa/s. Holmes et al. [2]. With kind permission of John Wiley and Sons

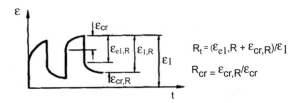

Fig. 9.7 Definition of the total strain recovery ratio (R_t) and creep strain recovery ratio (R_{cr}) used to quantify the amount of strain recovery during the cyclic-creep experiments. Holmes et al. [2]. With kind permission of John Wiley and Sons

The creep strain recovery ratio for a specific cycle is defined as

$$R_{cr} = \varepsilon_{cr.R}/\varepsilon_{cr}. \tag{9.3}$$

In Eq. (9.3), the elastic strain in loading and unloading is not included. Nearly linear behavior was obtained (see Fig. 9.8) in a composite at 1200 °C under a loading rate of 100 MPa/s.

Tensile-creep curves at various stresses from 0 to 200 MPa are shown in Fig. 9.9 and for 250 MPa in Fig. 9.10. Isothermal exposure at 1200 °C for 200 h under zero load produced no detectable change in specimen gage length (shown in Fig. 9.9) for the strain versus time curve, indicating that processing-related residual stresses were

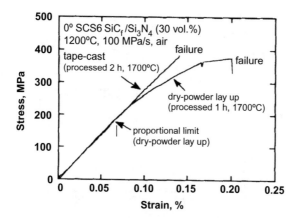

Fig. 9.8 Typical 1200 °C monotonic tensile behavior of the 0° SCS-6 SiC$_f$/Si$_3$N4 composite used in the present investigation (the composite was hot-pressed at 1700 °C for 2 h at 30 MPa). For comparison, the monotonic tensile behavior of an earlier 0° SCS-6 SiC$_f$/Si$_3$N$_4$ composite, processed by a dry-powder layup approach is shown (hot-pressed at 1700 °C for 1 h at 70 MPa). To minimize time-dependent phenomena, the monotonic tension experiments were conducted in air at a constant loading rate of 100 MPa/s. The UTS and Young's modulus were 385 MPa and 280 ± 10 GPa of the composite, respectively. Holmes et al. [2]. With kind permission of John Wiley and Sons

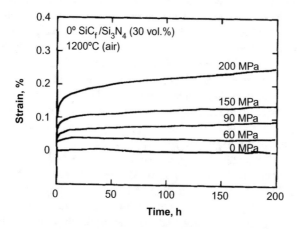

Fig. 9.9 Typical tensile-creep curves for sustained loading of 0° SCS-6 SiC$_f$/Si$_3$N$_4$ at stresses from 0 to 200 MPa. In all cases, specimens were loaded to the creep stress at a rate of 100 MPa/s. The creep curves include the instantaneous elastic strain that occurred during specimen loading (at 200 MPa, the elastic strain was approximately 0.07%). At 60 MPa, the creep rate was below 10^{-12} s^{-1}. Holmes et al. [2]. With kind permission of John Wiley and Sons

low. This means that the strain measured under the experimental test conditions is a consequence only of creep and creep recovery. As shown in Fig. 9.9, transient creep is present at all the stress levels, but its extent increases with stress becoming more pronounced (see, for example, the 200 MPa line and the 250 MPa plot in

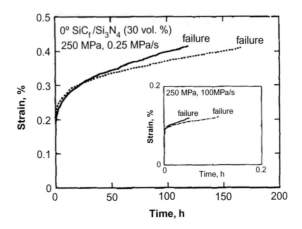

Fig. 9.10 Tensile-creep curves found for sustained loading of 0° SCS-6 SiC$_f$/Si$_3$N$_4$ at 230 MPa. To examine the influence of initial stress relaxation in the matrix on creep life, specimens were loaded to 250 MPa at either 0.25 or 100 MPa/s (see *Inset*). Loading slowly to the creep stress allows the matrix stress to relax, decreasing the likelihood of matrix fracture and dramatically increasing creep life. Holmes et al. [2]. With kind permission of John Wiley and Sons

Fig. 9.10). The strain following transient creep also increases with the stress level. One can assume that the parts of the curves following transient creep represent steady-state creep, referred to by the authors as a 'quasi-steady-state region.'

The creep lives of composites with brittle matrices increase with slower loading rates (as seen in Fig. 9.10). Thus, their lifetimes were 118 and 167 h for the faster and slower loading rates, respectively. The failure mechanism occurs by periodic fiber fracture during long-duration tensile loading. Load transfer from the matrix to the fibers (which are more creep-resistant) causes a progressive increase in fiber stress, causing fiber fracture. This is expected to occur in composites, in which the creep rate of the matrix is greater than that of the fiber.

The extent of recovery is appreciable in cyclic-creep loading. In Fig. 9.11, cyclic-creep behavior is shown for various loading and unloading histories (see the summary in Table 9.2). The introduction of a rapid unloading and reloading cycle, every 50 h (Fig. 9.11a), without a recovery hold period, did not introduce transient creep on reloading; the prior creep rate was immediately re-established and continued from the place of unloading (see Fig. 9.11b).

A significant change in primary-creep behavior was found for the shorter duration 300-s-creep/300-s-recovery cycles (Fig. 9.11c). For this loading history, the duration of transient creep was reduced to less than 20 h, versus roughly 70 h for sustained creep at 200 MPa. The insets in Fig. 9.11c show the creep behavior of the composite at selected times.

The amount of strain recovery was influenced by cycle's duration. For example, creep strain recovery ratios approaching 80% were observed during long-duration (50-h-creep/50-h-recovery) creep cycles between stress limits of 200 and 2 MPa (Fig. 9.12).

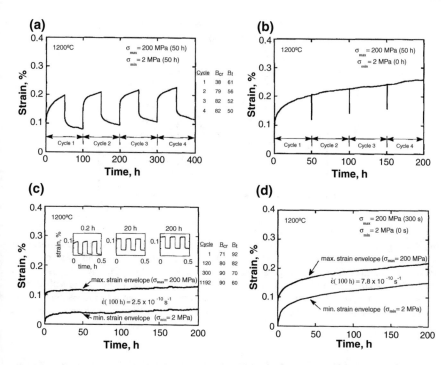

Fig. 9.11 Isothermal (1200 °C) cyclic-creep behavior of 0° SCS-6 Si_f/Si_3N_4. Specimens were cycled between stress limits of 200 and 2 MPa. For loading histories with a finite recovery hold time, the total strain recovery and creep strain recovery ratios ($R_t = (\varepsilon_{el.R} + \varepsilon_{cr.R})/\varepsilon_t$, and $R_{cr} = \varepsilon_{cr.R}/\varepsilon_{cr}$, respectively) are shown adjacent to the creep curves. **a** 50-h creep/50-h recovery. **b** 50-h creep/0-s recovery. **c** 300-s creep/300-s recovery. **d** 300-s creep/0-s recovery. Holmes et al. [2]. With kind permission of John Wiley and Sons

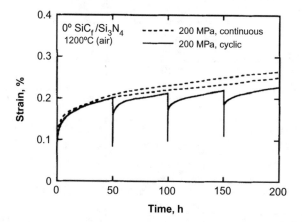

Fig. 9.12 Comparison of the accumulated creep strain and tensile-creep rate for sustained loading at 200 MPa and long-duration cyclic loading (50-h creep/50-h recovery) between stress limits of 200 and 2 MPa. Only the loading portions of the cyclic-creep curve are shown (the recovery segments were deleted, and the resulting curves were shifted to the left to allow a comparison of creep strain accumulation to be made for an equivalent time at the creep stress of 200 MPa). Holmes et al. [2]. With kind permission of John Wiley and Sons

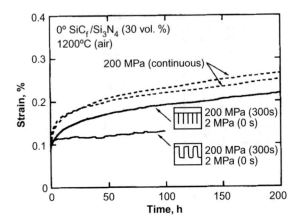

Fig. 9.13 Comparison of the accumulated creep strain and creep rate for sustained loading at 200 MPa and for short-duration cyclic loading (300-s creep/300-s recovery and 300-s creep/0-s recovery). For the cyclic-creep experiments, only the traces of the strain versus time curves obtained at the creep stress of 200 MPa are shown. As in Fig. 9.12, the recovery segments of each cycle have been removed to allow a comparison of accumulated creep strain to be made for an equivalent time at the creep stress of 200 MPa. Holmes et al. [2]. With kind permission of John Wiley and Sons

Table 9.3 Summary of loading histories and experimental results[a]. Holmes et al. [2]. With kind permission of John Wiley and Sons

Loading history	Total cycles	Avg creep rate[b] (s^{-1})	Total strain (%) (at 200 h)	R_t (%) (first/last cycle)
60 MPa, 200 h	1	$<10^{-12}$	0.04	
75 MPa, 200 h	1	2.8×10^{-10}	0.06	
90 MPa, 200 h	1	3.8×10^{-10}	0.09	
90 MPa, 200 h	1	3.9×10^{-10} (at 100 h)[c]	0.07 (at 100 h)	
150 MPa, 200 h	1	5.4×10^{-10}	0.14	
200 MPa, 200 h (+25-h recovery)	1	8.6×10^{-10}	0.27	45 (1 cycle)
200 MPa, 200 h (+25-h recovery)	1	8.3×10^{-10}	0.26	46 (1 cycle)
250 MPa, (100 MPa/s)	1	Failed at $\approx$210 s	ε (200 s) ≈ 0.12	
250 MPa (100 MPa/s)	1	Failed at $\approx$402 s	ε (400 s) ≈ 0.12	
250 MPa (0.25 MPa/s)	1	2.9×10^{-9} (failed at 118 h)	ε (118 h) ≈ 0.42	
250 MPa (0.25 MPa/s)	1	1.8×10^{-9} (failed at 167 h)	ε (167 h) ≈ 0.41	
Cyclic 300 s/0 s	2384	7.8×10^{-10}	0.30	$\approx$0/0
Cyclic 300 s/300 s	1192	2.5×10^{-10}	0.13	90/58
Cyclic 300 s/300 s	1192	2.3×10^{-10}	0.12	92/60

(continued)

Table 9.3 (continued)

Loading history	Total cycles	Avg creep rate[b] (s^{-1})	Total strain (%) (at 200 h)	R_t (%) (first/last cycle)
Cyclic 180,000 s/0 s	4	8.5×10^{-10}	0.26	$\approx 0/0$
Cyclic 180,000 s/180,000 s	4		$\varepsilon(200\ h) = 0.20$	61/50
			$\varepsilon(400\ h) = 0.26$	

[a]All cyclic-creep experiments were conducted in air at 1200 °C between stress limits of 200 and 2 MPa. The creep rates for the stress-increment experiment conducted between 60 and 120 MPa are given in Fig. 9.8
[b]Average creep rate between 100 and 200 h: at 100 h if failure occurred in under 200 h.
[c]Equipment failure at 100 h

For the same stress limits, the creep strain recovery ratio increased to 90% during shorter (300-s-creep/300-s-recovery) creep cycles (Fig. 9.13). In fiber-reinforced composites, the recovery process is assisted by the residual-stress state that develops in the composite upon specimen unloading. The extent of primary creep was significantly reduced during cyclic loading with a finite recovery hold time. Under sustained creep loading at 200 MPa, primary creep persisted for approximately 70 h. In the case of cyclic loading with a 300-s hold at 200 MPa, followed by rapid unloading and reloading (without a recovery hold time), the duration of primary creep was, again, roughly 70 h. These results may be compared with less than 20 h of primary creep, observed for cyclic loading with a 300-s hold at 200 MPa, followed by 300 s of recovery per cycle (Table 9.3).

A knowledge of creep strain recovery behavior may be used to increase the lifetimes of components subjected to sustained- and cyclic-creep load. More detailed information on the SiC_f/Si_3N_4 system may be found in the work of Holmes et al.

References

1. Wu X, Holmes JW (1993) J Am Ceram Soc 76:2695
2. Holmes JW, Park YH, Jones JW (1993) J Am Ceram Soc 76:1281

Chapter 10
Empirical Relations

Abstract Stress rupture and creep life are particularly important for preventing catastrophic failure. The knowledge on this important subject is based on empirical concepts. Even Andrade's creep relations are based on empirical observations. Several empirical relations are considered in this chapter. These are the Larson-Miller parametric method, the Monkman–Grant relationship, the Sherby-Dorn parametric method, the Orr-Sherby-Dorn approach, and the Manson-Haferd parameter. Appropriate relations and illustration of these empirical methods to evaluate creep life are the subject of this chapter.

Most of the creep relations, even those of Andrade, are basically empirical relations, including the MG creep concept and its modifications (as previously discussed in Chap. 7). As such, the major empirical relations relating to creep phenomena deserve a brief, dedicated discussion. Stress rupture and creep life are particularly important for preventing catastrophic failure. To this end, the emphasis here is on experimentally supposed and well-analyzed empirical concepts, rather than on fundamental physical principles.

10.1 The Larson-Miller Parameter (LMP)

The "Larson-Miller relation" or "Larson-Miller parameter" (LMP) is basically an extrapolation of creep experimental data. In essence, it originates in their idea that the creep rate, r, may be described by any Arrhenius-type equation expressed as

$$r = A\exp\left(-\frac{\Delta H}{RT}\right),$$

(10.1)

where A is a constant and ΔH is the activation energy of the creep. Expressing Eq. (10.1) in logarithmic form gives

$$\ln r = \ln A - \frac{\Delta H}{RT}$$

(10.2)

J. Pelleg, *Creep in Ceramics*, Solid Mechanics and Its Applications 241,
DOI 10.1007/978-3-319-50826-9_10

Rearrange the above to obtain

$$\frac{\Delta H}{R} = T (\ln A - \ln r) \tag{10.3}$$

Since creep is an inverse time-dependent process, another equation may be written in terms of time as

$$\frac{\Delta l}{\Delta t} = A' \exp\left(-\frac{\Delta H}{RT}\right) \tag{10.4}$$

And again, taking the logarithm on both sides results in

$$\ln\left(\frac{\Delta l}{\Delta t}\right) = \ln A' - \frac{\Delta H}{RT} \tag{10.5}$$

Equation (10.5) may then be rewritten as

$$\begin{aligned}\frac{\Delta H}{R} &= \left[\ln A' - \ln\frac{\Delta l}{\Delta t}\right] T \\ &= [\ln A' - \ln(\Delta l) + \ln(\Delta t)]T\end{aligned} \tag{10.6}$$

or:

$$\frac{\Delta H}{R} = \left[\ln\frac{A'}{\Delta l} + \ln(\Delta t)\right] T \tag{10.7}$$

Write $\ln\frac{A'}{\Delta l} = B$, so that Eq. (10.7) becomes

$$\frac{\Delta H}{R} = T[B + \ln(\Delta t)] \tag{10.8}$$

This is now the same as the LMP, which is a parametric relation used to extrapolate experimental data on creep and creep rupture, given by

$$LMP = T (C + \log t) \tag{10.9}$$

This is significant, since rupture lifetime is impractical to assess in a laboratory. C was found to be in the range of 20–22, typically ~ 20. This value of C seems to be applicable to many cases and materials, but deviations from this value have been observed. Selecting a proper value, which may be determined for a material of interest, can narrow-down the scatter problem (quite common in ceramic experiments). Equation (10.9) is a stress-dependent, temperature-compensated rupture-life function. Here, t is the stress rupture time (in hours).

When using this relation, it is assumed that the activation energy is independent of the applied stress, and it may be used to evaluate the difference in rupture life

Fig. 10.1 Tensile stress versus Larson-Miller parameter at different stresses in argon at 1000, 1100, 1200, and 1300 °C. Zhu et al. [14]. With kind permission of Elsevier

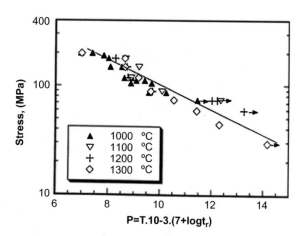

with the difference in temperature for a given stress. Note that, in LMP, time is expressed in hours and T in Rankine degrees. The LMP, P, is one of the useful parameters used for predicting creep life in metallic materials. Its basic assumptions are that $m = 1$ and Q is a function of stress. It has also been applied to ceramics. One such example is the SiC-fiber-reinforced SiC. The LMP was used to correlate the stress–temperature-life relationship in SiC/SiC composites, by means of the expression given in Eq. (10.10), reproduced as

$$P = (C + \log\ t_r) \tag{10.10}$$

where T is in K, and t_r is the time to rupture (as indicated earlier in hours). It was found that data at different temperatures fall on the same line with the best fit, when the constant, C, is between 5 and 10 (in metals C $\sim$ 20). Figure 10.1 shows the relation of stress to the LMP with C being 7. Note that, for monolithic silicon nitride, $C = 30$–40, depending on the grade of the silicon nitride. For oxides, it ranges from 10 to 22. In order to use the LMP method, a family of stress-rupture curves is required, representing different test temperatures for a give material. These then re-plotted on a revised temperature-compensated time axis, i.e., the LMP. The family of curves chosen is superimposed on a single, master curve, as illustrated in Fig. 10.1 for SiC/SiC.

10.2   The Monkman-Grant Relationship (MGR)

The Monkman-Grant relationship (MGR) uses an exponential relationship, given as:

$$t_r \dot{\varepsilon}^m = C_{M-G} \tag{10.11}$$

The MGR relates the minimum creep rate or steady-state strain rate and time to fracture, t_r, m, and C are material constants, m being the strain-rate exponent. (The subscript was dropped). Rewrite Eq. (10.11) on a logarithmic scale as

$$log\, t_r = -m \log \dot{\varepsilon} + \log C \qquad (10.12)$$

The MGR provides a method for creep-life prediction not only for metals and alloys, but also for ceramics. Figure 10.2 shows the MGR straight line for the SiC/SiC composite.

In this figure, the steady-state creep-strain rate versus time to rupture is shown. The data fall into a straight line, i.e., fit the MGR, indicating that $m = 0.72$. Using the MGR, the creep life can also be calculated, according to Eq. (10.11), if the steady-state creep rate is known. The exemplar for creep-life prediction is Si_3N_4 (Fig. 10.3).

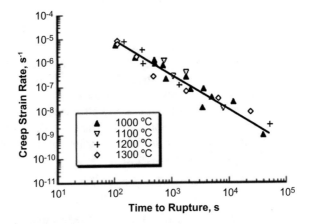

Fig. 10.2 Tensile minimum creep-strain rate versus time-to-rupture at different stresses in argon at 1000, 1100, 1200, and 1300. Zhu et al. [14]. With kind permission of Elsevier

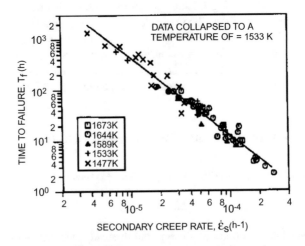

Fig. 10.3 Temperature-dependence of the Monkman-Grant lines correlated with an additional temperature term (Eq. 10.10). Menon et al. [8]. With kind permission of John Wiley and Sons

As may be inferred from the above, the advantage of using the MGR is that, once the relationship is established from short-term lab tests, all that is needed for the estimation of failure time for a new set of conditions is the value of the secondary-creep rate. The tests are relatively short, because the secondary-creep stage, in general, comprises only a short portion of the creep-failure time. For a detailed and extensive discussion on stress rupture and more information on the MGR, the reader is referred to Sect. 7.2 above.

10.3 Sherby-Dorn Parametric Method

Lifetime predictions are essential for industrial applications. The Sherby-Dorn parametric method is basically an empirical Norton power law (see Chap. 5, Eq. 5.22) for obtaining the dependence of σ on temperature-normalized lifetime, Θ, as

$$\log \sigma = A - \frac{1}{n} \log \Theta \tag{10.13}$$

One immediately recognizes that the Norton Eq. (5.22), reproduced here, can be easily brought to the form of Eq. (10.13):

$$\varepsilon_{ss} = A\sigma^n \exp\left(\frac{-Q}{RT}\right) \tag{5.22}$$

Θ from Eq. (10.13) is given as

$$\Theta = t_f \exp\left(-\frac{Q}{RT}\right) \tag{10.14}$$

A and Q are parameters, the latter being an activation energy of creep and the former is a material constant; t_f is the failure time.

A logarithmic plot of Eq. (10.13), of $\log \sigma$ versus $\log \Theta$, yields a curve with an intercept of A and a slope of the inverse exponent, i.e., $1/n$. One example of such a plot is for two oxide-dispersion-strengthened Nickel-base superalloys–MA 754 and MA 6000 (Fig. 10.4).

Often, the Sherby-Dorn method appears in the literature as

$$P_{SD} = \log \theta = \log t_f - \frac{\log e}{R} \frac{Q}{T} \tag{10.15}$$

where P_{SD} is the Sherby-Dorn parameter and Q is assumed to be independent of stress and temperature. According to this method, a number of tests are run at various temperatures and stresses to determine the time to failure and the activation energy, Q. A universal plot is then constructed of the stresses as a function of P_{SD}. An allowable stress to failure combined with temperature, namely P_{SD}, may be

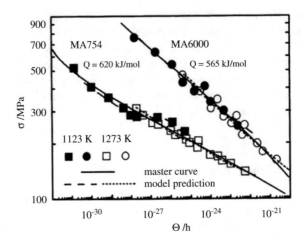

Fig. 10.4 Temperature-normalized creep rupture lifetime diagram according to Sherby-Dorn. Ref. [3]

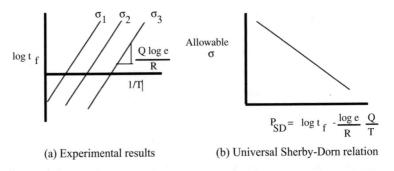

(a) Experimental results   (b) Universal Sherby-Dorn relation

Fig. 10.5 Summary of Sherby-Dorn relation **a** experimental; **b** Sherby Dorn relation

determined from the curve. Schematic experimental results and the universal curve are shown in Fig. 10.5.

It would be illuminating to show the actual use of the Sherby-Dorn technique, regardless of whether the example is for an alloy, such as a certain Al–Mg alloy. The stress-P_{SD} relation is found to be

$$\sigma = f(P_{SD}) = -11.3P_{SD} - 124 \quad (25 \leq \sigma \leq 85 \text{ MPa}) \quad (10.16)$$

The objective is to determine the allowable stress to give, say 2000 h lifetime at 200 °C (473 K). With the values of activation energy 150,500 Jmol^{-1}, R = 8.314 J/mol K, t_f = 2000 h and T 473 K. P_{SD} is calculated as -13.21. Substituting this value into (10.16) provides the allowable stress as 25 MPa.

10.4   The Orr-Sherby-Dorn (OSD) Approach

The Orr-Sherby-Dorn (OSD) parameter, P_{OSD}, is given as

$$P_{OSD} = t_r \exp\left(-\frac{Q}{RT}\right) \qquad (10.17)$$

which can be basically formulated differently in a logarithmic manner as

$$\ln P_{OSD} = \ln t_r - \frac{Q}{RT} \qquad (10.18)$$

or rearranging the logarithmic form as

$$\log t = \log P_{OSD} + 0.434\frac{Q}{RT} \qquad (10.19)$$

with the symbols having their usual meanings. Note, however, t_r, which expresses the time to rupture, may also indicate the time needed to reach a given creep rate, just before stress rupture occurs. This relation has been mostly applied to metals and alloys, and much less to ceramics, although the literature frequently indicates that it may be applied to ceramics and polymers, as well.

   Like the LMP, the OSD method assumes that log t is a linear function of $1/T$, as schematically illustrated in Fig. 10.6 [Eqs. (10.18) and (10.19) or (10.20)]:

$$\theta = t \exp\left(-\frac{\Delta H}{RT}\right) \qquad (10.20)$$

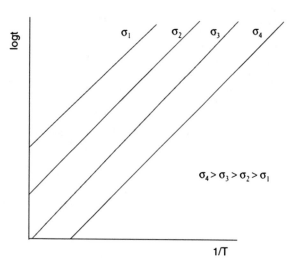

Fig. 10.6 A schematic presentation of OSD creep method for obtaining the parameter

Equation (10.20) is the same as Eq. (10.17), with θ being P_{OSD} and ΔH being Q. Also note that in Eqs. (10.1) and (10.2), expressing the LMPs r and A is equivalent to P_{OSD} and t, respectively. The expressions (10.1) and (10.20) represent the LM parameters. In Fig. 10.6, the $\log t$ versus $1/T$ plots are parallel isostress lines each curve for one specific stress value, like Dorn's concept, who assumes that only one creep process is actually operating at some high temperature.

Note that the OSD technique is based on Norton's empirical power law for obtaining the dependence of σ on the lifetime, θ, expressed as

$$\log \sigma = A - \frac{1}{n}\log \theta \qquad (10.21)$$

The OSD parameter is stress dependent, and a master curve for this relation has to be created. In order to determine an empirical model for this relation, creep data are used to find the best fit between stress and the OSD parameter. For instance, a Ni-based superalloy, often used at high temperatures, in which creep conditions prevail, may be considered. Figure 10.7 (following OSD parametric creep-life prediction) presents a temperature-compensated, creep rupture lifetime diagram for Ni-based alloys. The master curves shown in this figure are based on the experimental results (shown in Fig. 10.8).

In Fig. 10.8 the variation of stress with rupture time for creep and hot tensile tests are presented. Note the excellent compatibility of the tensile data with those of the creep. Also the variation of creep rupture time with temperature is required. Such illustrations are shown in the work of dos Reis Sobrinho and de Oliveira Bueno and the reader of interest can turn to their paper.

Fig. 10.7 The parameterization curves for creep and CSR tensile data based on the Log (tr) versus 1/T diagram: dos Reis Sobrinho and de Bueno Oliveira [13]. With kind permission of Dr. Levi de Oliveira Bueno

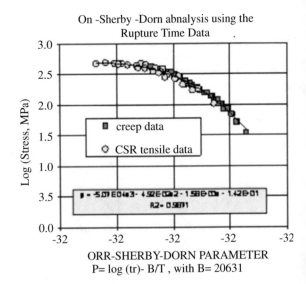

The minimum creep rate and also the tensile test data as a function of OSD parameter are shown in Fig. 10.9. The results are very similar to those obtained with the analysis considering the rupture times. The authors dos Reis Sobrinho and de Oliveira Bueno claim on the basis of their experimental results that hot tensile data can also be used as a very helpful complement in the determination of parameterization curves for creep data.

The OSD approach involves a time-temperature constant, P_{OSD}, based on the linear relationship of $\log t$ versus $1/T$. In this approach, the initial premise of the

Fig. 10.8 Variation of the stress with rupture time in creep tests, plotted together with the ultimate tensile stress, and the time for its occurrence in the tensile tests. dos Reis Sobrinho and de Bueno Oliveira [13]. With kind permission of Dr. Levi de Oliveira Bueno

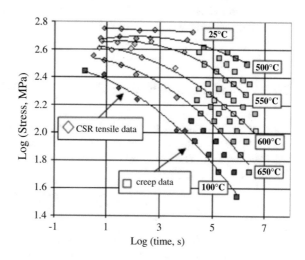

Fig. 10.9 The parameterization curves for minimum creep rate using creep and CSR data. based on the log(tr) versus 1/T diagram. dos Reis Sobrinho and de Bueno Oliveira [13]. With kind permission of Dr. Levi de Oliveira Bueno

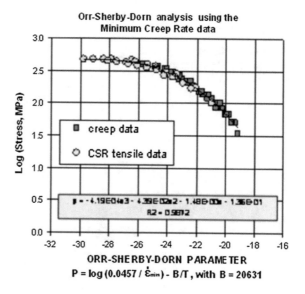

LMP has been modified such that the constant C_{LM} (C in Eq. 10.9) becomes a function of stress, and P_{LM} (in Eq. 10.9 LMP) becomes a constant. Based on these assumptions, the LMP Eq. (10.9) can be re-arranged to yield the Orr-Sherby-Dorn equation as given in Eq. (10.22):

$$P_{OSD} = f(\sigma) = \log t_f - C_{OSD}/T \tag{10.22}$$

where P_{OSD} and C_{OSD} are the Orr-Sherby-Dorn parameter and constant, respectively. In Eq. 10.19 $C_{OSD} \equiv 0.434\ Q/R$.

This method is claimed to be a better approach for providing a creep rupture prediction method than the other parametric method.

10.5   The Manson-Haferd Parameter (MFP)

Another modification of the LMP, the Manson-Haferd parameter (MFP) has been proven to be more applicable to some materials than the LMP. One such example is the 9% Cr steels. The MHP may be expressed as

$$P_{MH} = f(\sigma) = \frac{\log t_r - \log t_a}{T - T_a} \tag{10.23}$$

P_{MH} is the MHP, and t_r is the rupture duration (in hours). Constructing a family of curves at different stresses indicates that they terminate at the same point, which determines $\log t_a$ and T_a. A schematic illustration of $\log t_r$ versus T is shown in Fig. 10.10.

Fig. 10.10 Schematic diagram of the MHP model. Individual plots for each stress

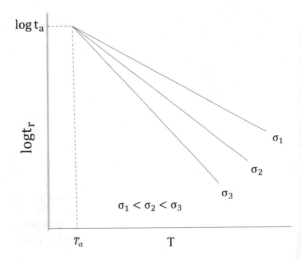

Fig. 10.11 The parameterization curves for creep and CSR tensile data based on the log tr versus T diagram: Manson-Haferd model. dos Reis Sobrinho and de Bueno Oliveira [13]. With kind permission of Dr. Levi de Oliveira Bueno

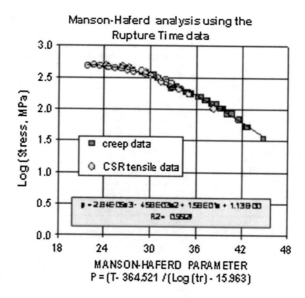

As seen in Eq. (10.23), unlike the LMP approach, two constants are required in the MHP method. It seems that the MHP approach is more accurate, since the variation of $C = 20$, given in the LMP is eliminated in the MHP technique. Furthermore, when plotting experimental creep data on the P_{MH} versus stress curve, they seem to fall on a single curve, a master curve, as seen in Fig. 10.11. The MHP method was specifically developed to improve the LMP method (especially in regard to the fixed value of the constant C).

Note that, in the above figures, CSR stands for "constant-strain rate." Note that the MHP method relates time to temperature for a given stress, assuming that the algorithm of time varies linearly with temperature at a constant initial-stress condition. Recall that the LMP method plots time against $1/T$ (Eq. 10.9), rather than versus T, as in the MHP approach. P_{MH} is composed of two constants relating time to temperature, thus, possibly giving this model better sensitivity to the time–temperature relationship.

Regarding the schematic curves in Fig. 10.10, the experimental creep data of HK 40 steel, 25Cr-20Ni alloy (an austenitic Fe–Cr–Ni alloy) are collected and listed in Table 10.1 and then plotted at each stress, as shown in Fig. 10.12. The MHP linear time–temperature relationship was developed not only for stress rupture, but also to mark the time to a given creep-strain level, specifically the second stage and minimum creep rate. A minimum creep-rate curve appears in Fig. 10.13. The MHP approach, like the LMP method, assumes that steady-state creep is dominated by power-law behavior.

Thus, the MHP intends to remedy the uncertainty found in the LMP approach. The parameters of P_{MH} (Eq. 10.23) and LMP (Eq. 10.9) are graphically derived from the intersection point of the extrapolated isostress lines, when plotted on a log of minimum creep rate versus absolute temperature (the MHP method), instead of

Table 10.1 Results from rupture time in steel. Latorre [5]. With kind permission of CT&F—Ciencia, Tecnología y Futuro

Sample	Service time Hr,000	Stress-applied Mpa	Temperature K	Rupture time, tr hours	Final deformation mm/mm	Minimum creep hr^{-1}
7S–1	74	27	1198	337.9		
7S–2	74	27	1223	149		
7S–4	74	27	1273	37.3		
2–21	74	27	1148	2429.0	0.060	0.0000027
14–9	88	27	1173	534.4	0.296	0.000197
7N–5	74	27	1198	347.9		
7N–6	74	27	1223	126.9		
7N–7	74	27	1248	109.4		
7N–8	74	27	1273	40.3		
2–28	74	30	1098	7323.8		
2–27	74	30	1148	1516.1	0.067	0.00000114
2–26	74	30	1198	237.2	0.109	−0.00156
2–23	74	30	1248	36.8	0.034	0.0001438
2–25	74	30	1273	23.9	0.144	0.0016
2–20	74	35	1123	991.2	0.158	0.00000888
14–11	88	35	1148	350.4	0.328	−0.000026
14–12	88	35	1198	56.6	0.246	0.00241
14–14	88	35	1223	27.8		0.0136
14–16	88	35	1248	11.8	0.290	0.00734
2–24	74	35	1273	9.0	0.187	0.00009859
2–30	74	38	1173	117.6	0.041	0.0006434
2–29	74	38	1223	30.6	0.042	0.0017435
2–32	74	38	1248	12.6	0.066	0.000542
14–10	88	42	1123	263.7	0.328	0.015928
2–19	74	42	1148	213.1	0.149	0.003354
14–17	88	42	1173	46.3	0.242	0.00747
14–13	88	42	1198	20.5	0.301	0.0173
14–15	88	42	1223	10.7	0.341	0.00359
2–22	74	42	1248	14.4	0.132	0.0001247
2–37	74	50	1098	327.1	0.070	0.0010142
2–36	74	50	1148	70.2	0.160	0.00521
2–33	74	50	1198	21.8	0.219	0.0114
2–34	74	50	1223	8.4	0.222	0.0176
2–35	74	50	1248	4.8	0.160	0.022
14–18	88	60	1173	7.0	0.384	0.000372
2–31	74	60	1098	168.7	0.180	

the inverse temperature (LMP approach), respectively. The intersection point then determines the constants t_a and T_a in Eq. (10.23). It is claimed that the MHP method may be used for a variety of materials.

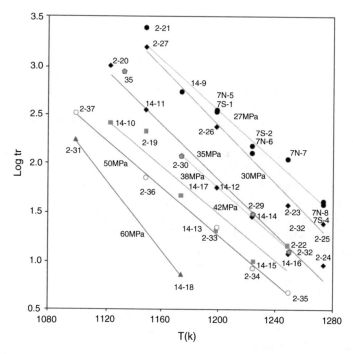

Fig. 10.12 Manson-Haferd model, PMH. Calculation, data and nomenclature in Table 10.1. Each curve corresponds to a stress applied on the sample. Latorre [5]. With kind permission of CT&F—Ciencia, Tecnología y Futuro

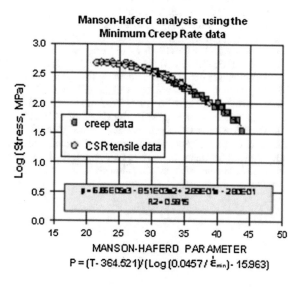

Fig. 10.13 The parameterization curve for minimum creep rate using creep and CSR data based on the log (tr) versus T diagram. dos Reis Sobrinho and de Bueno Oliveira L [13]. With kind permission of Dr. Levi de Oliveira Bueno

10.6 Summary

This chapter reviewed the several most commonly used empirical and parametric approaches to creep. Most of the experimental data are taken from metals and alloys, which have been extensively used for creep resistance. Much less experimental data has been reported in the open literature, despite the fact that the tendency in modern technology is to use ceramic materials, which are expected to exhibit excellent creep resistance up to elevated temperatures. Even so, the researchers in the field, and particularly those who were proposing their parametric and empirical relations, claim that the relationships are valid for a large variety of materials, including ceramics and polymers. The obvious interest in creep-resistant ceramics comes from the turbine production industries and the field of aviation. Although the primary intent of this chapter was to indicate the parametric relationships in ceramic materials, to discuss their use, and to compare the merits and difficulties in predicting component lifetimes under creep deformation, it was convenient to provide some classic examples from metallic materials. This is justified by the realization that there is no universal approach, based on theoretical grounds, so one must rely heavily on experimental results in order to understand creep behavior under various conditions, and especially at high temperatures. In light of the absence of a theoretically-founded approach, the aforementioned parametric methods become even more significant for creep-resistant design and manufacturing.

References

1. Chokshi AH, Porter JR (1985) J Am Ceram Soc 68:C-144
2. Ferber MK, Jenkins MG (1992) J Am Ceram Soc 75:2453
3. Heilmaier M, Reppich B (1996) Metall Mater Trans A 27:3861
4. Larson FR, Miller J (1952) Trans ASME 74:765
5. Latorre G (2000) C. T. F. - Cienc Tecnol Futuro 2(1), Bucaramanga, Jan./Dec. 2000
6. Lin H-T, Becher PF (1990) J Am Ceram Soc 73:1378
7. Manson SS, Haferd AM (1953) A linear time-temperature relation extrapolation of creep and stress-rupture data. National Advisory Committee for Aeronautics, Hampton, VA
8. Menon MN, Ho TF, Wu DC, Jenkins MG, Ferber MK (1994) J Am Ceram Soc 77:1255
9. Monkman FC, Grant NJ (1956) Proc Am Soc Test Mater 56:593
10. Ohji T, Yamauchi Y (1995) J Am Ceram Soc 76:3105
11. Orr RL, Sherby OD, Dorn JE (1954) Trans ASM 46:113
12. Pelleg J (2014) Mechanical properties of ceramics. Springer, Berlin, pp 519–528
13. dos Reis Sobrinho JF, de Bueno Oliveira L (2005) Revista Matéria 10:463
14. Zhu S, Mizuno M, Kagawa Y, Mutoh Y (1999) Compos Sci Technol 59:833

Chapter 11
Design for Creep Resistance

Abstract In the absence of theoretical fundamentals for creep, one has to rely on experimental creep data in order to develop creep-resistant alloys and design components for service-lifetime evaluation. New designs must be based on extensive experimentation performed on a variety of materials, while applying all manner of possible work conditions while keeping in mind that creep consists of three stages.

In the absence of theoretical fundamentals for creep, one has to rely on experimental creep data in order to develop creep-resistant alloys and design components for service-lifetime evaluation. Again, general and universal designs for creep-resistant material development do not exist yet. Therefore, new designs must be based on extensive experimentation performed on a variety of materials, while applying all manner of possible work conditions. This objective is very difficult to obtain for several reasons, the main obstacles being the facts that: (a) creep has three stages, (b) the test materials may be metallic, ceramic or polymeric, and (c) there many potential work conditions to be tested: different temperatures, service durations, environments, structures, types of deformation, applied loads, etc. in varying combinations. As such, this short chapter will present several proven design concepts.

11.1 The Stages of Creep

From the descriptions of creep in the earlier chapters, particularly of the different creep stages, it is clear that one should avoid using components in which tertiary creep is active. At this stage, the strain rate increases exponentially with stress. The necking of specimens occur with fracture setting in. A desirable component design should favor primary creep, with its relatively high initial creep rate that diminishes with increasing exposure time, leading to second-stage creep, in which the creep

© Springer International Publishing AG 2017
J. Pelleg, *Creep in Ceramics*, Solid Mechanics and Its Applications 241,
DOI 10.1007/978-3-319-50826-9_11

rate in the specimen is balanced by work hardening and reaches its minimum creep rate. The minimum creep rate is a constant creep rate, which is an important design parameter; its magnitude is temperature-and stress-dependent. Two criteria of minimum creep rate are commonly applied to alloys: (a) the stress needed to produce a creep rate of $0.1 \times 10^{-3}\%/h$ (or 1% in 1×104 h) and (b) the stress needed to produce a creep rate of $0.1 \times 10^{-4}\%/h$, namely 1% in 100×103 h, which is about 11.5 years. The first criterion is used for turbine blades, while the second is usually applied to steam turbines. (For more discussion on minimum creep rate see Chap. 1). To the best of this author's knowledge, no such criteria are given for ceramics in the literature. But it is logical to expect an almost unlimitedly high lifetime from ceramics, the very reason for choosing ceramics to operate at high temperatures under loads. The objectives of proper creep tests are to determine the minimum creep rate at stage II creep, on the one hand, and to evaluate the time at which failure sets in, on the other. Such information is essential, so that the proper ceramics will be selected to prevent failure during service and to evaluate the time period of safe use in high-temperature applications, in which structural stability is essential. By making the proper choice, good ceramic components may be selected, capable of operating at various conditions of high temperature and creep deformation.

11.2 The Material

In general, the structures of materials differ. Metallic materials are not the same as ceramic or polymeric ones. Even within the same class of materials, there are different structures that may be changed at will by certain treatments. There are differences between grains—their sizes, shapes, distributions may differ—as well as their nature (crystalline or amorphous), and even their dislocation content may be altered from its natural state after modification following deformation. Nonetheless, since creep is a slow, time-dependent deformation process, dislocation motion must be involved in some of the creep processes. The vacancy content in a material is also associated with creep; even the contents of these vacancies are different in the various materials. In short, creep is different for each material.

As such, material designs must take into account the effects of dislocations on creep. Low-dislocation-content materials should be selected for creep resistance. In particular, the occurrence of dislocation slip and climb may be reduced, if the proper materials are chosen. Ceramics make a good choice. In fact, vacancy content is not only dependent on the material chosen, but also on the component's service temperature. Vacancy-assisted creep occurs in Coble creep (grain-boundary diffusion) and in the Nabarro-Herring creep, as well, which is vacancy-diffusion-controlled. Designers must be familiar with the properties of dislocations and vacancy contents and with their distributions in the structure, as well as the expected changes that may occur in the wake of long-term exposure to stress and temperature. A good designer is an expert regarding the roles of dislocations, vacancies, and the various obstacles that hinder dislocation motion, such as: grain boundaries (in polycrystalline materials), precipitates, solutes, and impurity particles, and strain fields, resulting from other dislocations or pile-ups, that also retard dislocation motion. Recall that dislocations may glide (leading to slip in their planes), climb, and cross-slip; by successfully hindering or overcoming such motions—the creep-resistance and the lifetime of a material are increased. Furthermore, different types of vacancies in ceramics carry different charges; each existing or formed vacancy must be balanced to preserve neutrality. For example, even in oxide ceramics, oxygen vacancies may exist. There is a correlation between dislocation climb and the vacancy content. The ability of a dislocation to climb requires that vacancies be present in its immediate vicinity. Thus, for effective, ideal creep design, developed materials should have low-dislocation and vacancy contents.

11.3 Various Conditions

Materials exposed to creep operate under various conditions of temperature, time, structure, and environment, as well as being affected by the type of deformation process in action. Some of these factors are briefly and generally considered below.

11.3.1 Temperature

Generally, creep is related to a material's melting point, T_m. The potential exposure temperature applied, to avoid creep failure, depends on the melting point of the tested material. The higher the melting point, the longer the expected lifetime. Ceramics are characterized by high-melting points; thus, ceramic components are frequently used for design purposes. The idea of choosing high-melting point materials is a consequence of the fact that diffusion processes are related to temperature-dependent vacancy concentrations. The rate of diffusion (or self-diffusion) is slower in high-melting materials. As a general rule, the demarcation temperature for creep is $0.5T_m$; however, a safe temperature for avoiding creep is usually stated as $0.3T_m$. Creep below or at $0.5T_m$ is termed 'low temperature creep,' and since diffusion is relatively lower at such temperatures, the creep occurring at this temperature interval is not diffusion, but rather to other mechanisms associated with creep. Despite the extreme importance of avoiding creep, creep-related failure, and lifetime-extending material design for creep-resistance still rely mainly on parametric methods or on short experimental tests, since no theoretical functions are available yet due to the complex nature of creep.

11.3.2 Time

As indicated many times earlier, creep is a time-dependent deformation. Fortunately for designers, failure in brittle materials does not occur suddenly (as it does under tension and other types of deformation). Over time, creep strain develops in a material exposed to some stress at the temperature of application, depending on the duration of exposure. In order to better visualize the creep process, remember that stress and temperature determine the creep rate. The creep rate may be given by the known function, $\dot{\varepsilon} = f(t, T, \sigma)$, which tells the designer that time, temperature, and stress must be considered as acting in concert, if a successful creep-life is to be attained.

11.3.3 Structure

Creep is a complex type of deformation that depends on many parameters, thereby complicating the prediction of a material's lifetime. One such parameter is the structure of the material. The same material may begin as an amorphous material yet end up as a single crystal. The wide spectrum of structural variation within the same material imposes a challenge for designers, not only in choosing the design material, but also when determining its best structural type for the intended application.

Microstructural changes in a material can occur during exposure to certain temperatures, especially in the long-term, as in the case of creep, even in the absence of an applied stress. Such changes are accelerated and magnified when stress acts simultaneously with temperature over time. Designers must be aware of these potential, microstructural changes occurring under creep conditions (due to temperature, time, and stress) when selecting their creep-resistant material. Structural stability may be improved by introducing certain additives to ceramic materials, which a priori strengthens them even before exposure to creep conditions. These additives, such as solutes, have a stress field associated with them, thus hindering dislocation slide and climb motions. Additionally, the dispersion of insoluble, hard constituents (mostly as precipitates) throughout the material structure, such as oxides, nitrides, or carbides (usually high-melting) also hinder dislocation motion.

Moreover, one should not forget the importance of grain size in polycrystalline structures. Designers must be familiar with the effects of grain size, in general, and regarding creep, in particular. While small-grained structures are known to strengthen materials, effective creep processes prefer large-grain sizes. Extrapolating the grain size to such an extent that only one grain is present—the single-crystal case—represents, therefore, the culmination of creep resistance. The contradictory effect of grain size, regarding strength and creep resistance, results from the fact that the very large number of grain boundaries of the small grains are sources of vacancies. Vacancies generated at grain boundaries are responsible for climb, which is a major creep mechanism. Reducing the grain size diminishes the generation of vacancy formation by climb, thus producing more effective creep resistance. At the extreme, one-grained materials (single crystals) are the best performing structures. Thus, when polycrystalline structures are chosen (perhaps due to cost considerations), a compromise must be struck between the strengthening effect of small-grain size and the desire to improve creep resistance by the use of a large-grained structure (to reduce the number of vacancies required for climb). Nevertheless, and in spite of cost considerations, single crystals are used for special applications, as in turbine blades. Wherever possible, directionally solidified structures replace single-crystal structures. More consideration should be given to the orientation of the single crystals.

11.3.4 Environment

A material designer should be thoroughly familiar with each specific ceramics of interest; there are no general shortcuts. Roughly speaking, ceramic materials can be oxide and non-oxide ceramics. Non-oxide ceramics include technologically important carbides and nitrides, each responding differently to environmental effects. Materials operating at ambient conditions are influenced mainly by oxygen and moisture that depend on the operational temperature. The only common feature of a wide range of ceramics, each of which responds uniquely, is that their strength properties and corrosion resistance are environmentally sensitive. One of the most widely investigated effects on ceramics is the deleterious influence of H_2O on their properties; H_2O is harmful to all ceramics, irrespective of whether they are oxide or non-oxide ceramics. Crack propagation in ceramics is influenced by environmental phenomena and, therefore, one of the many techniques for understanding those environmental effects is by the investigation of crack propagation, which is environmentally sensitive for predicting lifetimes (the time-to-failure). The adverse effect of oxygen contamination in non-oxide ceramics, such as in carbides and borides, is well documented. An example to be cited is TiB_2. In this case, even during the processing and densification of the TiB_2 [1], oxygen is very harmful. Thus:

(i) oxygen promotes grain coarsening in hot pressing at 1400–1700 °C and limits the maximum attainable density. Be_2O_3 is formed, which can be reduced by the addition of C;

(ii) in case of pressureless sintering at 1700–2050 °C, oxygen remains as titanium oxides, increasing grain and pore coarsening, and limiting the attainable density. To inhibit abnormal grain growth, the total oxygen content of the powder should be limited to less than 0.5 wt% O, or strong reducing additives must be added to remove the TiO_x below 1600 °C.

The adverse effects of oxygen contamination on densification in other non-oxide ceramics (SiC, B_4C, etc.) are similar. When, for example, SiC is oxidized, a layer of SiO_2 is formed.

Oxide ceramics, such as YTZ [2] are susceptible to humid environments and water vapor over a temperature range of 65–500 °C, but, in aqueous solutions, the effect is more catastrophic at lower temperatures and over shorter times.

Hydrothermal corrosion over the 65–500 °C temperature range has been observed in silicon–nitride ceramics. The experimental observations on aging indicate that water or water vapor enhance the tetragonal-monoclinic transformation.

In summary, some examples of the effects of major ambient components (oxygen, water vapor) have been briefly presented. Designers should become familiar with their effects on creep resistance and utilize them in their designs for material longevity. Even ceramic materials that are very important for high-temperature applications are likely to be influenced by the harmful effects of the environment during their long-term exposures when loaded under creep conditions. Furthermore, care must be taken not to expose these ceramic components to harmful environmental effects during their manufacture from their respective powders.

References

1. Baik B, Becher PF (1987) J Am Ceram Soc 70:527
2. Lawson S (1995) J Eur Ceram Soc 15:485

Part II
Creep in Technologically Important Ceramics

In light of the information currently available in the literature, most of the general observations about creep will now be discussed in regard to a few significant, selected ceramics. As such, Chaps. 12–14 are devoted to the oxide ceramics: Chap. 12 discusses creep in alumina; Chap. 13—magnesia; and Chap. 14—zirconia. Then, chosen from among the carbide ceramics, SiC and BC are the topics of Chaps. 15 and 16. Finally, Si_3N_4 is considered in Chap. 17.

Chapter 12
Creep in Alumina (Al$_2$O$_3$)

Abstract Creep in alumina and alumina composites are discussed in this chapter. Relevant relations for creep specifically related to this technologically important ceramic are presented. Both polycrystalline and single crystal experimental details are considered. Creep tests in tension and compression are noted. Steady state creep rate is indicated and the parameters Q (activation energy), p (grain size exponent), and n (stress exponent) were evaluated.

12.1 Creep in Alumina

As indicated earlier in Chap. 4, the steady-state creep rate was given by Eqs. (4.1) and (4.2). This relation was also used by Chokshi and Porter and presented in Eq. (12.1), with G as the shear modulus, instead of the μ found in Eqs. (4.1) and (4.2):

$$\dot{\varepsilon} = \frac{AGb}{kT} D_0 \exp\left(-\frac{Q}{RT}\right)\left(\frac{b}{d}\right)^p \left(\frac{\sigma}{G}\right)^n \tag{12.1}$$

A is a dimensionless material constant, b is Burger's vector, $D_0\exp(-Q/RT)$ represents the diffusion coefficient, p is the inverse grain size exponent, d is the grain size, and n is the stress exponent, σ being the stress. Equation (12.1) suggests a diffusion-controlled creep mechanism. The creep mechanism and rate of its control are usually determined experimentally by evaluating the values of Q, p and n, and by comparing the resulting values with the theoretical one. It has been claimed that there are indications in fine-grained alumina that, under typical experimental conditions, $n \sim 1$ to 2 and $p \sim 2$ to 3. Various rate-controlling mechanisms have been suggested for creep, among them a diffusion-controlled creep process (Folweiler), GBS (Fryer and Roberts), and interface-controlled diffusional creep mechanisms (Cannon et al.), and so on. In Eq. (12.1), diffusion-controlled creep is indicated.

Figure 12.1 shows several creep rate curves at various stresses. These curves relate to polycrystalline alumina specimens doped with 0.25% MgO. The

© Springer International Publishing AG 2017
J. Pelleg, *Creep in Ceramics*, Solid Mechanics and Its Applications 241,
DOI 10.1007/978-3-319-50826-9_12

as-received material had a uniform grain size and no glassy phase was observed by TEM. The determined grain size was 1.6 ± 0.15 µm. The specimens were deformed by four-point flexure with outer and inner spans of 19 and 6.4 mm, respectively. In addition to the measurements made at 1673 K, creep measurements were performed at other temperatures as well, in order to evaluate the activation energy for creep in alumina. To evaluate the stress exponent strain rate versus stress, plots were constructed from the creep curves in Fig. 12.1. These plots are on a logarithmic scale and are exhibited in Fig. 12.2 for 1673 K. In Fig. 12.1, second-stage creep for the stresses 19.8 and 6.8 MPa are not observed.

Equation (12.1) indicates that, for a given stress and temperature, the creep rate decreases with an increase in grain size. This explains why large grain size materials are preferentially used to decrease the creep rate. In fact, single crystals make the most desirable applications for better creep resistance, thought they are costly.

The apparent activation energy for creep was evaluated from experiments performed at a constant stress of 75 MPa at three temperatures of 1623, 1673, and 1723 K, respectively. A typical plot was constructed for the steady-state creep rate versus the inverse temperature, as seen in Fig. 12.3.

From the Arrhenius plot of Fig. 12.3, the apparent activation energy is expressed as:

Fig. 12.1 The variation of strain rate with strain for polycrystalline alumina deformed at a temperature of 1673 K, $d = 1.6$ µm. Values of strain rate, σ, (o) 135 MPa, (□) 75.5 MPa, (Δ) 50.7 MPa, (◊) 19.8 MPa and (▽) 6.20 MPa. Chokshi and Porter [5]. With kind permission of Dr. Chokshi

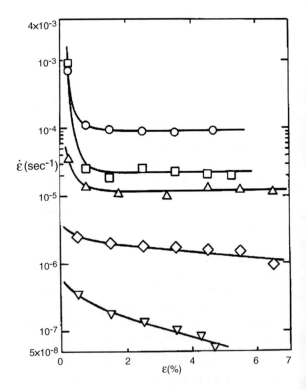

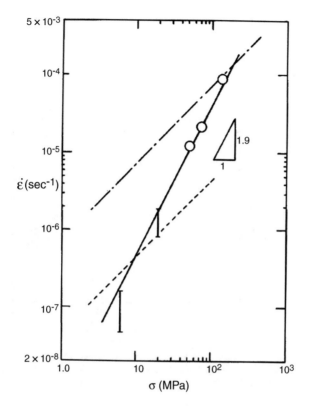

Fig. 12.2 The stress–strain rate relationship for polycrystalline alumina deformed at a temperature of 1673 K, $d = 1.6$ μm. (- - -) Nabarro-Herring [14, 20], (—·—) Coble [22]. Chokshi and Porter [5]. With kind permission of Dr. Chokshi

$$Q_a = -R\frac{\partial \ln \dot{\varepsilon}}{\partial(1/T)} \qquad (12.2)$$

The true activation energy is evaluated from:

$$Q_t = -R\frac{\partial \ln D}{\partial(1/T)} \qquad (12.3)$$

D is given as $D_0\exp(-Q/RT)$, as expressed in Eq. (12.1). The true activation energy evaluated is 635 kJ mol^{-1}. It is apparent from Fig. 12.1 that transient creep extends to ~1% strain, after which the creep rate decreases. Following primary stage creep, a well-developed second-stage creep is observed at high stresses. The stress exponent determined is 1.9. The experimental creep rate was higher than that of the Coble creep, as seen in Fig. 12.2. Thus, the creep mechanism in fine-grained alumina is diffusion controlled.

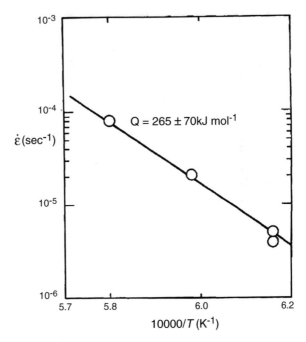

Fig. 12.3 The variation in steady-state strain rate with reciprocal temperature for polycrystalline alumina deformed at 75.5 MPa. $Q = 625 \pm 70$ kJ mol^{-1}. Chokshi and Porter [5]. With kind permission of Dr. Chokshi

12.2 Creep in Alumina Composite

As previously stated certain selected materials are often toughened (improving their general and mechanical properties and performance) for use as composites. Alumina is no exception and various alumina-based composites are in practical use, such as zirconia-toughened alumina (ZTA) composite. Equation (12.1) was also used to analyze the creep data of alumina-toughened zirconia (ATZ), the only difference being in the grain size exponent, indicated by the symbol m, rather than p. The experimental values of m, p, and Q (Eq. 12.2) are:

$$0 \leq m \leq 3, \quad 0.7 \leq n \leq 3.30 \, Q \leq 295 \leq 840 \, \text{kJ mol}^{-1}$$

It was found that rate-controlling creep mechanisms are either the lattice mechanism or the boundary mechanism. The lattice mechanism is intergranular, grain size independent, meaning that the exponent m equals 0, while the boundary mechanism involves grain size-dependent grain boundaries, where m is 1–3. Theoretically, it is possible to identify the active creep mechanism by analyzing the creep data for m, n, Q and D_0; however, in practice, this is difficult to do, since the mechanism also depends on composition and microstructure, making additional

Table 12.1 Designation, composition and flexural strength of studied materials, and creep parameters Q and n. Chevalier et al. [4]. With kind permission of Elsevier

Ceramic system	Nomenclature, preparation conditions	Composition	σ_f at RT (MPa)	Q (kJ/mol)	n
Alumina	A, sintered 1600 °C A1g, sintered 1650 °C A3	Purity > 99.98% Purity > 99.98% Purity C99 98% 1000 ppm MgO	415 ± 10 470 ± 20 380 ± 20	630 ± 10 650 ± 20 630 ± 10	2.5 ± 0.2
Zirconia	Z1, sintered 1750 °C	Mg PSZ 3% wt MgO, purity > 99.9% Mg PSZ	410 ± 40		
	Z5, sintered 1750 °C	Mg PSZ 3% wt MgO, purity > 99.6% Mg PSZ	620 ± 20		
	ZFME, sintered 1750 °C	Mg PSZ 4% wt MgO, purity > 99.6% Mg PSZ	375 ± 5		1.4 ± 0.2
	ZFYT	TZP 3% mol Y_2O_3	1000 ± 20		
Zirconia-toughened alumina	Al-10Z1, sintered 1600 °C	Al_2O_3: Al 10% vol Zro_2:Z1	540 ± 20	760 ± 20	2.5 ± 0.2
	Al-10ZY3, sintered 1600 °C	Al_2O_3:Al 10% vol finer ZrO_2:Zl	430 ± 20		
	Al-10Z4, sintered 1600 °C	Al_2O_3: Al 10% vol finer ZrO_2 (Z4)	550 ± 20		
	A2-10Zl, sintered 1600 °C	Al_2O_3:A2 purity > 99.6% 1000 ppm Mgo 10% vol Zro_2Zl	480 ± 5		

mechanisms possible. Such mechanisms were discussed in Chap. 3 and were also observed in alumina polycrystals.

Alumina and ATZ composite are compared below and Table 12.1 summarizes the details of the specimens discussed below. Information on zirconia is included as well. Creep tests were performed under varying stresses in the range of 50–200 MPa at temperatures up to 1400 °C in air. Flexural strengths were measured at room temperature with a four-point bending fixture having a 36 mm outer span and an 18 mm inner span. In order to obtain dimensional stability of all the parts of the creep testing machine, the system was maintained for one hour at the operating temperature before the creep tests were performed. Then, the load was increased uniformly to the nominal value. The applied load and specimen temperature were recorded as functions of time.

The flexure stress on the tensile face was evaluated by the expression:

$$\sigma = \frac{3P(L-l)}{2Bw^2} \tag{12.4}$$

P is the applied load, L the outer span, l the inner span, B the specimen width, and w is the specimen thickness. The creep rate was measured from the deflection, y_c, at the center of the beam using Hollenberg's method, given as:

$$\varepsilon = K(n)y_c \tag{12.5}$$

with:

$$K(n) = \frac{2w(n+2)}{(L-1)[n+1]+l^2(n+2)/2} \tag{12.6}$$

The constant $K(n)$, in addition to its dependence on n, is also a function of the spans, L and l. Hollenberg et al. have shown that, for the ratio (L/l) close to 2, $K(n)$ is almost insensitive to the value of n. Thus, Eq. (12.6) may be used to evaluate ε by means of Eq. (12.5), with an approximate value of n. By iteration, a proper value can be obtained for n, without large divergences between the final and initially used n values. In Fig. 12.4, the steady-state creep rate is shown at 1200 °C under a stress of 100 MPa.

The less pure alumina, A3 (see Table 12.1) indicates a higher steady-state creep rate compared with A1 and A2, as shown in Fig. 12.4. Similarly, larger grained alumina, A1g has a higher creep rate resistance (see Fig. 12.4). The activation energies for the various alumina (listed in Table 12.1) were determined from the slopes in Fig. 12.5 along with Eq. (12.2). The stress exponent, n, (Eq. 12.1) may also be determined from the variation in the creep rate as a function of the applied stress at the given temperature, 1200 °C (listed in Table 12.1 and shown in Fig. 12.6).

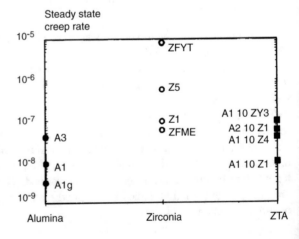

Fig. 12.4 Steady-state creep rates of the 11 ceramics at 1200 °C 100 MPa. Chevalier et al. [4]. With kind permission of Elsevier

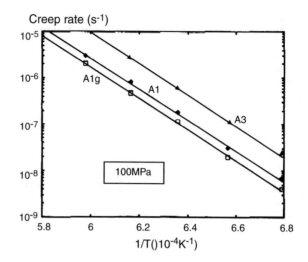

Fig. 12.5 Steady-state creep rate versus 104/*T* with σ = 100 MPa for Al, Alg and A3 alumina. Chevalier et al. [4]. With kind permission of Elsevier

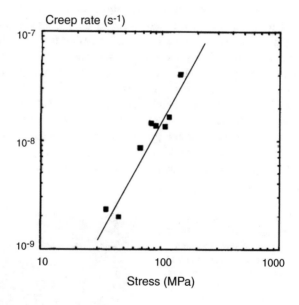

Fig. 12.6 Steady-state creep rate as a function of stress at 1200 °C for A1 alumina. Chevalier et al. [4]. With kind permission of Elsevier

The stress exponent, *n*, is calculated from the slope of the straight line in Fig. 12.6 by using a least-squares method. A value of *n* = 2.5 ± 0.2 for A1 alumina at 1200 °C is determined (listed in Table 12.1). This data on alumina is presented for the purpose of comparing the creep properties of pure alumina with those of ZTA composite, using data originating from tests performed in the same laboratory, applying the same technique, by the same authors. In this case, a different creep behavior is also expected, as seen from the comparison of A1 alumina

Fig. 12.7 Creep curves of Al
alumina, Zl zirconia and
Al-1021 zirconia-toughened
alumina at 1200 °C,
110 MPa. Chevalier et al. [4].
With kind permission of
Elsevier

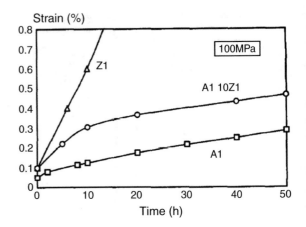

with the ZTA composite, A1-10Z1, exhibited in Fig. 12.7. The creep plot of zirconia is included.

This author believes that the higher creep rate of A1-10Z1, compared with that
of pure alumina, is a consequence of the higher creep rate of zirconia (as seen in
Figs. 12.4 and 12.7). Rather than increasing the creep rate of alumina, the addition
of zirconia reduces the creep resistance of the alumina. Thus, ZTA does not
improve the creep resistance of alumina ceramics. It is of interest to compare the
microstructures of the aluminas listed in Table 12.1 and illustrated in the above
Figures. The microstructures of A1 and A1-10Z1 are compared in Figs. 12.8 and
12.9. In A1 alumina, small cavities (A) appear at triple grain junctions and along
grain boundaries; cavity growth (B) and facet-sized cavities (C) may be observed.
The presence of a thin film of glassy phase at the grain boundaries and as pockets at
triple junctions are shown by TEM, as illustrated in Fig. 12.10.

Fig. 12.8 Microstructure of
Al alumina crept at 1400 °C
($\varepsilon = 2.6\%$) showing small
cavities at grain boundary (A),
cavity growth (B) and cavity
facets (C). Chevalier et al. [4].
With kind permission of
Elsevier

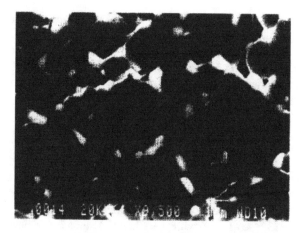

Fig. 12.9 Microstructure of Al-l0Z1 composite crept at 1200 °C showing the presence of numerous cavities. Chevalier et al. [4]. With kind permission of Elsevier

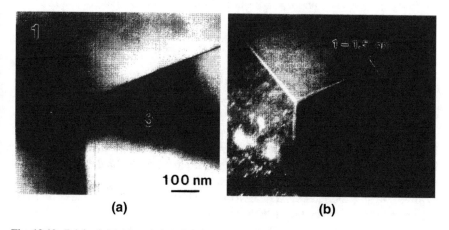

Fig. 12.10 Bright field (**a**) and dark field (**b**) TEM pictures, indicating the presence of a thin vitreous phase at grain boundary. Chevalier et al. [4]. With kind permission of Elsevier

In the ZTA, the cavities and the facet-sized flaws are already increased at 1200 °C creep, as seen in Fig. 12.9 and there were more than were found in pure alumina. The higher creep strain indicated earlier is probably associated with the higher degree of cavitation in the ZTA, appearing in the primary creep stage. The addition of yttria to the ceramic (see Table 12.1, as in partially stabilized zirconia, PSZ) did significantly deteriorate creep resistance and many long cracks appeared on all the faces of the sample.

SEM observations showed that cracking occurs along the grain boundaries and that the damage was homogeneously distributed within the sample. The tetragonal zirconia polycrystal (TZP) ceramics do not show macroscopic damage, although holes, cavities, and cracks clearly appear in SEM observations. Cracks were preferentially perpendicular to the traction axis. Apparently, a general cavitation mechanism occurs with the nucleation, growth and coalescence of cavities, which first leads to the formation of holes and then of cracks. It is of interest to indicate the creep rate used by Riedel and Rice for the effect of cavitation on creep, as indicated by Chevalier et al. given by:

$$\varepsilon = \varepsilon_s + N\left(\frac{2\pi r^2}{X}\right)\frac{2\Omega\delta D_b\sigma}{kTq(w)} \qquad (12.7)$$

where ε_s is the creep rate without cavitation, N—the volume density of cavities, r is the cavity-tip radius, δ—the grain boundary thickness, λ is the distance between cavities, D_b—the grain boundary diffusion coefficient, Ω is the atomic volume and q $(w) = -2\,lnw - (3 - w)(l - w)$ with $w = (2\,r/\lambda)^2$. Estimation of the contribution of cavitation on creep for the A1 alumina was done by taking the following parameters: $D = 2.7 \times 10^{-19}$ m^2/s; $\Omega = 4.2 \times 10^{-29}$ m^3; $T = 1673$ K; $\sigma = 96$ MPa; $N = 6 \times 10^{13}$ m^{-3} (measured by TEM observations); $\tau = 0.5$ μm; $\lambda = 6$ μm; $w = 1/36$; $q(w) = 4.28$; and $\delta = 10^{-3}$ μm. With these values, the cavitation contribution to the steady-state creep rate is $\sim 60\%$.

12.3   Creep in Single-Crystal Alumina

12.3.1   Compression Creep in Single-Crystal Alumina

In general, creep in single crystals is important and preferred (despite the cost) for increasing creep resistance and eliminating GBS, in regard to the deformation mode and the slip plane, and to determine whether a twinning mechanism is also involved. In an α-alumina basal plane (0001) $\langle 11\bar{2}0\rangle$, slip and rhombohedral $(10\bar{1}1)$ deformation twinning are the main deformation modes above 1000 °C. Clearly, slip deformation involves dislocation glide, which is also the deformation mode in compressive creep. Constant load was applied by compression and, since small total strains are associated with deformation, this experiment may be considered as having almost constant stress. A vacuum furnace was used for this creep test and the creep rate was measured continuously. For the experimental details, one is directed to the original work of Bertolotti and Scott.

Since basal slip is a thermally activated process, an Arrhenius-type relation may be used, given as:

$$\dot{\gamma} = v\,\exp-\frac{H(\tau)}{RT} \qquad (12.8)$$

$\dot{\gamma}$ is the strain rate, v is a constant, and $H(\tau)$ is the activation energy, which is a function of the shear stress, τ. The activation energy for basal slip in the $\langle 11\bar{2}0\rangle$

direction was evaluated in the steady-state creep region and the basal plane normal
was at 45° angles with the compression axis. The constant creep load was applied at
decreasing temperatures from 1700 to 1400 °C. The activation energy was derived
from the slope of a plot of log strain (ln $\dot{\gamma}$) versus $1/T$, applying a least-squares fit.
This test was performed at decreasing temperatures, since the analysis was done
assuming a constant number of dislocations. Activation energies were determined
up to a 7.5% true strain and were independent of strain over the range of experiments. The relevant plot is illustrated in Fig. 12.11 and the results appear in
Table 12.2.

Figure 12.11 shows that activation energy decreases with increasing applied
stress. Three specimens indicated an agreement of slopes at 2000 psi stress.

The effect of shear stress on the measured activation energy is shown in
Fig. 12.12. The effect of resolved shear stress on the activation volume is seen in
Fig. 12.13. For the determination of the activation volume, the following relations
were used. First, a polynomial was used to fit a curve to the data:

$$\ln \dot{\gamma} = A + B(\ln \tau) + C(\ln \tau)^2 \tag{12.9}$$

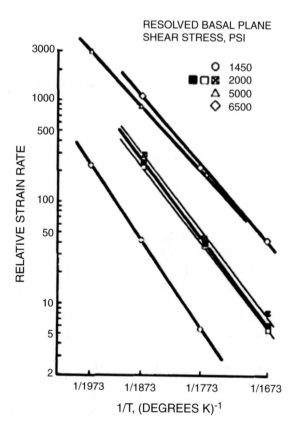

Fig. 12.11 Effect of
temperature and resolved
shear stress on rate of creep
by basal slip. Bertolotti and
Scott [1]. With kind
permission of John Wiley and
Sons

Table 12.2 Activation energy for basal slip in Al₂O₃. Bertolotti and Scott [1]. With kind permission of John Wiley and Sons

Specimen	Resolved shear stress (psi)	Activation energy (kcal/mol)
ST-45	1450	128
ST-43	2000	117
ST-45A	5000	94
ST-46	6500	102

Fig. 12.12 Effect of shear stress on activation energy for basal slip in α-alumina (Data of previous investigation taken from Conrad et al. Bertolotti and Scott [1]. With kind permission of John Wiley and Sons

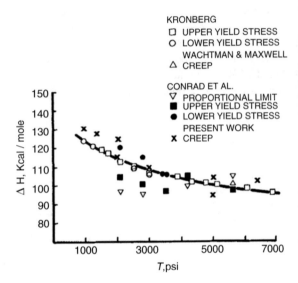

Then, the activation volume, v^*, was expressed by (Conrad et al.) as:

$$v^* = RT\left(\frac{\partial \ln \dot{\gamma}}{\partial \tau}\right)_T \qquad (12.10)$$

The magnitude of the activation volume may be explained either by cross-slip or by the overcoming of the Peierls–Nabarro stress–both possible mechanisms for controlling creep. The TEM investigation does not indicate significant cross-slip and, as such, the overcoming of a large Peirels–Nabarro stress is the most probable mechanism in the creep of Al₂O₃. Large Peirels–Nabarro stresses are expected in ionic crystals, such as Al₂O₃. Al₂O₃, grown in basal plane dislocation networks, usually lies in the $\langle 11\bar{2}0 \rangle$ direction, suggesting the possibility of a large Peirels–Nabarro stress.

Rombohedral twinning by shear stress in the $(10\bar{1}1)$ plane of Al is considered to be an additional mechanism involved in alumina creep, when the basal planes are perpendicular to the applied compressive stress. The activation energy for rhombohedral twin growth was measured from 1500 to 1700 °C, with the compression

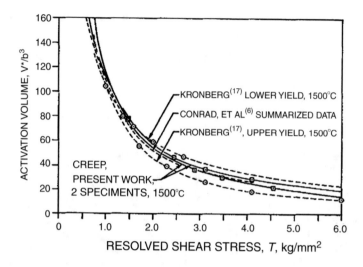

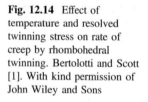

Fig. 12.13 Effect of resolved shear stress on activation volume. Present work and Kronberg's data. $b = 4.75$ Å. Bertolotti and Scott [1]. With kind permission of John Wiley and Sons

Fig. 12.14 Effect of temperature and resolved twinning stress on rate of creep by rhombohedral twinning. Bertolotti and Scott [1]. With kind permission of John Wiley and Sons

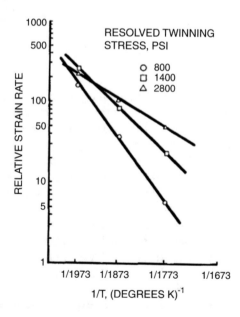

axis in the [0001] direction. For the chosen orientation, there are equal twinning stresses on all three rhombohedral planes and no shear stress on the basal planes. The strains used to eliminate fracture must be below 1.5% in the present case. The results of the successful tests (without fractures) are seen in Fig. 12.14 and listed in Table 12.3.

Table 12.3 Activation energy for rhombohedral twin growth in Al$_2$O$_3$. Bertolotti and Scott [1]. With kind permission of John Wiley and Sons

Specimen	Resolved twinning stress (psi)	Activation energy (kcal/mol)
ST-38X	800	114
ST-37	1400	84
ST-28	3800	52

Fig. 12.15 Creep curves for identically oriented Al$_2$O$_3$ crystals at 1500 °C. Numbers in parentheses are resolved basal plane shear stresses in psi; c refers to as cut and p to polished surfaces. Bertolotti and Scott [1]. With kind permission of John Wiley and Sons

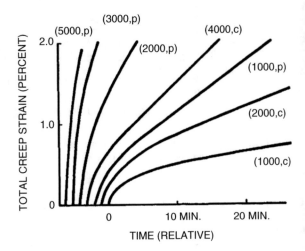

Surface damage resulting from specimen preparation may effect twin nucleation even at room temperature. Therefore, careful specimen preparation must be done. Figure 12.15 shows creep curves of identically oriented single crystals differing only in surface preparation. As-cut specimens showed the much slower creep than the polished ones. The orientation of these specimens was midway between the [0001] and $\langle 20\bar{2}5 \rangle$ directions. Only one rhombohedral plane was under large shear stress for twinning, whereas the basal planes were under moderate shear stress.

The measured activation energies determine the rate-controlling mechanism when both twinning and basal plane deformation occur. Activation energies were measured from 1400 to 1700 °C on cut specimens oriented for concurrent twinning and basal slip; the results are listed in Table 12.4 and the plots are shown in Fig. 12.16.

When reviewing these experiments, it became clear that the specimen preparation induces twin nuclei, which grow during creep. As such, in specimens in which the preparation effects were removed (by polishing) before compression, most of the twin nuclei were also removed.

If the measured activation energy agrees with the value for basal slip and not with twin growth (see Tables 12.3 and 12.4), then the controlling mechanism is thought to be slip in the basal planes. The difference between unprepared and

Table 12.4 Activation energy for creep in Al_2O_3 oriented for basal slip and rhombohedral twin. Bertolotti and Scott [1]. With kind permission of John Wiley and Sons

Specimen	Resolved basal plane shear stress (psi)	Activation energy (kcal/mol)
ST-55	1000	131
ST-54	2000	125
ST-57X	3000	114
ST-54A	5000	104

Fig. 12.16 Effect of temperature and resolved basal plane shear stress on rate of creep when both twinning and basal slip occur. Bertolotti and Scott [1]. With kind permission of John Wiley and Sons

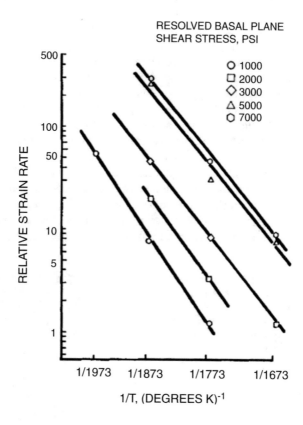

prepared specimens (i.e., polished) is explained on the basis of the relation between the basal planes and the twin-matrix interface. There are three possible basal plane slip vectors of $\langle 11\bar{2}0 \rangle$, but only one is common to both the twin and the matrix. Screw dislocations having this Burgers vector can cross-slip through the interface, but the other $\langle 11\bar{2}0 \rangle$ dislocations are effectively blocked. Therefore, the slower creep rate might be expected for the twinned crystals. The resolved twinning stress is considerably larger than the resolved basal plane shear stress in this orientation, yet the basal plane is rate controlling. Thus, basal slip is expected to be the rate-controlling mechanism at high temperatures for all but a few orientations in which the shear stress is very small or zero.

Appreciable deformation can occur by twinning. Crack formation and fracture in Al$_2$O$_3$ is associated with twinning. Nucleation of cracks can occur by intersecting twins. The controlling factor is whether the concentrated stress at the twin intersection is large enough to propagate the crack. A microstructure showing cracks at a twin intersection is visible in Fig. 12.17. Twins produced in as-cut specimens are usually formed after minor compression and grow thicker during continued loading.

When only one set of parallel twin planes is operating, twins may grow to several hundred micrometers. It was observed that twins formed after fracture looked different than those formed during creep. The former tend to be narrow and tapered (incoherent), while the latter are thicker and not tapered. An example of both these types of twins are shown in Fig. 12.18.

The fracture-produced twins are blocked by the creep-produced twins present before fracture. The experimental information indicates that 1700 °C anneal does not eliminate the surface damage that initiates twinning (twins were observed in creep of as-cut specimens annealed at 1700 °C after diamond sawing).

To summarize this section on compressive creep—both dislocation glide and rhombohedral twinning contribute to creep in Al$_2$O$_3$. In the 1400–1700 °C temperature range, basal-slip-controlled creep occurred whenever there was significant shear stress on these planes. Overcoming the Peirels–Nabarro stress is the most probable rate-controlling mechanism. In specimens properly oriented for twinning, only rhombohedral twinning may also be a significant mode of deformation. Specimen surface preparation plays a key role in determining creep behavior, since twins nucleated by surface abrasion grow under applied shear stress. As indicated,

Fig. 12.17 Cracks formed by intersecting twins at 1500 °C. Compression axis is vertical. Bertolotti and Scott [1]. With kind permission of John Wiley and Sons

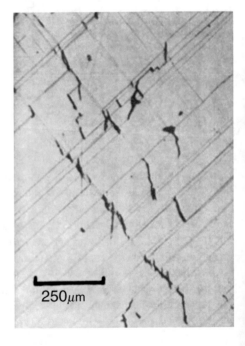

Fig. 12.18 Tapered twins observed after fracture in compressive creep at 1500 °C. The tapered twins resulting from fracture intersect a thick twin formed during creep. Bertolotti and Scott [1]. With kind permission of John Wiley and Sons

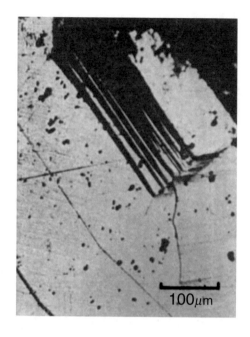

Fig. 12.19 Shear sense for $\langle 10\bar{1}1 \rangle$ twinning in α-Al$_2$O$_3$. Plane of paper is plane of shear. Scott [21]. With kind permission of John Wiley and Sons

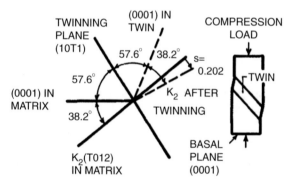

the intersection of twins on two rhombohedral planes may cause catastrophic failure due to crack nucleation. When basal slip and rhombohedral twinning occur concurrently, the creep rate is controlled by basal plane slip, but the presence of twins may substantially reduce the creep rate.

It may be of interest to note the illustration of the stereographic projection and the shear sense for $\langle 10\bar{1}1 \rangle$ twinning. Figure 12.19 illustrates the shear sense for rhombohedral twinning.

The shear sense for rhombohedral $\langle 10\bar{1}1 \rangle$ twinning is such that twinning in alumina should occur when the basal planes are perpendicular to an applied compressive stress. Symmetry conditions require that the second undistorted plane, K_2, be rotated clockwise while twinning.

Fig. 12.20 Stereographic
projection of crystal
orientation for concurrent
twinning and basal slip. Scott
[21]. With kind permission of
John Wiley and Sons

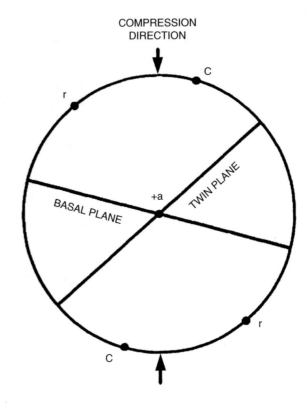

The geometries of the twin and basal planes for this orientation are shown in the stereographic projection in Fig. 12.20.

12.3.2   Tensile Creep in Single-Crystal Al$_2$O$_3$

Generally creep is more critical at high temperatures than at relatively low or room temperatures. Clearly, this depends on the material's melting point, the magnitude of the load and the expected lifetime. Therefore, conventionally, there is an interest in high-temperature creep. In this section, the subject is creep under tensile load. Creep tests at high temperatures encounter some problems: (a) maintaining the sample at the test temperature; (b) applying the stress; and (c) measuring the extension (strain) of the specimen exposed to the high-temperature creep test. As such, it is important to either have on hand or to design proper apparatus for performing these creep tests under tension. Figures 12.21 and 12.22 show tensile machines that are appropriate for creep testing and have efficient tensile grips. For a detailed description of a relevant experimental setup, consult the original work of Wachtman and Maxwell.

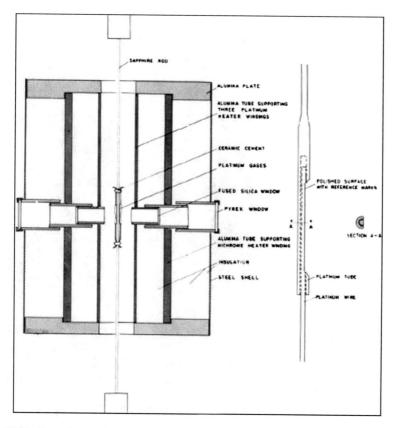

Fig. 12.21 Creep furnace for tests on sapphire rods in tension. The *top* and *bottom* of the heating chamber are closed with Fiberfrax, which effectively retards heat loss without constricting movement of the sapphire rod. Details of platinum gauge are shown at *right*. Wachtman and Maxwell [22]. With kind permission of John Wiley and Sons

A sapphire (α-Al$_2$O$_3$) creep curve obtained under constant load is illustrated in Fig. 12.23. Under 300 kg/cm^2 resolved shear stress, after a 142 h test, the rod did not deform plastically. However, when the load was increased to 400 kg/cm^2, creep deformation began after 50 h and a total elongation of 1.96% occured after 435 h. All three creep stages are seen in Fig. 12.23. Successive creep curves at 1100 °C are shown in Fig. 12.24. A set of selective creep curves are shown in Fig. 12.25, to show the extremes of behavior at 1000 °C. All these figures indicate increasing creep rates initially, and all those samples that did not break within the first 100 h showed eventually decreasing creep rates.

Studies of slip lines and orientation changes during deformation indicated that the geometry of plastic deformation in ceramic oxide single crystals is the same as that found in metals. It has also been shown that slip deformation in sapphire takes place on the (0001) plane in the $\langle 11\bar{2}0 \rangle$ direction, as indicated in the case of α-alumina under compression on the basal plane (in Sect. 12.3.1 above).

Fig. 12.22 Grips for tensile
tests on sapphire rods.
Wachtman and Maxwell [22].
With kind permission of John
Wiley and Sons

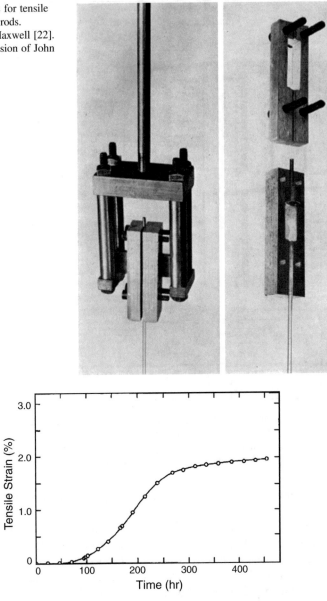

Fig. 12.23 Creep curve for sapphire rod SR69 in tension at 1000 °C (400 kg/cm^2 resolved shear
stress). Wachtman and Maxwell [22]. With kind permission of John Wiley and Sons

Moving on to tensile creep tests of single-crystal creep under tension, it is of
interest to focus on the steady-state creep in alumina. At high temperatures, plastic
deformation is accompanied by recovery. If the recovery rate is sufficiently fast, it
will balance the work-hardening rate and a steady state develops, providing the

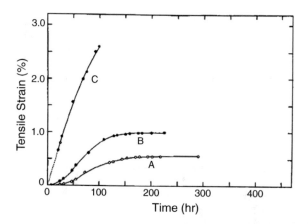

Fig. 12.24 Successive creep curves for sapphire rod SR55 in tension at 1100 °C.
A 0.69 × 108 kg/cm², **B** 0.93 × 10³ kg/cm², **C** 1.23 × 10³ kg/cm² (sample broke in cool
portion). Wachtman and Maxwell [22]. With kind permission of John Wiley and Sons

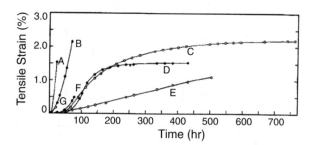

Fig. 12.25 Creep curves for saphire in tension at 1000 °C. **A** SR51, 0.92 × 10³ kg/cm², **B** SR49,
0.80 × 10³ kg/cm², **C** SR44, 1.01 × 10³ kg/cm², **D** SR40, 0.84 × 10³ kg/cm², **E** SR44,
1.20 × 10³ kg/cm, **F** SR38, 1.06 × 10³ kg/cm², **G** SR44, 1.50 × 10³ kg/cm². Wachtman and
Maxwell [22]. With kind permission of John Wiley and Sons

conditions for steady-state creep. Currently, the belief is that dislocation generation
and motion are associated with plastic deformation in steady-state
creep. (a) Dislocation climb involves the transport of material and, thus, the activation energy for steady-state creep is approximately equal to that of self-diffusion.
When dislocation climb becomes so rapid that it is no longer rate controlling, other
mechanisms of steady-state creep are proposed. Two important mechanisms under
consideration are (b) a microcreep mechanism and (c) the motion of dislocation
lines over Peierls stress hills. In (b), dislocation motion is hindered by an atmosphere of imperfections or impurities, and the velocity of dislocation motion is
proportional to the force exerted on them. The activation energy for steady-state
creep, according to this mechanism, is usually the same as that required for the
diffusion of the relevant impurities or imperfections. Here, the dislocation motion
over the Peierls stress hills is the rate-controlling mechanism. The activation energy

Fig. 12.26 Typical creep curve, sapphire (1.4 × 10^8 d/cm^2, 1823 K). Chang [3]. With kind permission of AIP Publishing LLC

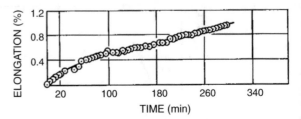

Fig. 12.27 'Steady-state' creep rate versus applied resolved shear stress, sapphire and ruby (1823 K). Chang [3]. With kind permission of AIP Publishing LLC

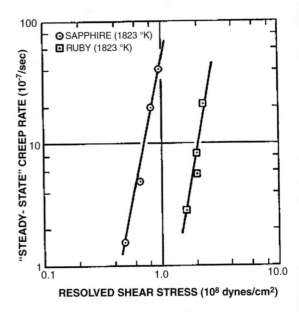

for steady-state creep is determined by the Peierls force of the material. In non-metals, such as alumina, the Peierls stress is high and it may be expected that the Peierls stress mechanism is the rate-controlling mechanism.

A typical creep curve for sapphire is shown in Fig. 12.26. The steady-state creep rate depends on the stress at a constant temperature and on the temperature at constant stress. Such dependencies appear in Figs. 12.27 and 12.28. Figure 12.27 is a plot of a constant temperature, 1823 K, as a function of resolved shear stress, while Fig. 12.26 shows the steady-state creep rate as a function of temperature.

Having already referred to steady-state mechanisms, now they may be inspected

(a) The Dislocation Climb Mechanism

One may assume that during high-temperature creep by means of a dislocation climb mechanism, a dislocation can climb perpendicular to the slip plane via vacancy diffusion. A further assumption is that equilibrium is established along the dislocation lines during climb. As such, (according to Chang following Weertman) for the steady-state creep rate one obtains:

Fig. 12.28 'Steady-state' creep rate versus temperature, sapphire and ruby (1823 K). Chang [3]. With kind permission of AIP Publishing LLC

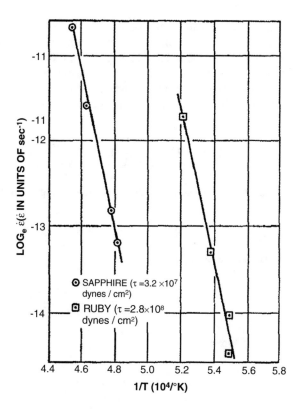

$$\dot{\varepsilon} = \left(\frac{6\pi^2 v\tau^2}{\mu^2}\right) \sinh\left[\frac{3\tau^5 b^3}{2k^2 T^2 \mu^3 M}\right]^{1/2}$$
$$x\exp\left(\frac{\Delta S}{k}\right)\exp\left(-\frac{\Delta H}{kT}\right) \tag{12.11}$$

where v is the vibrational frequency of a vacancy, τ is the resolved shear stress, μ is the rigidity modulus, **b** is the Burgers vector length of a dislocation, ΔS is the entropy, ΔH is the activation energy for self-diffusion, and M is the average density of the Frank-Read sources. T and k have their usual meanings of 'absolute temperature' and the 'Boltzman's constant,' respectively.

When the jog formation energy is high, so that equilibrium vacancy concentration cannot be maintained along the dislocation lines (due to the presence of only a few jogs), steady-state creep is given by:

$$\dot{\varepsilon} \cong \left(\frac{L}{b}\right)\left(\frac{6\pi^2 v\tau^2}{\mu^2}\right) \sinh\left[\frac{3\tau^5 b^3}{2k^2 T^2 \mu^3 M}\right]^{1/2}$$
$$x\exp\left(\frac{\Delta S^*}{k}\right)\exp\left(-\frac{\Delta H^*}{kT}\right) \tag{12.12}$$

L is the half-width between the Frank-Read sources, ΔS^* and ΔH^* are the sums of the entropies and the activation energies for self-diffusion and jog formation, respectively. Equation (12.11) is applicable when the probability that a jog exists in a dislocation segment of length L is greater than unity. Expression sinhx $\cong$ x is applicable when $(x \ll 1)$ is satisfied and when the applied stress is on the order of 10^8 d/cm^2. Then, Eq. (12.11) may be simplified as:

$$\dot{\varepsilon} = 72.4 \left(\frac{v^2 b^3 \tau^9}{k^2 T^2 \mu^7 M} \right)^{1/2} \exp\left(\frac{\Delta S}{k} \right) \exp\left(-\frac{\Delta H}{kT} \right) \tag{12.11.1}$$

According to Eq. (12.11.1) at constant temperature:

$$\dot{\varepsilon} = \text{cons} \tan t (\tau)^{9/2} \tag{12.11.2}$$

The constant being:

$$\text{cons} \tan t = 72.4 \left(\frac{v^2 b^3}{k^2 T^2 v^7 M} \right)^{1/2} \exp\left(\frac{\Delta S}{k} \right)$$

while at constant stress:

$$\dot{\varepsilon} = (\text{cons} \tan t) \frac{1}{T\mu^{7/2}} \exp\left(-\frac{\Delta H}{kT} \right) \tag{12.11.3}$$

The constant is clearly given by:

$$\text{cons} \tan t = 72.4 \left(\frac{v^2 b^3 \tau^9}{k^2 M} \right)^{1/2} \exp\left(\frac{\Delta S}{k} \right)$$

The plot of Eq. (12.11.2) appears in Fig. 12.27. The slope derived for sapphire (single crystal alumina is 4.5, which agrees well with the expected value of 9/2. Regarding Eq. (12.11.3), throughout these experiments, T and τ are negligibly small and, as such, the equation becomes:

$$\dot{\varepsilon} \cong (\text{cons} \tan t) \exp\left(-\frac{\Delta H}{kT} \right) \tag{12.11.4}$$

The plot of this function was shown in Fig. 12.28, yielding a value of $\Delta H = 7.8$ eV for alumina. The activation energy for self-diffusion, measured indirectly from sintering data is 7.0–7.8 eV. The activation energy for the diffusion of oxygen in Al$_2$O$_3$ is 7.5 eV. Thus, it appears that the activation energy for steady-state creep in Al$_2$O$_3$ single crystals is almost equal to that of self-diffusion. This suggests that the steady-state creep mechanism in Al$_2$O$_3$ single crystals is probably the dislocation climb mechanism. In order to use

Eq. (12.11.1), the density of the Frank-Read sources, M, and the entropy term, $\Delta s/k$, are required. Assuming the respective values of $M \sim 10^{11}$ cm^{-3} and $\Delta s/k \sim 20$, one obtains a reasonable agreement between the experiments and the theories.

(b) The Viscous-Drag Mechanism

The viscous drag mechanism represents the hindering effect of imperfections on dislocation motion. Since the activation energy of steady-state creep is high, a viscous drag mechanism in dislocations versus impurities is unlikely. It is more likely that lattice imperfections act as dragging points, hindering dislocation motion. An example of such a mechanism is given in Fig. 12.29, which illustrates a moving dislocation.

The mean distance between the dragging points is l, **b** is the Burgers vector, and τ is the resolved shear stress. The steady-state creep rate may be expressed as:

$$\dot{\varepsilon} \cong 2\lambda\Lambda b^2 v_0 \exp\left(-\frac{\Delta F}{kT}\right) \sinh\left(\frac{lb^2\tau}{kT}\right) \tag{12.13}$$

Λ is the total free length of dislocation per unit volume (10^6 em^{-3}); **b** is the Burgers vector (8.22×10^{-8} cm); ΔF is the free energy of activation; v_0 is the vibrational frequency of an atom (10^{13}/s); k is Boltzmann's constant (1.38×10^{-16} ergs per deg); and λ is a correction factor converting shear strain to tensile strain ($((3/2))^{1/2}$ for rods with a basal plane inclined at 30° from the rod axis. Expressing ΔF as $\Delta H - T\Delta S$ and sinh ($lb^2\tau/kT$) as $\cong 1/2\exp(lb^2\tau/kT)$ for $lb^2\tau \geq kT$, Eq. (12.13) to be written as:

$$\dot{\varepsilon} \cong \lambda\Lambda b^2 v_0 \exp\left(\frac{\Delta S}{k}\right) \exp\left(-\frac{\Delta H + lb^2\tau}{kT}\right) \tag{12.14}$$

A plot of Eq. (12.14), as log $\dot{\varepsilon}$ versus τ at constant temperature, should provide the slope:

Fig. 12.29 Moving dislocations with dragging points. This schematic drawing follows Chang [3]

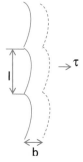

$$lb^2/kT \qquad (12.14.1)$$

and an intercept (extrapolated to zero) of:

$$\log\left(\lambda \Lambda b^2 v_0\right) + \left(\frac{\Delta S}{k}\right) - \left(\frac{\Delta H}{kT}\right) \qquad (12.14.2)$$

A plot of $\log \dot{\varepsilon}$ versus $1/T$ at a given stress, instead of τ, yields a slope of:

$$-\frac{\Delta H + lb^2 \tau}{kT} \qquad (12.13.1)$$

and an intercept (extrapolated to $1/T = 0$) of:

$$\log\left(\lambda \Lambda b^2 v_0\right) + \frac{\Delta S}{k} \qquad (12.13.2)$$

The mean dragging distance, l, may be obtained from Eq. (12.14.1). Quantity ΔH may be evaluated using Eqs. (12.14.1) and (12.13.1) and also independently from the difference in Eqs. (12.13.2) and (12.14.2). Knowing ΔH enables the determination of $\Delta S/k$, either from Eq. (12.14.2) or (12.13.2).

The facts that the main distance between the dragging points is on the order of hundreds of angstroms and that the activation energy should be the same as for self-diffusion makes this mechanism doubtful at present, until future research proves the validity of findings based on this mechanism.

(c). The Peierls Stress Mechanism

According to Weertman, in cases where the Peierls mechanism is controlling creep, steady-state creep may be expressed as:

$$\dot{\varepsilon} \cong \left(\frac{12\tau^5 \gamma^2 a^2}{\mu^6 b^5 M}\right)^{/2} \exp\left(-\frac{Q}{kT}\right) \exp\left(\frac{\pi \tau Q}{2\tau_0 kT}\right) \qquad (12.15)$$

where a is the distance between the Peierls hills, Q is the barrier to dislocation climb over the Peierls hills, and τ_0 is the Peierls force. Seeger relates Q and τ_0 by:

$$Q \sim \left(\frac{4ab}{\pi}\right)\left(\frac{2\mu\tau_0 ab}{5\pi}\right)^{1/2} \qquad (12.16)$$

It is difficult to evaluate the Peierls force, τ_0. Assuming that Q is equal to ΔH (experimental: 7.8 eV), Eq. (12.16) provides for a Peierels force of $\sim 2 \times 10^9$ d cm^{-2}. This value is about a factor of 10^7–10^8 higher than the value calculated by means of Eq. (12.15), suggesting that the Peierls stress mechanism is not rate controlling in the steady-state creep of Al$_2$O$_3$ single crystals.

12.4   Creep Rupture in Alumina

Certainly, failure by rupture can occur when specimens are exposed to tensile or flexural loads. Such failures and the various responses of specimens to various stresses should be studied and understood. At the core of rupture, and/or other types of creep failure, are cracks which nucleate, grow and ultimately leading to fracture. The interest in these properties is a consequence of the desire to develop structural ceramics for high-temperature applications having lifetimes of many thousands of work hours. Tests for such extended periods are not practical and, therefore, creep components are designed for short-term tests (long-term tests are costly). Thus, the factors controlling the propensity for creep rupture are of great interest for an understanding of the basic creep process. It is known that Al_2O_3, by itself, is not ideal for high-temperature applications and that the use of various additives is common (examples are considered later on). It is easier to gain a fundamental understanding of creep, failure and even microstructure without involving complications caused, for instance, by alloying. To this end, the next section considers unalloyed Al_2O_3.

These experiments on hot-pressed alumina were performed at lower temperatures and strain rates than in earlier ones (Dalgleish et al.; Johnson) to eliminate short-time failure and to extend the creep test procedure. Two kinds of specimens were used, one with 0.3% MgO, to aid sintering (termed AVCO) and a second having only 200 ppm MgO. Both had a residual porosity of <0.05% and the grains were equiaxial, without detectable grain-boundary glass. Flexural and tensile tests were performed in air and deformation was monitored continuously by extensometers. Most of the experiments were performed by the application of a single stress to fracture but, in some tests, incremental stress changes were applied in order to determine the creep-test exponent. Some compressive tests were also performed. To eliminate swelling due to trapped gas, the tests were performed at 1250 °C and below (at 1350 °C gas trapping occurs). In Fig. 12.30, flexural creep results are shown for both alumina-type materials. Typical creep curves are seen for flexural creep tests at 1150 °C.

The outer tensile strain, ε, is calculated from the 3-point probe displacements using the analysis of Hollenberg et al. The stress at steady state is:

$$\sigma = \frac{2n+1}{3n}\sigma_{el} \qquad (12.17)$$

where n is the stress exponent: for ARCO it is 1.8 ± 0.2 and for the AVCO alumina it is 2.0 ± 0.2. The creep rate data are plotted in Fig. 12.31.

Power law creep was observed with the indicated exponents and the activation energies are 390 and 480 kJ mol^{-1} for the ARCO and AVCO aluminas, respectively. The failure strain, at about 175 MPa, sets in leading to fracture in flexure tests, as shown in Fig. 12.32.

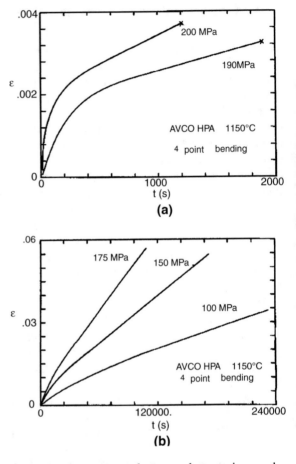

Fig. 12.30 Flexural creep curves in AVCO alumina at (**a**) high stresses and (**b**) intermediate stresses. The strain and steady-state stresses are calculated for a stable neutral axis and a stress exponent of 2.0. Robertson et al. [20]. With kind permission of John Wiley and Sons

Flexural stress rupture in both aluminas occurred at very low strains, under 1%, within very short times (typically less than 2 h); results are plotted in Fig. 12.33.

Furthermore, tests were done applying tensile stress at 1250 °C. As opposed to bend tests, in these experiments, the creep curves showed very little evidence of primary creep. This may be seen in Fig. 12.33. Also very little tertiary creep is observed before failure. The steady-state creep data are summarized in Fig. 12.35. These data were obtained from both fixed stress and stress change tests. The stress exponent for tensile creep is 1.8, the same as that obtained from the compression and flexural tests (Fig. 12.34).

The failure strains, both true failure strain and longitudinal strain, as functions of strain, are illustrated in Fig. 12.36. These strains are:

$$\varepsilon_f = \ln \frac{A_0}{A_f}$$

Fig. 12.31 Flexural creep rates in **a** AVCO and **b** ARCO alumina as a function of steady-state stress and temperature. *Solid squares* are for samples that failed before attaining steady-state creep. Robertson et al. [20]. With kind permission of John Wiley and Sons

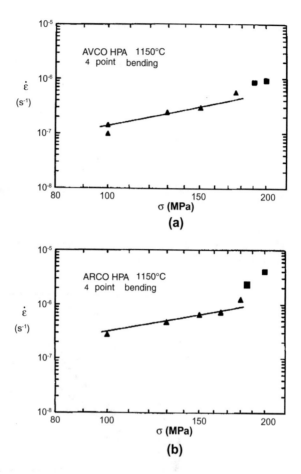

(a)

(b)

Fig. 12.32 Failure strain, ε_f, as a function of the steady-state stress flexure tests in AVCO alumina. Robertson et al. [20]. With kind permission of John Wiley and Sons

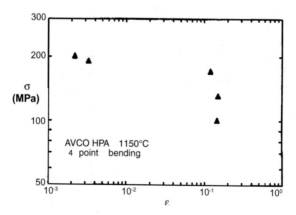

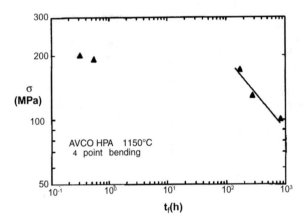

Fig. 12.33 Flexural stress rupture data for AVCO alumina. Robertson et al. [20]. With kind permission of John Wiley and Sons

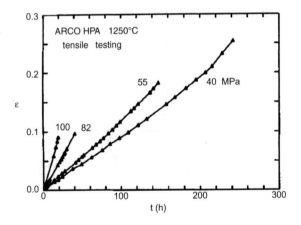

Fig. 12.34 Creep curves for ARCO alumina in tension. Robertson et al. [20]. With kind permission of John Wiley and Sons

Fig. 12.35 Steady-state creep data for ARCO alumina in tension. Robertson et al. [20]. With kind permission of John Wiley and Sons

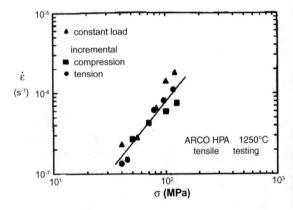

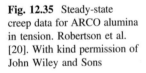

Fig. 12.36 Failure strain in ARCO alumina deformed in tension as a function of applied stress. Robertson et al. [20]. With kind permission of John Wiley and Sons

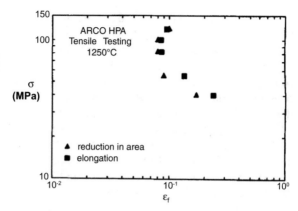

and:

$$e_f = \ln \frac{l_f}{l_0}$$

The failure strain is about 9% at high stresses but, at lower stresses, it increases to 17%.

The stress rupture results may be seen in Fig. 12.37.

Two regimes may be seen in Fig. 12.37. At the two highest stresses, there is a relation of the form $t_f \sim \sigma^{-m}$, where the stress exponent is $m \sim 2.5$, while at the lowest stress range, it is ~ 1.8 and the time to failure is longer. The stress levels to failure in these two regimes are 80 and 55 MPa, respectively, and between them lies a region in which the time-to-failure increases rapidly with decreasing stress. Stress rupture is a consequence of crack nucleation, its growth leading to fracture. A sequence of micrographs of high-temperature creep rupture appear in Fig. 12.38.

Fig. 12.37 Time to failure as a function of applied stress for ARCO alumina tested in tension. Robertson et al. [20]. With kind permission of John Wiley and Sons

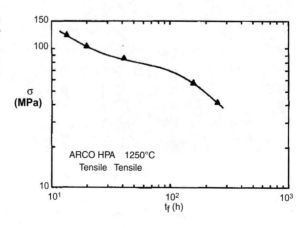

A ceramic material may contain pre-existing flaws of sufficient size, so that the initial stress intensity factor, K_i, is greater than the threshold intensity factor, K_{th}, and when cracks tend to blunt, they lead to thickening without significant growth. In this case, creep rupture is considered to be controlled by crack growth with a consequent short life to rupture. However, when the initial flaws are small and blunting occurs, rupture becomes controlled by creep damage. This creep rupture duality is illustrated schematically in Fig. 12.39.

The concept of duality in stress rupture was also indicated in a hot-pressed Al$_2$O$_3$/1/4MgO tested at 1250–1300 °C by flexure at a constant peak stress in air. This is shown in Fig. 12.40, where the stress, σ, is plotted against the rupture strain, ε_f. The separation of the results indicated in Fig. 12.40 into two groups is obvious: the first group exhibits failure strains of $\sim \leq 1\%$, while in the second group shows $\sim \geq 8\%$.

It has been also observed that the rupture strain in the specimens that exhibit large deformation before failure varies with the stress as a simple product, as illustrated in Fig. 12.41.

SEM micrographs of the failure modes discussed above are seen in Fig. 12.42.

Damage observation by a series of SEM micrographs are illustrated in Figs. 12.43, 12.44, 12.45, 12.46, 12.47 and 12.48. In Fig. 12.43 the flaw consisted

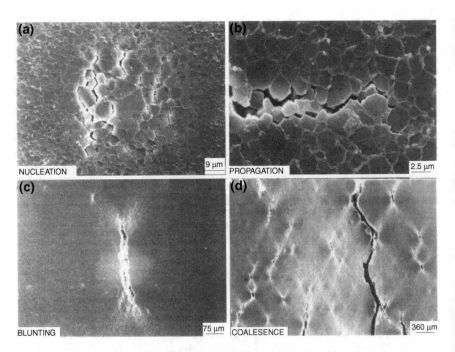

Fig. 12.38 Scanning electron micrographs of sequence involved in high-temperature rupture, showing **a** crack initiation at large-grained heterogeneity, **b** coplanar crack propagation into damage zone, **c** shear band formation at arrested crack, and **d** failure by damage coalescence across shear bands. Dalgleish et al. [9]. With kind permission of John Wiley and Sons

Fig. 12.39 Schematic indicating crack growth and creep damage controlled regimes of creep rupture and associated blunting threshold (a refers to length of pre-existing cracks, and K_{th} is threshold stress intensity factor). Dalgleish et al. [9]. With kind permission of John Wiley and Sons

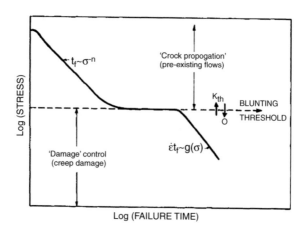

Fig. 12.40 Plot of failure strain, $\dot{\varepsilon}t_f$, as a function of stress, σ_f, indicating that the data exhibit two distinct regimes of behavior: one at small failure strains and the other at large failure strains. Dalgleish et al. [9]. With kind permission of John Wiley and Sons

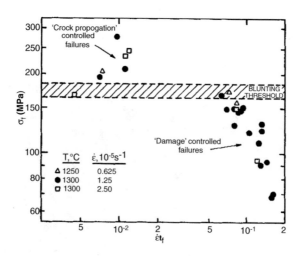

of either a large-grained region (A) or a region beneath the tensile surface. A region of cavity damage is seen around the failure origin in (B), which contains isolated cavities (on the two grain interfaces). This may indicate creep crack growth prior to rupture. However, shear bands were observed on the fracture surface seen in Fig. 12.45. These nucleated on both the tensile side and side surfaces at a strain of $\sim\geq 3\%$. The number of shear bands increased with strain. Most of the bands were found to initiate at large-grained microstructural heterogeneities as seen in Fig. 12.44.

Once the shear bands were formed, they extended rapidly to a distance of $\approx 3R$, R being the heterogeneity radius. Thereafter, further growth was minimal. With more increase in strain beyond $\sim 4\%$, cracks were propagated across the bands (see Fig. 12.45). It was also observed that the large-grained heterogeneities were subject to cracking at very small strains ($\leq 1\%$). Suggestions were put forward that prior

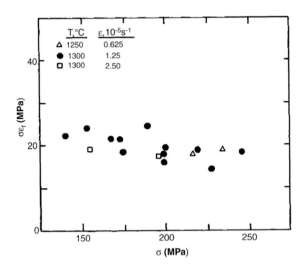

Fig. 12.41 Plot of product of stress and failure strain, $\sigma\varepsilon_f$, as a function of stress, σ, for large-strain failures, revealing that the product is approximately constant in this failure regime. As seen in the figure, $\sigma\varepsilon_f$ is ≈ 20 MPa. Dalgleish et al. [9]. With kind permission of John Wiley and Sons

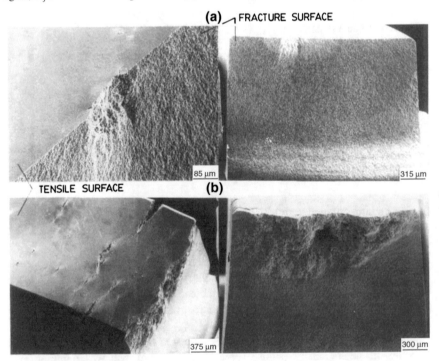

Fig. 12.42 Scanning electron micrographs of failures in two regimes of behavior: **a** short-term failure, showing that rupture occurs from a single flaw with no evidence of creep damage, and **b** long-term failure, showing rupture by coalescence of creep damage across shear bands. Dalgleish et al. [9]. With kind permission of John Wiley and Sons

Fig. 12.43 Scanning electron micrographs of a large-grained failure origin in stress-controlled rupture regime: **a** large-grained zone; **b** slow-growth regime outside large-grained zone, showing cavities on two-grain interfaces. Dalgleish et al. [9]. With kind permission of John Wiley and Sons

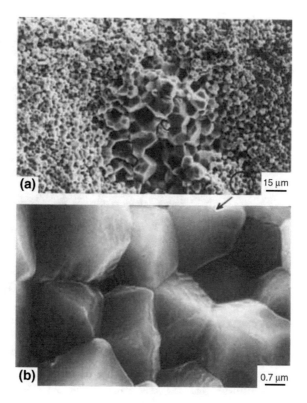

cracking might be involved in the shear-band initiation process. Appreciable coalescence of the damaged shear bands was observed, as seen in Fig. 12.47. This coalescence occurred either as interband coalescence (Fig. 12.47a) or as coalescence with an edge crack (Fig. 12.47b). Failure surfaces indicate that the shear bands that caused rupture are approximately normal to the tensile surface, visible in Fig. 12.48.

Concerning rupture, crack growth may be described in the crack propagation region, where $K_i > K_{th}$, given as:

$$\frac{da}{dt} = v_0 \left(\frac{K}{K_c}\right)^n \quad (12.18)$$

v_0 and n are material and temperature-dependent parameters, respectively (as discussed earlier). Creep-crack growth in linear material predicts that $n \approx 1$ and v_0 is given by:

$$v_0 = F(z, \lambda/l)/\eta Z \quad (12.19)$$

η is the viscosity of the material, z is the grain facet in the damage zone, λ is the spacing between cavities, l is the grain facet length, and F is a function plotted in

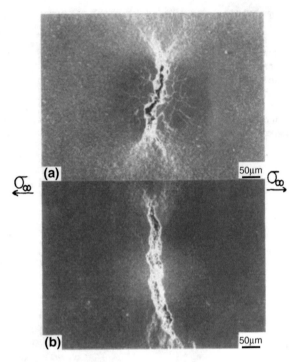

Fig. 12.44 Scanning electron micrographs of shear band nuclei, showing **a** large-grained region and **b** nickel-rich zone. Dalgleish et al. [9]. With kind permission of John Wiley and Sons

Fig. 12.49. At a test temperature of 1300 °C, the v_0 for alumina, when $z \approx 1$ and $\lambda/l = 1/10$, was determined to be 2×10^{-8} m s^{-1}.

$$K = (2/\sqrt{\pi})\sigma\sqrt{a} \tag{12.20}$$

The lifetime is predicted to vary according to:

$$t_f = \sqrt{\pi}K_c\left[\left(\frac{\sqrt{\pi}K_c}{2\sigma}\right) - \sqrt{a_0}\right]/\sigma v_0 \tag{12.21}$$

or:

$$\frac{t_f\sigma}{\sqrt{a_0}} = (A/\sigma\sqrt{a_0}) - B \tag{12.22}$$

with $A = \pi/K_c^2/2v_0$ and $B = \sqrt{\pi}K_c/v_0$ which are material parameters. The short life failure data are shown in Fig. 12.50 and are compared with predicted rupture times (2×10^{-8} m s^{-1}) obtained from crack growth measurements. The predicted rupture time consistently exceeds the measured values.

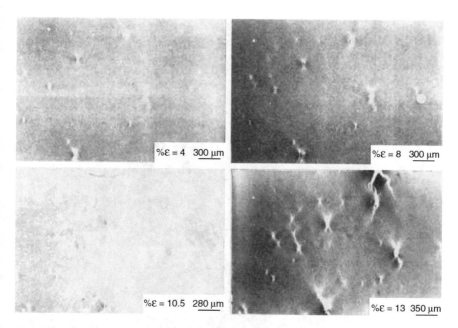

$\%\varepsilon = 4$ 300 μm $\%\varepsilon = 8$ 300 μm $\%\varepsilon = 10.5$ 280 μm $\%\varepsilon = 13$ 350 μm

Fig. 12.45 Note that number density of bands increases with increase in strain. Sequence of scanning electron micrographs indicating development of shear bands with strain. Dalgleish et al. [9]. With kind permission of John Wiley and Sons

Fig. 12.46 Cavities on two-grain interfaces within large-grained region. Dalgleish et al. [9]. With kind permission of John Wiley and Sons

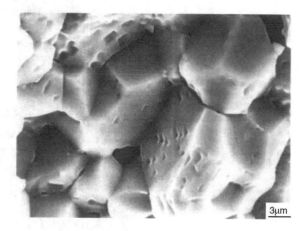

3μm

It can thus be concluded, in regard to the dual creep rupture concept, that creep-damage-controlled creep rupture at stress intensities below the crack-blunting threshold is characterized by a simple failure law of $\sigma\varepsilon_f =$ constant. At stress intensities above the blunting threshold, creep rupture is determined by the growth of pre-existing flaws in the material.

Fig. 12.47 Coalescence of
shear bands just before
rupture, showing **a** interband
coalescence and
b coalescence with specimen
edge. Dalgleish et al. [9].
With kind permission of John
Wiley and Sons

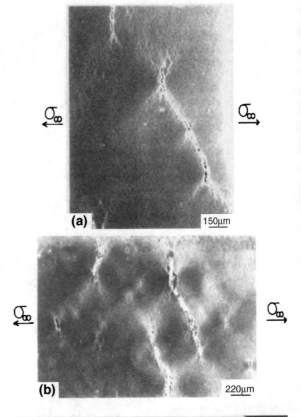

Fig. 12.48 Failure surface
indicating that shear bands
form approximately normal to
tensile surface and at $\approx \pi/3$ to
strain axis. Dalgleish et al.
[9]. With kind permission of
John Wiley and Sons

12.5   Superplasticity in Al$_2$O$_3$

Before discussing superplasticity specifically in Al$_2$O$_3$, one may review the
essential structural prerequisites for superplasticity as summarized by Nieh et al.
Due to dynamic grain growth, it is unlikely to obtain superplasticity in pure Al$_3$O$_3$

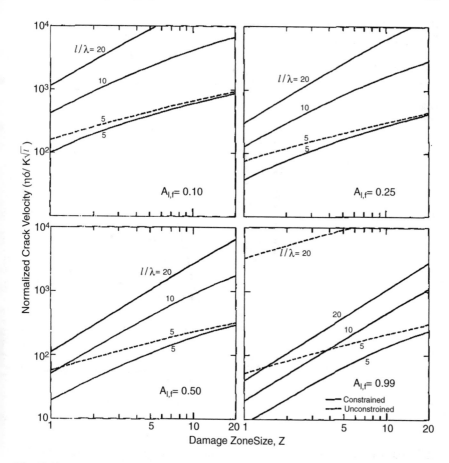

Fig. 12.49 Plot of predicted trends in crack velocity with damage zone size. Dalgleish et al. [9]. With kind permission of John Wiley and Sons

without dopants or alloying. Various charge-carrying dopants, such as Ti^{4+}, Mn^{2+}, Zr^{4+}, etc., are aded to the basic alumina. Often, ZrO_2 is used as an additive to alumna to obtain superplasticity by hindering grain growth and, thus, imparting microstructural stability. Superplastic deformation may be expressed as:

$$\dot{\varepsilon} = A\frac{\sigma^n}{d^p} \tag{12.23}$$

As indicated in earlier equations [for example Eq. (8.9)], $\dot{\varepsilon}$ is the strain rate, σ is the stress, d is the grain size, and n and p are the stress and the grain size exponents, respectively. A is a temperature-dependent and diffusion-related coefficient. For superplastic ceramics, n an p are between 1 and 3. According to Eq. (12.23), superplasticity is encouraged by small grain size and an increasing

Fig. 12.50 Plot of failure
time in stress-controlled
regime compared with
predictions based on creep
crack growth in homogeneous
material. Dalgleish et al. [9].
With kind permission of John
Wiley and Sons

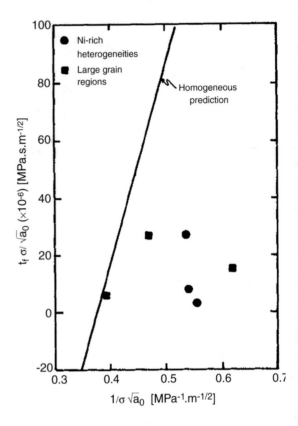

diffusion rate (A is diffusion-related). Although low-temperature sintering does not cause significant grain growth, the accepted method is by the use of various additives. MgO and ZrO$_2$ are successfully applied in alumina. Mg ions and zirconia particles effectively pin the grains, thus inducing superplasticity in the alumina ceramics up to a level of 100% engineering strain at 1450 °C. Zirconia, however, is the best additive for alumina, achieving very high superplasticity of up to a 550% elongation. So far, even an elongation of 850% has been reported in alumina-based ceramics (Kim et al.). Superplastic alumina is compared with its undeformed state in Fig. 12.51.

12.5.1 High-Temperature Superplasticity

The composition of the (alumina-based) spinel after sintering was Al$_2$O$_3$-10 vol% ZrO$_2$-10 vol% spinel. During these experiments, tensile tests were carried out in the 1400–1500 °C temperature range in a vacuum. The as-sintered ceramic is shown in a SEM micrograph in Fig. 12.52.

Fig. 12.51 Undeformed and superplastically deformed specimens. Kim et al. [16]. With kind permission of Elsevier

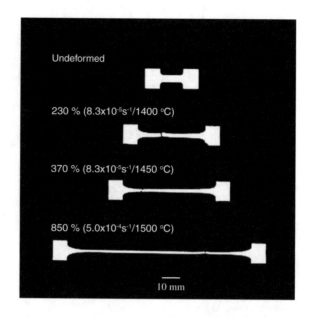

Fig. 12.52 As-sintered microstructure. "A", "Z" and "S" represent Al$_2$O$_3$, ZrO$_2$ and spinel grains, respectively. The dihedral angle between Al$_2$O$_3$/spinel interphase boundaries is smaller than that between Al$_2$O$_3$/Al$_2$O$_3$ grain boundary and spinel/Al$_2$O$_3$ interphase boundary, as indicated by *arrows*. Kim et al. [16]. With kind permission of Elsevier

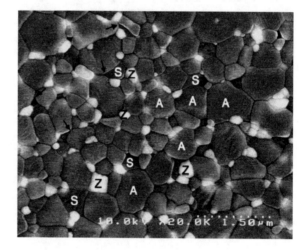

The gray grains are Al$_2$O$_3$ and the ZrO$_2$ and spinel phases are white and dark particles, respectively. The ZrO$_2$ particles are located at the quadruple junctions of the matrix grains, while the spinel particles among the matrix grains. The average radius of the Al$_2$O$_3$ grains is 0.25 mm, and those of the ZrO$_2$ and spinel particles are 0.09 and 0.18 mm, respectively. HRTEM observation revealed (Fig. 12.53) that no amorphous phases exist along the Al$_2$O$_3$/Al$_2$O$_3$ grain boundaries or Al$_2$O$_3$/ZrO$_2$ and Al$_2$O$_3$/spinel interphase boundaries in the as-sintered material.

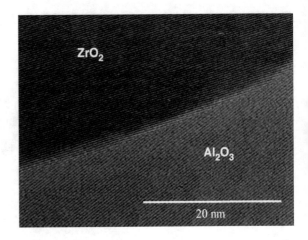

Fig. 12.54 Static grain
growth behavior. The open
markers are for Al$_2$O$_3$-10 vol
% ZrO$_2$. Kim et al. [16]. With
kind permission of Elsevier

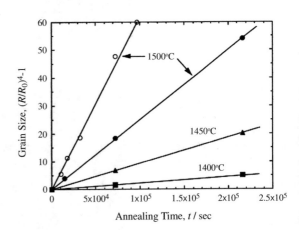

The results of various experiments on static grain growth at 1400–1500 °C are plotted in Fig. 12.54. Static grain growth may be expressed as:

$$R_m - R_0^m = kt \tag{12.24}$$

Here, R is the average grain radius, R_0 is the initial radius, m is the grain growth exponent, k is the rate constant (proportional to grain-boundary mobility and grain-boundary energy), and t is the annealing time. The value of m for alumina is 4. The slopes of the lines in Fig. 12.54 are k/R_0^4 and their temperature dependence give the activation energy as $Q_g = 588$ kJ mol^{-1}.

Tensile stress–strain curves of the alumina spinel are shown in Fig. 12.55 for three initial strain rates at 1500 °C. Also shown for comparison is a curve of Al$_2$O$_3$-10 vol% ZrO$_2$. At the lower strain rates, almost no strain hardening is observed after the yield stress. Also note that with increasing strain rate (and flow stress

Fig. 12.55 Typical stress-strain curves at 1500 °C. The *dotted curve* is for Al$_2$O$_3$-10 vol% ZrO$_2$. Kim et al. [16]. With kind permission of Elsevier

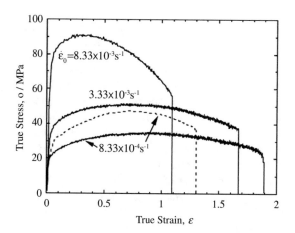

increase), the strain decreases. Moreover, it seems that the alumina spinel has a lower flow stress than Al$_2$O$_3$-10 vol% ZrO$_2$ at the same strain rate (8.33 × 10^{-3} s^{-1}), but the elongation is higher.

The strain-rate dependence of tensile elongation is shown for 1400–1500 °C in Fig. 12.56. The dispersion of ZrO$_2$ and spinel particles leads to enhanced superplasticity The maximum tensile elongation reached 850%, which is the largest elongation ever reported in Al$_2$O$_3$-based ceramics. The dispersion of these phases is very effective in suppressing both static and dynamic grain growth. In the present material, the kinetic constant for static grain growth and the rate constant of dynamic grain growth were lower than those in Al$_2$O$_3$-10% ZrO$_2$ by 40 and 27%, respectively. α depends on several factors, such as grain shape and grain size distribution. A value for $\alpha = 0.9$ was determined for Al$_2$O$_3$-10% ZrO$_2$. Dynamic grain growth is given by Eq. (12.25):

Fig. 12.56 Tensile elongation vs. initial strain rate for Al$_2$O$_3$-10 vol% ZrO$_2$-10 vol% spinel (*filled markers*). The *open markers* are for Al$_2$O$_3$-10 vol% ZrO$_2$. Kim et al. [16]. With kind permission of Elsevier

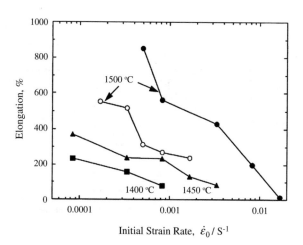

$$dR = \frac{k}{m}R^{1-m}dr + \alpha R d\varepsilon \tag{12.25}$$

where ε is the strain and α is the rate constant for dynamic grain growth. By knowing the parameters in Eq. (12.25) and integrating, one can predict dynamic grain growth. By using the measured Al$_2$O$_3$ grain sizes and fitting Eq. (12.25) to these values, a value of $\alpha = 0.43$ was obtained. Taking this value, a plot for dynamic grain growth may be constructed as indicated in Fig. 12.57. The values of $m = 4$ and $k = 9.9 \times 10^{-31}$ at 1500 °C are reasonable for the grain sizes, regardless of the strain rate variations.

For smaller values of α and k from Eqs. (12.24) and (12.25), for a given value of m, the dynamic and static components (grain size) decrease. This is the objective of the spinel particle dispersion (10 vol%), with the consequent decrease of α by 27% and of k by 40%. The suppression of dynamic grain growth enhances superplasticity

The creep deformation (as described in one of the earlier chapters) is described by:

$$\dot{\varepsilon} = A\frac{\sigma^n}{R^p}\exp\left(\frac{Q}{R_g T}\right) \tag{12.26}$$

where $\dot{\varepsilon}$ is the strain rate, σ is the stress, Q is the apparent activation energy, R_g is the gas constant, T is the absolute temperature, A is a proportional constant, and n and p are the stress and grain size exponents, respectively. The stress exponent may be obtained by the slope of the $\log(\varepsilon) - \log(\sigma)$ plot for the same grain size. By using Eq. (12.26), the grain size exponent may be evaluated from the slope of a plot of $\log\left(\frac{\sigma^n}{\varepsilon}\right)$ versus $\log(R)$, as illustrated in Fig. 12.58.

For $n = 2.2$, evaluated from the plot in Fig. 12.58 and for $\sigma < 30$ MPa and $\dot{\varepsilon} < 10^{-3}\text{s}^{-1}$, one gets Fig. 12.59 for the ratio $\frac{\sigma^{22}}{\varepsilon}$ as a function of R.

Fig. 12.57 Dynamic grain growth behavior at 1500 °C. The theoretical fitting from Eq. (12.25) is represented by solid lines for $\alpha = 0.43$. Kim et al. [16]. With kind permission of Elsevier

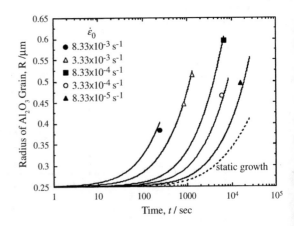

Fig. 12.58 Strain rate versus stress at $R = 0.28$ μm. Kim et al. [16]. With kind permission of Elsevier

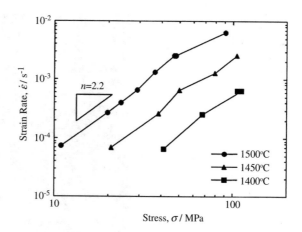

The slope in the linear part of the curves is 3.2 for the aforementioned conditions, indicating that the choice of $n = 2.2$ is reasonable. The deviation from linearity is due to growing cavitation damage. Note that for $\dot{\varepsilon} > 10^{-23} \text{s}^{-1}$, the shape and location of the $\frac{\sigma^{22}}{\dot{\varepsilon}} - R$ curve is different than those of $\dot{\varepsilon} < 10^{-3} \text{s}^{-1}$, and the linear portion does not exist, as in Fig. 12.59.

Superplastic deformation is associated with GBS, simultaneous grain growth and grain elongation. However, grain boundary elongation is regarded as having a small or negligible effect in alumina, since the aspect ratio remains small (less than 1.5). The limited grain elongation during superplastic deformation is related to grain boundary migration and grain growth. High grain boundary mobility restricts grain elongation. The microstructure of a deformed alumina-based ceramic may be seen in Fig. 12.60.

The elongation occurring in grain boundaries is diffusion controlled during GBS in order to maintain microstructural continuity. The major role of the dispersed ZrO₂ and spinel particles is the inhibition of grain-boundary migration. The shape of the ZrO₂

Fig. 12.59 Relationships between $\frac{\sigma^{22}}{\dot{\varepsilon}}$ and R at 1500 °C. Kim et al. [16]. With kind permission of Elsevier

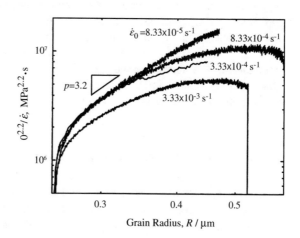

Fig. 12.60 Deformed
microstructure (560%) at
8.33×10^{-4} s^{-1} and at
1500 °C. The stress axis is
vertical. Kim et al. [16]. With
kind permission of Elsevier

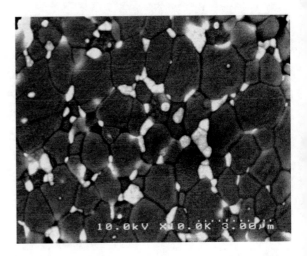

particles depends on their location, as seen in Fig. 12.60. Those grains that are along
the Al$_2$O$_3$/Al$_2$O$_3$ boundaries are elongated (Al$_2$O$_3$ matrix phase is gray grains, while
the ZrO$_2$ and spinel phases appear as white and dark particles, respectively), while
those that are embedded in the Al$_2$O$_3$ grains are equiaxed. The observed elongation of
Al$_2$O$_3$ grains along the stress axis seems to be associated with grain-boundary diffusion. Particles of ZrO$_2$ along grain boundaries are associated with GBS, contrary to
those ZrO$_2$ particles within the Al$_2$O$_3$ grains. The relative stability of the ZrO$_2$ particles in the equiaxed Al$_2$O$_3$ grains is an indication that neither lattice diffusion nor
intragranular dislocation motion in the Al$_2$O$_3$ grains is the primary deformation
mechanism. Indeed, grain-boundary diffusion is the primary deformation mechanism.
The small or negligible dislocation concentration within the Al$_2$O$_3$, confirmed by
TEM observations, experimentally support the above determination.

Experiments have shown that concurrent cavitation, particularly cavity interlinkage in the direction normal to the stress axis, limits the tensile ductility of
superplastic ceramics. The role of the simultaneous dispersion of ZrO$_2$ and spinel
particles is to suppress cavity formation during tensile deformation. Suppressed
cavity formation retards cavity interlinkage in the direction perpendicular to the
stress axis, resulting in elongated cavities and, thereby, leading to large tensile
elongation of the alumina-based 10 vol% zirconia and 10 vol% spinel ceramics.
Thus, grain-boundary diffusion and cavity elongation are associated with superplasticity in these alumina-based ceramics.

12.5.2   Low-Temperature Superplasticity

Again, Eq. (12.23) also describes superplasticity at the lower temperatures. The
low-temperature creep reported here is for two alumina-based composites: (a) 1 mol%
Ti^{4+} and 1 mol% Mg^{2+}, with additional dopants of 1000 ppm ZrO$_2$, 1000 ppm MgO

and 500 ppm Y$_2$O$_3$, and (b) is the same as (a) except with 3 mol% ZrO$_2$. The purpose of the MgO and Y$_2$O$_3$ additives is to improve microstructural homogeneity and to retard grain growth. The 0.1% zirconia (1000 ppm) is intended to establish a baseline to account for the very strong hardening effect in this system, while the 3% zirconia is meant to pin grain-boundary particles. SEM micrographs of the Ti^{4+}- and Mg^{2+}-doped alumina ceramics appear in Fig. 12.61 (for experimental details see Xue and Chan).

Cavity formation is observed in the alumina containing 0.1% ZrO$_2$, but not in the 3% ZrO$_2$, as seen in the above figure, in (c) and (d), respectively. Grain growth is

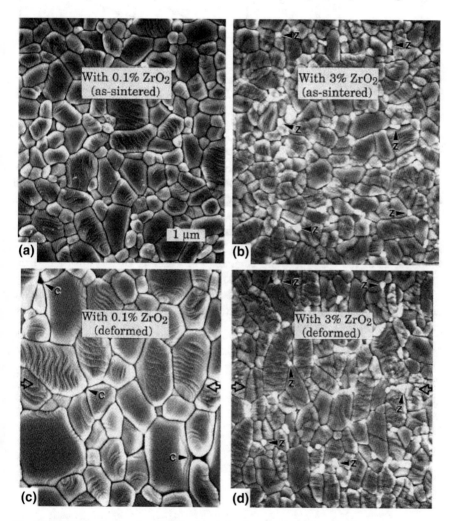

Fig. 12.61 SEM micrographs of Ti/Mn-doped alumina. As-sintered specimens **a** 0.1% ZrO$_2$ and **b** 3% ZrO$_2$. Deformed specimens with **c** 0.1% ZrO$_2$, $\varepsilon = -0.64$. The compression axis is marked by *hollow arrows* in **b** and **d** and representative ZrO$_2$ particles in **b** and **d** are indicated by small *solid arrows* (z). Examples of cavities in **c** are also marked by *small solid arrows*. Xue and Chan [24]. With kind permission of John Wiley and Sons

Fig. 12.62 Stress-strain
curves at 1330 °C, at constant
strain rate of 3 × 10^{-4}/s, for
0.1%-ZrO$_2$-added and 3%-
ZrO2-added Ti/Mn-doped
alumina. Xue and Chan [24].
With kind permission of John
Wiley and Sons

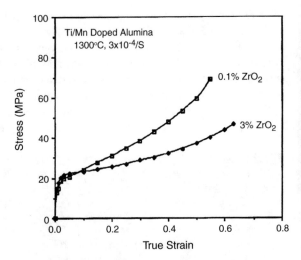

much more pronounced in the specimen containing 0.1% ZrO$_2$. This difference may
account for the more pronounced strain hardening and also for the cavity formation.
In Fig. 12.62, the stress–strain curves obtained under compression at 1300 °C and
at a strain rate of 3 × 10^{-4} s^{-1} for the two alumina-based composites are shown.

It seems that this grain growth occurs during deformation—it is dynamic grain
growth—which is more pronounced than static grain growth, as a result of
annealing. Strain hardening is observed in both specimens, but it is particularly
pronounced in the 0.1% zirconia ceramic. The addition of 3% zirconia to the
ceramic effectively retards both static and dynamic grain growth.

The strain rate and the flow stress are related in Fig. 12.63 on a logarithmic
scale. Despite the difference indicated in Fig. 12.62, the deformation behavior of

Fig. 12.63 Strain rate–stress
relationship at 1300 °C for
0.1%-ZrO$_2$ added and 3%-
ZrO$_2$ added Ti/Mn-doped
alumina. Xue and Chan [24].
With kind permission of John
Wiley and Sons

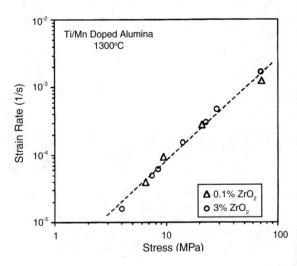

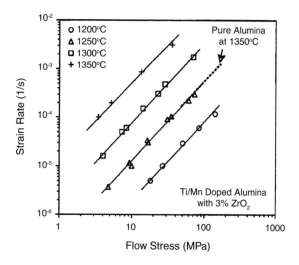

Fig. 12.64 Relationship between strain rate and flow stress at various temperatures for the Ti/Mn-doped alumina (0.66 μm) with 3% ZrO₂. The *dashed line* is for pure alumina (0.5 μm) at 1350 °C. Xue and Chan [24]. With kind permission of John Wiley and Sons

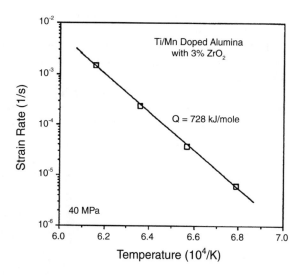

Fig. 12.65 Strain rate of the Ti/Mn-doped alumina with 3% ZrO₂ at 40 MPa as a function of reciprocal temperature. Xue and Chan [24]. With kind permission of John Wiley and Sons

the two composites is identical, as observed in Fig. 12.63—the single line representing both composites. The effect of temperature on the stress rate versus stress is shown in Fig. 12.64.

The inverse temperature dependence of the strain rate at 40 MPa for the 3% ZrO₂ containing alumina is illustrated in Fig. 12.65. The activation energy evaluated from the slope in this figure is 728 kJ/mol, which is much higher than the one obtained for pure alumina (460 kJ/mol). The superplastic behavior of this ceramic is illustrated in Fig. 12.66, by stretching the 3% Zr-alumina ceramic by means of punch displacement.

Fig. 12.66 Forming load
versus punch displacement for
the Ti/Mn-doped alumina
with 3% ZrO$_2$ during
superplastic stretching at
1280 °C. The average strain
rate is 2.3 × 10^{-4}s^{-1}. Xue
and Chan [24]. With kind
permission of John Wiley and
Sons

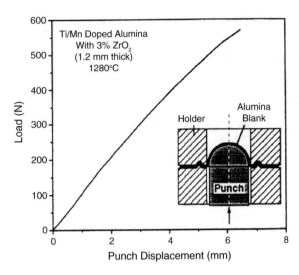

The ceramics under consideration still possessed a fine microstructure with a grain size of ∼0.8 μm after the superplastic stretching. To improve its high-temperature creep resistance, postforming annealing was performed to coarsen its microstructure. Annealing at 1650 °C (for 4 h) increased the grain size to 23 μm (Fig. 12.67).

The creep rate of a specimen annealed in 1400 °C is compared with that of the as-sintered one at 1350 °C (shown in Fig. 12.68). It is known that the charge-compensating dopants, Ti^{4+} and Mn^{2+}, have a higher solubility in alumina

Fig. 12.67 SEM micrograph
of a polished and thermally
etched Ti/Mn-doped alumina
specimen with intergranular
and intragranular ZrO$_2$
particles (annealed at 1650 °C
for 4 h). Xue and Chan [24].
With kind permission of John
Wiley and Sons

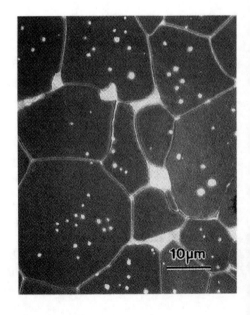

Fig. 12.68 Comparison of deformation rate for the Ti/Mn-doped alumina with 3% ZrO$_2$ before and after annealing. Xue and Chan [24]. With kind permission of John Wiley and Sons

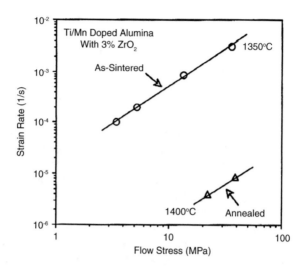

and that they significantly enhance the diffusion/deformation process during sintering and deformation. Codoping of these constituents increases the strain rate by a factor of 70. In addition, their effect on grain growth may be controlled by adding 3 mol% ZrO$_2$, which acts as pinning agent. By the addition of these constituents, superplastic alumina may be obtained with a forming capability of 100% strain by biaxial tension at temperatures as low as 1250 °C. Furthermore, it was found that annealing increases creep resistance by a factor of 2000 (seen from the extrapolation of the as-sintered specimen to 1400 °C, Fig. 12.64). This stretchability demonstrates the feasibility of the superplastic forming of alumina ceramics, even at low temperatures.

12.6 Creep in Nano-Alumina

Technological interest lies in composite alumina, rather than in monolothic alumina, due to the desire to obtain improved properties, such as improved creep resistance. Nowadays, nanostructures are of great technological interest. Often, nanostructured or nano-alumina is also strengthened by incorporating various additives into the monolithic alumina. To this end, one often used additive is SiC, in various forms. The addition of SiC to nano-alumina improves its overall properties, especially creep resistance, since ceramics are intended for high-temperature applications. Now, the tensile creep behavior of nano-alumina with SiC additives will be discussed and compared with that of monolithic alumina.

The tested nanocomposite had 17 vol% SiC and was prepared by hot pressing the powders at 1800 °C in a nitrogen atmosphere under an applied pressure of 30 MPa for 1 h. A high temperature was required for the sintering, since the presence of SiC nanoparticles suppresses densification and grain growth.

The monolithic alumina was hot pressed at 1500 °C, so that the grain sizes of the two components were equal. Tensile creep tests were performed for both the alumina-7 vol% SiC composite and for the monolithic alumina at 1200–1330 °C at 50–150 MPa. Excellent creep resistance was obtained in the nanocomposite, compared to the monolithic alumina. The improvement in creep resistance, in terms of the minimum creep rate, was manifested by a lower creep rate by ~ three orders and an increased creep lifetime of about 10 times that of the monolithic alumina. Compared to the accelerated creep observed in the monolith, the composite showed only transient creep to failure. Rotating and plunging the SiC nanoparticles into the alumina matrix increased the creep resistance with GBS.

Flexural creep tests were also performed in air at 1200 °C at 100, 150 and 200 MPa. The specimens for the flexure tests were 2 and 3 mm thick and wide, respectively. They were loaded in a four-point flexure fixture with inner and outer span lengths of 10 and 30 mm, respectively. The applied stress and the resulting strain were calculated from the load and displacement relation. The tensile creep curves tested at 1200 °C for both the monolith and the composite are compared in Fig. 12.69.

The curve of the monolith consists of primary, steady-state, and a very small tertiary creep regimes (while very little was observed in the composite). The monolith's lifetime was ~150 h and ~4% creep strain at fracture, with a large number of observed microcracks. The composite, however, had very good creep resistance and its creep strain at fracture was only 0.5%. No microcracks were detected by optical microscopy. The composite also achieved better creep resistance in the flexural creep test.

The stress dependencies of the monolith and composite are compared in Fig. 12.70. In this figure below, note that the flexural creep test indicates that the flexural creep rate is smaller than the tensile creep rate under the same applied stress. The TEM microstructure of the crept and fractured nanocomposite is shown

Fig. 12.69 Tensile creep curves of the monolith and nanocomposite at 1200 °C and 50 MPa. Slight accelerated creep and steady-state creep were present in the monolith, while they were little observed in the nanocomposite. Ohji et al. [19]. With kind permission of John Wiley and Sons

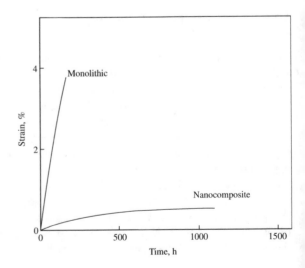

Fig. 12.70 Dependencies of steady-state or minimum creep rates in the tension (*closed symbol*) and the flexure (*open symbol*) for the monolith and the nanocomposite. The temperature is 1200 °C. The stress exponent for creep rate is 2.2 for the monolith and 3.1 for the nanocomposite in tension, and 2.9 for the monolith and 2.2 for the nanocomposite in flexure. Ohji et al. [19]. With kind permission of John Wiley and Sons

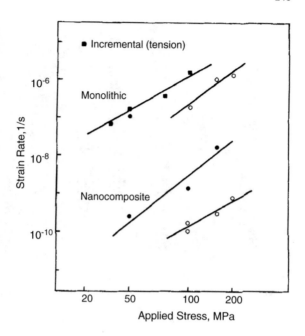

in Fig. 12.71. In (a), it shows the rotation of SiC particles accompanied by GBS and the formation of small cavities around the particles. The strain-contrast contours formed may be observed at the corners (top right and top left) of the particle and also small cavities produced between a particle and an upper grain. The SiC particles seem to penetrate into the other grains, consequently producing a greater pinning effect. Thus, larger creep resistance results and the creep remains in its transient stage. More evidence of rotating and plunging intergranular SiC particles and associated cavity formation is given in Fig. 12.71b. Traces of intergranular crack propagation are seen in Fig. 12.72. The crack formed propagates at the alumina–alumina grain boundary, where small cavities formed around the SiC particles as a consequence of GBS. Note that some dislocations are also seen, as are small transgranular cracks in the crept specimens around the transgranular nanoparticles. It was suggested that these features form during the cooling down from the sintering temperature, rather than during creep.

In conclusion, one can summarize the effect of SiC additive as follows. The creep resistance of the nanocomposite was excellent, compared to that of the monolithic nano-alumina, as attested by Figs. 12.69 and 12.70. In a more recent work, the significantly improved mechanical properties of this nano-alumina composite at high temperatures has been substantiated, far outdoing the monolithic nano-alumina. The SiC content was only 5%, but MgO was also added as a sintering aid. Fabrication was done by pressureless sintering and post hot isostatic pressure (HIP). The density of the composite after sintering and with a different MgO content is found in Fig. 12.73. In the Al_2O_3/SiC composite (unlike the monolithic alumina), the open pores disappear at a bulk density of $\sim 90\%$, as

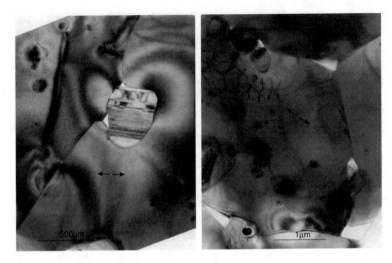

Fig. 12.71 Transmission electron micrographs of microstructures of the nanocomposite tested at 1300 °C and 50 MPa in tension, showing examples of rotating and plunging of intergranular silicon carbide particles and associated cavity formation. The stress direction is indicated by *arrows*. Ohji et al. [19]. With kind permission of John Wiley and Sons

Fig. 12.72 Transmission electron micrograph of a trace of intergranular crack propagation. The sample was tested at 1200 °C and 100 MPa in tension. Note the transgranular-fractured nanoparticle The stress direction is indicated by an *arrow*. Ohji et al. [19]. With kind permission of John Wiley and Sons

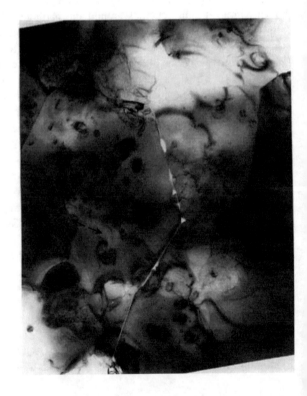

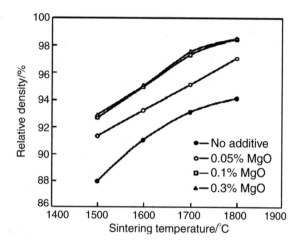

Fig. 12.73 Relative density as function of sintering temperature for Al$_2$O$_3$/5%SiC composites with different MgO contents. Jeong and Niihara [14]. With kind permission of John Wiley and Sons

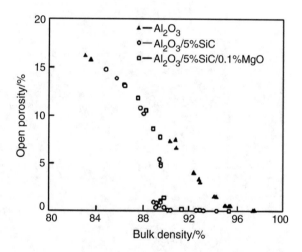

Fig. 12.74 Open porosity as function of bulk density for monolithic Al$_2$O$_3$ and Al$_2$O$_3$/5%SiC composites without and with 0.1% MgO. Jeong and Niihara [14]. With kind permission of John Wiley and Sons

shown in Fig. 12.74. Therefore, the densification for before its use was >90%, and could be achieved by the combined fabrication by pressureless sintering and HIP (see Fig. 12.75). The microstructure of the Al$_2$O$_3$/SiC composite, before and after HIP) is illustrated in Fig. 12.76. The effect of the MgO additive may also be seen in Fig. 12.76a and b. It is evident that MgO sintering promotes densification. A number of closed pores remained although the sintering was done at a high temperature (1800 °C).

Following HIP, complete densification occurred, whether MgO was present or not, and the grain sizes were the same as seen in Fig. 12.76c and d. HIP treatment of Al$_2$O$_3$/SiC produces an homogeneous distribution of the SiC particles both within the Al$_2$O$_3$ grains and within the grain boundaries, as indicated in Fig. 12.77.

Fig. 12.75 Variations of HIP density with sintered density for monolithic Al$_2$O$_3$ and Al$_2$O$_3$/5%SiC composites without and with 0.1% MgO. Jeong and Niihara [14]. With kind permission of John Wiley and Sons

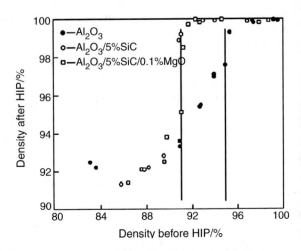

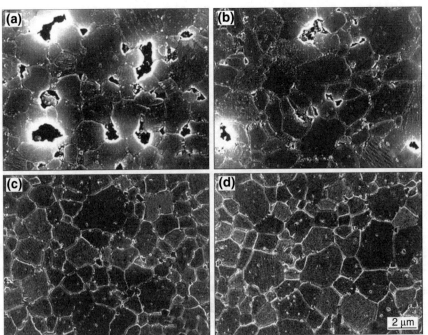

Fig. 12.76 SEM micrographs of thermally-etched Al$_2$O$_3$/5% SiC nanocomposites sintered at 1800 °C for 2 h: **a** without MgO, before HIP; **b** with 0.1% MgO, before HIP; **c** without MgO, after HIP; **d** with 0.1% MgO, after HIP. Jeong and Niihara [14]. With kind permission of John Wiley and Sons

The fracture strength, as a function of the MgO content, before and after HIP, is shown in Fig. 12.78. The pre-HIP fracture strength increased with the MgO content and sintering temperature, owing to the increase in the sintered density. Apparently, the HIP produces the significant effect, since a high fracture strength of 1 GPa was

Fig. 12.77 TEM image of Al$_2$O$_3$/5%SiC/0.1%MgO nanocomposite fabricated by pressureless sintering at 1800 °C for 2 h and subsequent HIP treatment at 1600 °C for 1 h under 150 MPa. Jeong and Niihara [14]. With kind permission of John Wiley and Sons

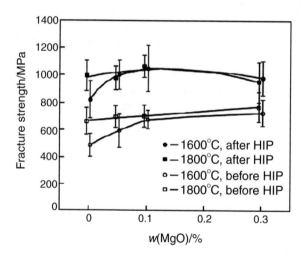

Fig. 12.78 Fracture strength as function of MgO content for Al$_2$O$_3$/SiC nanocomposites before and after HIP. Jeong and Niihara [14]. With kind permission of John Wiley and Sons

achieved, regardless of the presence or absence of MgO in the Al$_2$O$_3$/SiC nanocomposite (see Fig. 12.78).

The strength improvement due to the HIP treatment is also related to the fracture mode indicated in Fig. 12.79. Accordingly, the fracture surface of the sintered ceramics exhibit both intergranular and transgranular fracture. The HIP-treated Al$_2$O$_3$/SiC showed complete transgranular fracture. Bridging at a crack site is considered the primary strengthening mechanism in a ceramic nanocomposite. This mechanism leads to a very high crack-growth resistance curve (*R* curve). Crack

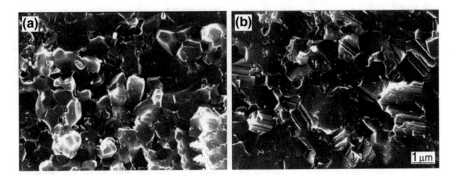

Fig. 12.79 SEM micrographs showing fracture surfaces of Al$_2$O$_3$/SiC/0.1% MgO nanocomposite sintered at 1600 °C for 2 h: **a** before HIP; **b** after HIP. Jeong and Niihara [14]. With kind permission of John Wiley and Sons

extension through the nearest SiC particles, induced by thermal residual tension, causes a bridging mechanism to operate effectively, even at a small SiC volume fraction of 5%, as in the current experimental results.

Thus, one can conclude that a composition of as low as 5% SiC (with a small amount of MgO as a sintering aid) and by the application of HIP, may be considered a favorable processing method resulting in high-strength nanocomposites. Clearly, the above composite is only one example of the many possible additives that may be effectively used with nanostructure alumina. It was chosen because it has been frequently used in creep resistance studies and there is ample proof that it improves nanostructured alumina.

References

1. Bertolotti RL, Scott WD (1971) J Am Ceram Soc 54:286
2. Cannon RM, RhodesWH, Heuer AH (1980) J Amer Ceram Soc 63:46
3. Chang R (1969) J Appl Phys 31:484
4. Chevalier J, Olagnon C, Fantozzi G, Gros H (1997) J Eur Ceram Soc 17:859
5. Chokshi AH, Porter JR (1986) J Mat Sci 21:705
6. Conrad H, Stone G, Janowski K (1965) Trans AIME 233:889
7. Dalgleish BJ, Evans AG (1985) J Am Ceram Soc 68:44
8. Dalgleish BJ, Johnson SM, Evans AG (1984) J Am Ceram Soc 67:741
9. Dalgleish BJ, Slamovich EB, Evans AG (1985) J Am Ceram Soc 68:575
10. Edington JW, Melton KN, Cutler CP (1976) Prog Mater Sci 21:61
11. Folweiler RC (1961) J Appl Phys 32:773
12. Fryer GM, Roberts JP (1966) Proc Br Ceram Soc 6:225
13. Hollenberg GW, Terwilliger GR, Gordon RS (1971) J Am Ceram Soc 54:196
14. Jeong Y-K, Niihara K (2011) Trans Nonferrous Met Soc China 21:s1
15. Johnson SM, Dalgleish BJ, Evans AG (1984) J Am Ceram Soc 67:759
16. Kim B-N, Hiraga K, Morita K, Sakka Y (2001) Acta Mater 49:887
17. Kronberg MK (1957) Acta Met 5:507
18. Nieh TG, McNally CM, Wadsworth J (1988) Scripta Metall 22:1297

19. Ohji T, Nakahira A, Hirano T, Niihara K (1994) J Am Ceram Soc 77:3259
20. Robertson AG, Wilkinson DS, Caceres CH (1991) J Am Ceram Soc 74:915
21. Scott WD (1971) J Am Ceram Soc 54:286
22. Wachtman GB, Jr, Maxwell LH (1954) J Am Ceram Soc 37:291
23. Weertman J (1955) J Appl Phys 26:1213; (1957): 28:362; (1957): 28:1185
24. Xue LA, Chan I-W (1996) J Am Ceram Soc 79:233

Chapter 13
Creep in MgO

Abstract MgO and MgO composite (MgO·Al$_2$O$_3$) are discussed in this chapter. The value of the stress exponent determined as $n = 3.3$ suggests a dislocation model as the rate-controlling creep mechanism. In the absence of glide the dislocation motion is that of climb, which is the rate-controlling mechanism. Knowledge of the structure is of great importance for understanding the creep deformation mechanism in the power law range. It is revealed that the typical dislocation structure of creep-deformed MgO is qualitatively very similar to that of creep-deformed metals and that the grains are divided into well-defined subgrains. Creep in polycrystalline and single-crystal MgO are considered in this chapter and the experiments were performed by tensile, compressive and flexural tests. Creep rupture, superplasticity, and nano-MgO are important sections of this chapter.

One of the early expressions for creep at high temperatures via a vacancy-controlled mechanism (supposedly determined by the stress-directed lattice diffusion of vacancies) was given by Nabarro-Herring as:

$$\dot{\varepsilon} = \frac{B \,\Omega\sigma}{d^2 \, kT} D_L \tag{13.1}$$

where B is a constant ~ 10 for equiaxed polycrystals, Ω is the atomic volume, D_L is the lattice self-diffusion coefficient, Q_L is the lattice diffusion activation energy, and σ and d have their usual meanings of applied stress and grain diameter, respectively. Recall that:

$$D_L = D_{0(L)} \exp\left(-\frac{Q_L}{kT}\right) \tag{13.2}$$

The diffusion path may be along the grain boundaries. For such a case, Coble gave the expression:

$$\dot{\varepsilon} = \frac{150 \,\sigma W\Omega}{\pi \, d^3 kT} D_{GB} \tag{13.3}$$

© Springer International Publishing AG 2017
J. Pelleg, *Creep in Ceramics*, Solid Mechanics and Its Applications 241,
DOI 10.1007/978-3-319-50826-9_13

where D_{GB} is the grain-boundary diffusion coefficient and W is the width of the grain boundary. A similar equation to Eq. (13.2) may be given for D_{GB} with the grain-boundary subscripts GB. Agreement has been found between the calculated values and those for cation diffusion in some oxide ceramics, such as Al_2O_3, Be oxide, MgO, etc. However, other creep studies on ceramic polycrystals have shown that the creep rate obeys a power law, rather than linear stress dependence, suggesting a glide-and-climb mechanism for dislocations. In that case, Weertman provided:

$$\dot{\varepsilon} = \frac{A\sigma^n}{kT} \exp\left(-\frac{Q_c}{kT}\right) \tag{13.4}$$

Exponent n is a constant–4.5, A is a constant that sometimes depends on temperature, and Q_c is the activation energy for creep. Several, contradictory results on creep in MgO appear in the literature, sometimes in regard to porous or impure MgO; therefore, in the experiments described below, only poreless, high-purity MgO is used to determine the creep rate-controlling mechanism at a temperature of $\sim 0.5 T_m$, i.e., 1200 °C. The creep curves tested at 1200 °C are shown in Fig. 13.1 at various, constant loads in the 500–20,000 psi range and the strain–time curves are recorded for each case. Figures 13.1, 13.2 and 13.3 refer to the smallest grain size—11.8 μm. Figure 13.1 shows primary and secondary creep, tested under compression. No accelerated creep stage is observed, since the test had to be terminated early to eliminate fracture. The steady-state creep was plotted versus stress as indicated in Fig. 13.2.

All the points of the small specimen lie on the same line, independent of grain size, with a slope of $n = 3.3$. In order to determine the activation energy according to Eq. (13.5), the temperature was cyclically changed by ~ 25 °C at strain increments of ~ 0.02, so that the structure could be assumed to remain constant:

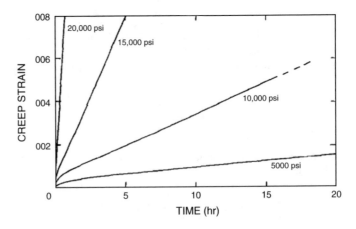

Fig. 13.1 Creep curves for MgO specimens of smallest grain size (11.8 μm), tested in compression at 1200 °C. Langdon and Pask [13]. With kind permission of Elsevier

Fig. 13.2 Steady-state creep
rate versus stress for MgO
specimens of grain size
11.8 μm (□: length l/width
w = 4; ∇: l/w = 1.52), 33 μm
(Δ: l/w = 1.52) and 52 μm
(○: l/w = 1.52). The lines of
slope n = 1 indicate
predictions arising from
theories of diffusional creep
via the lattice and grain
boundaries, respectively.
Langdon and Pask [13]. With
kind permission of Elsevier

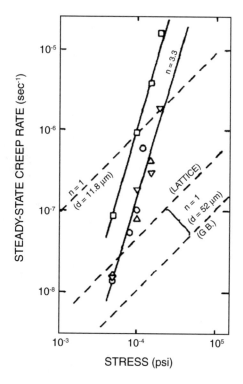

$$Q_c = \frac{\partial \ln \dot{\varepsilon}}{\partial(-1/RT)} \sim \frac{R\ln(\dot{\varepsilon}_2/\dot{\varepsilon}_1)}{(T_2 - T_1)/T_1 T_2} \tag{13.5}$$

$\dot{\varepsilon}_1$ and $\dot{\varepsilon}_2$ are the instantaneous creep strain rates immediately after each temperature
change from T_1 to T_2. The average activation energy obtained was 51 ± 5 kcal
mol^{-1}. The scatter of the experimental points is a consequence of the fact that
thermal equilibrium was not attained immediately after the ~ 25 °C temperature
changes (increase or decrease). It seems from Fig. 13.2 that diffusional creep is not
the mechanism acting in polycrystalline MgO (despite existing evidence to the
contrary), since the following criteria are not satisfied: (a) the slope of the line gives
$n = 3.3$ and not unity, and (b) there is no grain size dependence in the 11.8–52 μm
range. Calculations made (Langdon and Pask) at constant stress and temperature for
two strain rates and grain diameters are given for the lattice and grain-boundary
diffusion as:

$$\frac{\dot{\varepsilon}_1}{\dot{\varepsilon}_2} = \frac{d_2^2}{d_1^2} \quad \text{lattice diffusion} \tag{13.6}$$

$$\frac{\dot{\varepsilon}_1}{\dot{\varepsilon}_2} = \frac{d_2^3}{d_1^3} \quad \text{grain boundaey diffusion} \tag{13.7}$$

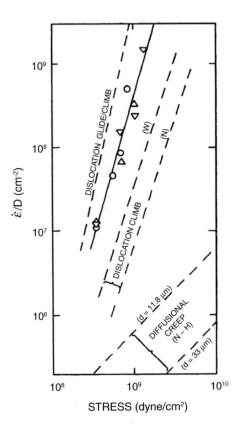

Fig. 13.3 Diffusion-compensated creep rate $\left(\frac{\dot{\varepsilon}}{D}\right)$ versus stress, taking D for extrinsic lattice diffusion of O^{2-} to calculate the experimental points. The *dashed lines* represent predictions arising from Nabarro-Herring diffusional creep (for $d = 11.8\ \mu m$ and 33 μm respectively), a dislocation glide/climb mechanism, and the dislocation climb model formulated by Nabarro (N) and reanalyzed by Weertman (W). Langdon and Pask [13]. With kind permission of Elsevier

Also, the use of experimentally observed values to draw the lines of slope $n = 1$ through the points indicates great discrepancy in the experimental data and the diffusional models. The value of $n = 3.3$ for the stress dependence (see Fig. 13.2) suggests a dislocation model as the rate-controlling creep mechanism. Dislocation climb and glide (Weertman) require $n = 4.5$. Using Weertman's model as:

$$\frac{\dot{\varepsilon}}{D} = \frac{3\pi^2\sigma^2}{2(2)^{0.5}G^2b^2}\sinh\left(\frac{(3)^{0.5}\sigma^{2.5}b^{1.5}}{8G^{1.5}M^{0.5}kT}\right) \tag{13.8}$$

G is the shear modulus ($\sim 0.4\ E$, where E is Young's modulus), b is the Burgers vector, and M is the number of Frank-Read sources cm^{-3} ($M^{0.5} = 0.526\rho^{0.76}$, where ρ is the dislocation density). Taking $\rho = 10^8$ dislocations cm^{-2} and $b = 3 \times 10^{-8}$ cm,

$\frac{\dot{\varepsilon}}{D}$ was calculated. Figure 13.3 shows the results with the experimental points calculated by taking the extrinsic lattice diffusion of O^{2-} and the predicted Nabarro-Herring creep for $d = 11.8$ and 33 μm. The authors indicate that the dislocation glide-and-climb model breaks down at high stresses because of the sinh term in Eq. (13.8). The breakdown stress may be estimated by setting the sinh term to unity. With this procedure, Weertman's model is valid up to $\sim 1.5 \times 10^9$ dynes cm^{-2}. By assuming that the lattice diffusion of O^{2-} is rate controlling, the experimental results show better agreement with dislocation mechanisms than with diffusional creep, as indicated in Fig. 13.3. Thus it may be concluded, based on the experiments of Langdon and Pask, that some form of dislocation motion, such as climb (in the absence of glide) is the rate-controlling mechanism for creep, rather than the stress-directed diffusion of vacancies.

Furthermore, regarding the dislocation concept as a rate-controlling mechanism of creep (where the creep rate dependence on stress follows a power law), the stress exponent varies in the 2.3–4.0 range. The measured activation energies were 46, 111 ± 12 and 51 ± 5 kcal mol^{-1}. The knowledge of the structure is of great importance for understanding the creep deformation mechanism in the power law range. TEM investigations provided this information, as shown in Figs. 13.4 and 13.5

It is revealed that the typical dislocation structure of creep-deformed MgO is qualitatively very similar to that of creep-deformed metals and that the grains are divided into well-defined subgrains. No entanglements or pile-ups are observed in the subgrains in the three-dimensional dislocation network (Fig. 13.4). The dislocation density inside the subgrains was measured. When a random dislocation distribution is assumed, the volume density cm/cm^3 is twice the area density (i.e., dislocations/cm^2). For creep-deformed metals, the relation found is:

$$\rho = \frac{\sigma^2}{b^2 G^2} \tag{13.9}$$

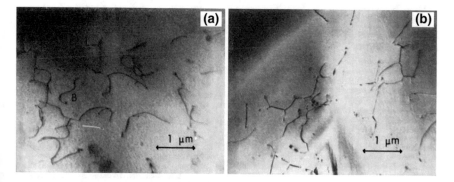

Fig. 13.4 Typical dislocation structure. In **a**, loops, L, and bowed-out dislocation, **b**, are visible. Bilde-Sörensen [5]. With kind permission of John Wiley and Sons

Fig. 13.5 Subgrain boundary
consisting of hexagonal
network. Bilde-Sörensen [5].
With kind permission of John
Wiley and Sons

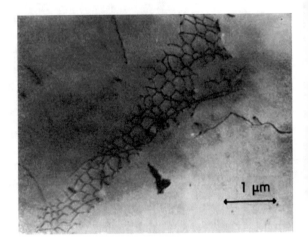

The plot of volume density versus load shows a reasonable agreement with Eq. (13.9), as seen in Fig. 13.6.

These creep experiments were conducted under compression at $<5 \times 10^{-5}$ torr. The load varied from 2.5 to 5.5 kgf/mm^2 and the temperature from 1300 to 1460 °C. The secondary creep rate was fitted to:

$$\dot{\varepsilon} = K\sigma^{\alpha} \tag{13.10}$$

The evaluated values of α are: 2.6 ± 0.6 at 1300 °C, 3.5 ± 0.6 at 1400 °C and 3.7 ± 0.7 at 1460 °C. The relation used to describe creep in MgO is given as:

Fig. 13.6 Dislocation
density versus load. Broken
lines show results from
$\rho = \sigma^2/b^2G^2$ for 1300 °C
(*lower line*) and 1460 °C
(*upper line*). Bilde-Sörensen
[5]. With kind permission of
John Wiley and Sons

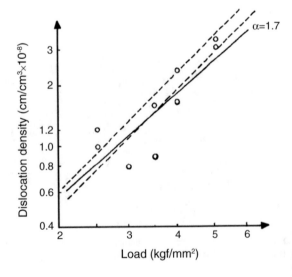

Fig. 13.7 Plot of ln $\left(\frac{\dot{\varepsilon}kT}{bDG}\right)$ versus ln $\left(\frac{\sigma}{G}\right)$. *Broken lines* show results calculated from *a* proposed model and *b* Nabarro model. Bilde-Sörensen [5]. With kind permission of John Wiley and Sons

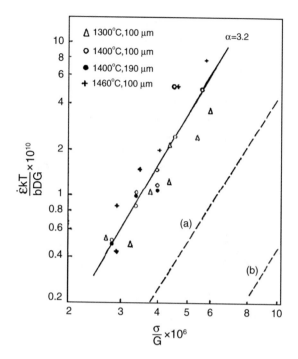

$$\left(\frac{\dot{\varepsilon}kT}{bDG}\right) = A\left(\frac{\sigma}{G}\right)^{\alpha} \tag{13.11}$$

And is plotted in Fig. 13.7. The self-diffusion coefficient of the O^{2-} ion was used for D. A stress exponent of 3.3 ± 0.3 was estimated from Fig. 13.7. This exponent did not vary with temperature. As such, an actual value of 3.2 for the stress exponent was used to calculate the activation energy for creep by this mechanism, giving a value of 76 ± 12 kcal/mol. The creep rates of the specimens with grain sizes of 100 and 190 μm did not vary significantly. This was surprising, since large grain-sized specimens are required for creep resistance. Consequently, no appreciable GBS is expected and processes of diffusion along grain boundaries do not contribute significantly to the creep rate, based on this model.

One can, thus, conclude that within the range of the investigated parameters, several processes may be operating simultaneously; however, the main cause for creep deformation is a combined climb-and-slip process.

13.1 Creep in Composite MgO

Relatively few data are available in the literature on this important subject. One such composite is the $MgAl_2O_4$ spinel, often written as $MgO \cdot Al_2O_3$, obtained from the reaction between MgO and Al_2O_3 at ~1200–2000 °C and pressures of

1.0–4.0 GPa. In a recent publication [18], MgO·Al$_2$O$_3$ was prepared by plasma-spark sintering at 1100–1200 °C with an applied pressure of 120–200 MPa and a dwell time of 2 h. Creep curves were obtained, as illustrated in Fig. 13.8. The plot shows the creep strain versus time. Each creep curve consists of three regions. At each temperature, the first segments are shallow, showing almost no significant creep strain. Softening, with increased applied stress, may be seen even at the lower temperature. The increase in slope (dashed line) is an indication of creep strain change (increase). From these slopes, the strain rates may be evaluated and their variations with the pressure may be seen in Fig. 13.9. As indicated earlier, the main interest is in the values of the stress exponent, which (as previously indicated) is usually obtained from:

Fig. 13.8 Creep curves for spinel under pressure of 120–200 MPa in the 1100–1200 °C range. The *dashed lines* indicate change of the slope. Ratzker et al. [18]. With kind permission of Mr. Ratzker for the authors

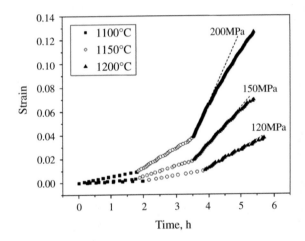

Fig. 13.9 Creep rates of spinel as a function of pressure, tested at various temperatures. Ratzker et al. [18]. With kind permission of Mr. Ratzker for the authors

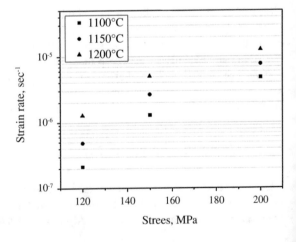

$$\dot{\varepsilon} = A\sigma^n \exp\left(-\frac{Q}{RT}\right) \tag{13.12}$$

Equation (13.12), in logarithmic form, yields:

$$\ln(\dot{\varepsilon}) = \ln A + n \ln(\sigma) - \frac{Q}{RT} \tag{13.13}$$

and a plot of this equation may be seen in Fig. 13.10. The values of the stress exponent at each temperature are listed in Table 13.1. These values are higher than $n \cong 1$, which may indicate that several processes operate concurrently in creep deformation; however, the main form of creep deformation here is a combined climb-and-slip process. The authors consider GBS to be the creep mechanism, however, they admit that dislocation slip-and-climb may be additional processes accommodating GBS.

Equation (13.12) or its logarithnic form also enables the determination of the activation energy for creep. A plot of the activation energy versus stress appears in Fig. 13.11 and Table 13.2 lists the activation energies at various stresses. Note that the activation energy increases with the decrease in the applied stress, as expected.

The microstructures of polished and thermally etched specimens before and after creep, obtained by high-resolution SEM (HRSEM), are shown in Fig. 13.12.

Fig. 13.10 ln(strain rate) versus ln(stress) for spinel tested under 120, 150 and 200 MPa. Ratzker et al. [18]. With kind permission of Mr. Ratzker for the authors

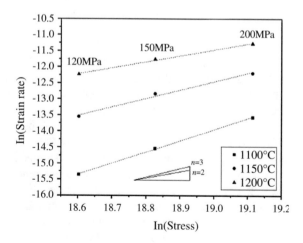

Table 13.1 Values of the stress exponent at various temperatures. Ratzker et al. [18]. With kind permission of Mr. Ratzker for the authors	Temperature (°C)	Stress exponent (n)
	1100	3.48 ± 0.1
	1150	2.64 ± 0.26
	1200	1.87 ± 0.15

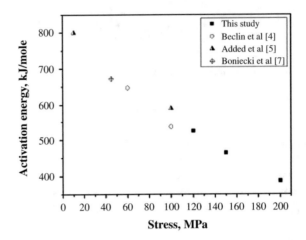

Fig. 13.11 Apparent activation energy as a function of stress. Ratzker et al. [18]. With kind permission of Mr. Ratzker for the authors

Table 13.2 Apparent activation energies for polycrystalline magnesium aluminate spinel. Ratzker et al. [18]. With kind permission of Mr. Ratzker for the authors

Applied stress (MPa)	Activation energy (Q) (kJ/mol)
120	526 ± 35
150	465 ± 50
200	387 ± 36

The equiaxed grain size increases with creep deformation from 250 to 400 nm, without any change in the equiaxed shape of the grains, possibly indicating GBS as a main creep mechanism. In order to obtain more data on creep in spinel, HRTEM structures were acquired (see Fig. 13.13).

The HRTEM images confirmed (attested by the formation of triple points) that GBS is one mechanism of creep in spinel. Dislocations may also be involved at higher stress levels (Fig. 13.13c, d), accomodating GBS.

Although title of this section is "Creep in Composite MgO," this is as good a place as any to consider an example of an additive that produces solid solutions. Recall that composite materials are usually mixtures of two or more components, each of which gets different properties than it had originally. In contrast, although the starting materials in a solid solution each have different properties, the final product is not a mixture, for example, as in the case of the MgO–Fe_2O_3 solid–solution system.

By vacuum hot-pressing solid solutions of polycrystalline MgO and MgO-Fe_2O_3 with 0.10–8.08 wt% Fe_2O_3, specimens were fabricated very close to the theoretical density [20]. Creep testing was performed in air and in the temperature 1000–1400 °C range at stresses of 50–550 kg cm^{-2}, but steady-state

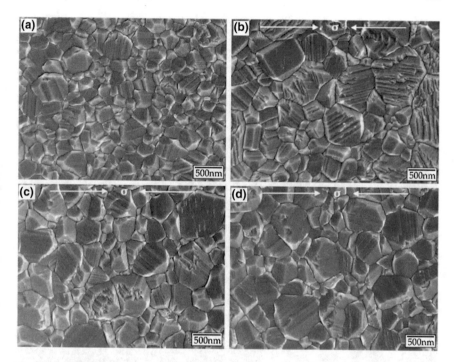

Fig. 13.12 HRSEM images of the spinel samples before creep (**a**); after creep at 1100–1200 °C (4% strain) under 120 (**b**); (7% strain) 150 (**c**) and (13% strain) 200 MPa (**d**). Compression direction is marked. Ratzker et al. [18]. With kind permission of Mr. Ratzker for the authors

creep was not reached even after 50 h creeping. Hot-pressed specimens of MgO, with and without doping by Fe_2O_3, are shown in Fig. 13.14 and after creep testing in Fig. 13.15.

The symbols of the specimens in Figs. 13.14 and 13.15 are indicated in Table 13.1. The initial grain sizes in the undoped specimens (18–19 µm) are considerably larger than in the doped specimens (1–5 µm). Subnormal grain growth and irregular distribution were encountered in the hot-pressed, undoped oxide (Fig. 13.14a). A nonuniform distribution of iron was observed in the heavily doped samples, namely, 0.94, 2.95 and 8.08% Fe_2O_3. At lower doping concentrations (0.10 and 0.48%), no segregation was observed. Grain growth occurred during annealing and creep, but the grain distribution became more uniform, as seen in the MgO (undoped) specimen in Fig. 13.15a after annealing and in (b) after the creep test. Porosity was not found in the hot-pressed specimens but, after annealing, pores coalesced to a size visible to optical microscopy. These pores are located at grain boundaries or triple points as seen in Fig. 13.15b.

Details of the stress change experiments, including densities, temperatures, grain sizes, and stress exponents are listed in Table 13.3. The maximum stress in the outer fiber, σ, of a beam is given for creep by the calculation:

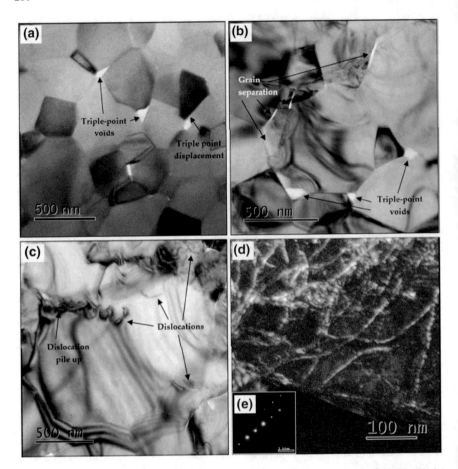

Fig. 13.13 HRTEM images of spinel samples after deformation at 1100–1200 °C. Under 120 MPa (4% strain) triple-point voids and displaced triple points are shown (**a**); under 200 MPa (13% strain) grain separation and sliding along the grain boundaries (**b**) and dislocations (**c**) are shown. A weak-beam dark field (WBDF) image for $g = 440$ shows the high dislocation density within the grain after creep in response to 200 MPa pressure (**d**); the selected area diffraction pattern is presented in the *insert* (**e**). The examined cross-sections were perpendicular to the compression axis. Ratzker et al. [18]. With kind permission of Mr. Ratzker for the authors

$$\sigma = \frac{3}{2}\frac{L-a}{bh^2}F\left(\frac{2N+1}{3N}\right) \tag{13.14}$$

Note that here $N (\equiv n)$ stands for the stress exponent. L is the distance between the supporting points; the distance between the load points, $F (\equiv P)$, gives applied load; and b and h are the width and height of the specimen, respectively. Recall that the applied stress is flexural. For $N = 1$, Eq. (13.14) reduces to:

(a)   **(b)**

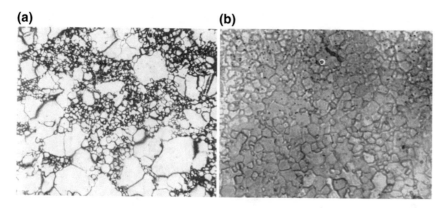

Fig. 13.14 Typical microstructures of hot-pressed specimens. **a** Undoped MgO (U-G) with $GS_{avg} \approx 19.4$ μm (× 250) and **b** 0.48% Fe_2O_3 (D-D) with $GS_{avg} \approx 4.7$ μm (× 690). Terwilliger et al. [20]. With kind permission of John Wiley and Sons

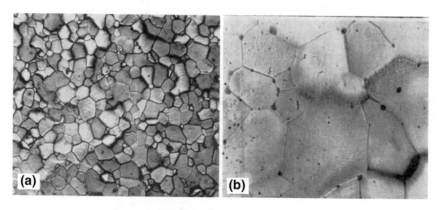

Fig. 13.15 Typical microstructures after annealing and creep **a** Undoped MgO (U-H) annealed 9 h at 1300 °C (×200) and **b** undoped MgO (U-G6) creep-tested ≈75 h at 1400 °C (×450). Terwilliger et al. [20]. With kind permission of John Wiley and Sons

$$\sigma = \frac{3}{2} \frac{L-a}{bh^2} F \tag{13.15}$$

When a beam under four-point loading is elastically deformed, the strain in the outer fiber is given by measuring the deflection at the load points (x) as:

$$\varepsilon = \frac{6h}{(L-a)(L+2a)} x \tag{13.16}$$

Differentiating Eq. (13.16) with respect to time, t, one gets the creep rate by:

Table 13.3 Stress exponents from stress change experiments. Terwilliger et al. [20]. With kind permission of John Wiley and Sons

Specimen	Density[a] (g/cm[1])	Stress range[b] (kg/cm[2])	Temp. (°C)	Stress exponent, N	Approx grain size at Stress change (µm)
Undoped					
U–G–1	3.571	342–492	1300	1.91	
U–G–2		379–440		1.22	23–29
		440–530		2.86	
U–H–1	3.580	365–389		1.74	40
U–H–2		387–439		2.55 Avg	40
U–H–3		396–450		2.81 2.38	20
U–H–4		433–491		3.26	40
		491–433		2.36	
U–H–5		380–431	1400	3.81	45
		331–476		3.76 Avg	
		476–431		3.32 3.63	55
0.10% Fe_2O_3					
D–A–1	3.558	394–552	1300	1.31 Avg	20
D–B–1	3.577	315–490		0.79 1.05	14
D–A–2		169–284	1400	1.19	36
		284–420		0.59	40
		420–535		1.17 Avg	50
		535–420		0.82 1.05	60
D–B–2		149–306		1.11	36
		306–413		1.43	50
0.48% Fe_2O_3					12
D–C–1	3.557	445–492	1300	2.48	12
D–C–2		149–241	1400	0.89	12
2.95% Fe_2O_3					
D–F–1	3.597	228–326	1200	0.86	≈10
D–F–2		148–258	1300	1.24	≈10

[a]Theoretical density of pure MgO assumed to be 3.584 g/cm^3

[b]First number is stress before weight change; second is stress after weight change

$$\dot{\varepsilon} = \frac{6h}{(L-a)(L+2a)}\dot{x} = Kh\dot{x} \tag{13.17}$$

Clearly, $\dot{x}$ is the deflection rate. Figure 13.16 shows the deflection rate versus the creep time of MgO, compared to that of the solid solution with 0.48% Fe_2O_3. Initially, the creep rate decreases rapidly and, later on, it slowly decays. An accelerating creep region was observed in specimens containing 2.95 and 8.08% Fe_2O_3. In such specimens, where there was straining to fracture, oriented voids were formed at the grain boundaries perpendicular to the tensile direction of the stress, thereby decreasing the cross-sectional area remaining for load bearing. There have been suggestions that voids form at grain boundaries during GBS.

The stress exponent, N ($\equiv n$), was determined using the dependence of the strain rate on the stress (load). By changing the load and measuring the change in creep rate, N was evaluated as follows:

$$N = \frac{\log \dot{x}_2 - \log \dot{x}_1}{\log \sigma_2 - \log \sigma_1} \tag{13.18}$$

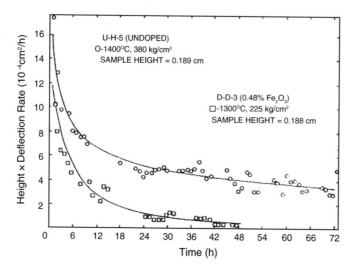

Fig. 13.16 Comparison of best-fit curves and deflection rate data for typical creep experiments. Terwilliger et al. [20]. With kind permission of John Wiley and Sons

The value of $\dot{x}_2$ was determined by extrapolation. The extrapolated value appears in Fig. 13.17. This procedure eliminated the effect of the transient and permitted the values of $\dot{x}_1$ and $\dot{x}_2$ to correspond to the same time. The resultant stress exponents are listed in Table 13.3.

Stress exponent, N, increased significantly as grain size increased in the undoped specimens, indicating that dislocation creep probably predominates in relatively large-grained (23–55 μm) MgO, although direct metallographic evidence for dislocations was not observed in the work reported here. The stress exponents of most

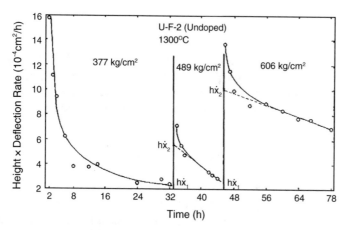

Fig. 13.17 Stress change experiments on undoped creep specimen at 1300 °C. Terwilliger et al. [20]. With kind permission of John Wiley and Sons

doped specimens is unity (see Table 13.3). This indicates the presence of a viscous- or diffusion-creep mechanism. The strain rate is reciprocally related to a power of the grain size. An equation for a viscous mechanism is:

$$\dot{\varepsilon} = \frac{k_1'\sigma}{GS^m} = \frac{k_1}{GS^m} \tag{13.19}$$

The grain size (GS) exponent is $m = 1$–3, depending on the acting creep mechanism. Grain growth is a given as:

$$GS^n - GS_0^n = k_n t \tag{13.20}$$

From Eqs. (13.19) and (13.20), one may derive an expression of the form:

$$\dot{\varepsilon} = \frac{C_1}{(t + C_2)^p} \tag{13.21}$$

with:

$$p = \frac{m}{n} \tag{13.21a}$$

$$C_1 = \frac{k_1}{(k_n)^p} \tag{13.21b}$$

$$C_2 = \frac{GS_0^n}{k_n} \tag{13.21c}$$

When C_2 in Eq. (13.21) is small, it reduces to:

$$\dot{\varepsilon} = \frac{C_1}{t^p} \tag{13.21d}$$

The values of the time exponents vary within the $p = 0.33$–1.5 range, as a function of the creep and grain growth mechanisms. Using the deflection data and Eq. (13.21), the values of C_1, C_2, and p may be calculated. An initial value of C_2 is assumed and then a linear regression is performed according to Eq. (13.22), given as:

$$\log(h\dot{x}_i) = -p\log(t_i + C_2) + \log C_1' \tag{13.22}$$

One immediately realizes that this equation is obtained by expressing the strain rate found in Eq. (13.21) as $\dot{\varepsilon} = Kh\dot{x}$ (from Eq. 13.17), then writing $\frac{C_1}{K} = C_1'$ and expressing the resulting equation in a logarithmic form. In the iteration, C_2 is varied until the standard deviation, $\hat{\sigma}$, becomes a minimum. This iteration is performed according to:

$$\hat{\sigma} = \sqrt{\frac{\sum_{i=1}^{n} \left[\log(h\dot{x}_i) - \log C_1' + p\log(t_i + C_2)\right]^2}{n}} \tag{13.23}$$

The deflection, x, may be obtained by integrating Eq. (13.22):

$$x = \left[\frac{C_1}{h}\right] \ln\left(\frac{t}{C_2} + 1\right) \quad \text{for } p = 1 \tag{13.24}$$

and:

$$x = \frac{C_1'/h}{(1-p)}\left[(t + C_2)^{1-p} + C_2^{1-p}\right] \quad \text{for } p \neq 1 \tag{13.24a}$$

Note that in Eq. (13.24) the logarithmic term is ln. The parameters determined by Eqs. (13.24) and (13.24a) are summarized in Table 13.4.

Observe that, in the undoped specimens, p is smaller than in those containing Fe_2O_3, except in 8.08% Fe_2O_3. The small value of p in the unalloyed specimens of MgO may indicate creep by dislocation motion, whereas, in the 8.08% Fe_2O_3

Table 13.4 Comparison of parameters calculated from deflection and deflection rate (best-fit) data. Terwilliger et al. [20]. With kind permission of John Wiley and Sons

Specimen	Stress (kg/cm²)	Temp. (°C)	p		C_1' (cm²h^{p-1})	
			Deflection rate (best-log fit) calculation	From log-deflection-time plots	Deflection rate (best-fit) calculation	From semi-log-deflection time plots
Undoped						
U-F-5	216	1200	0.61	0.55	1.18×10^{-3}	
U-G-3	277	1300	0.53	0.49	3.25×10^{-4}	
U-H-5	380	1400	0.35	0.38	1.58×10^{-3}	
0.10% Fe_2O_3						
D-A-3	178	1200	1.10		1.17×10^{-3}	9.20×10^{-4}
D-A-5	225	1300	0.97		1.98×10^{-3}	2.09×10^{-3}
D-A-6	294	1300	0.86	0.80	9.50×10^{-4}	
D-A-2	169	1400	0.61	0.60	4.94×10^{-4}	
0.48% Fe_2O_3						
D-C-3	291	1300	0.89		1.68×10^{-3}	2.54×10^{-3}
D-C-1	445	1300	0.98		8.50×10^{-3}	9.10×10^{-3}
0.94% Fe_2O_3						
D-E-1	137	1200	0.69	0.55	6.96×10^{-4}	
D-E-2	239	1300	0.93		3.17×10^{-3}	7.8×10^{-3}
D-E-3	316	1400	0.75	0.71	6.09×10^{-3}	
8.08% Fe_2O_3						
D-G-2	134	1100	0.64	0.58	1.19×10^{-3}	
D-G-3	162	1200	0.40	0.36	2.41×10^{-3}	

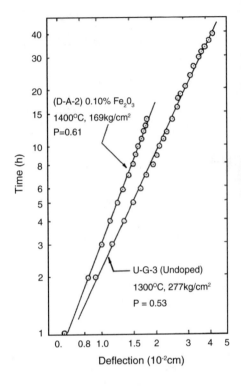

Fig. 13.18 Representative log–log plots of deflection versus time. Terwilliger et al. [20]. With kind permission of John Wiley and Sons

specimens, the formation of voids may explain the low p values. Figures 13.18 and 13.19 show the time-deflection relation.

Grain growth occurring during creep in MgO and in specimens containing Fe_2O_3 at 1300 and 1400 °C are shown in Fig. 13.20. The graph in Fig. 13.20 uses Eq. 13.20 with $n = 2$ at 1300 °C and 3 at 1400 °C. Thus, squared and cubic growth relations are indicated in the plots of Fig. 13.20. Growth in undoped MgO follows a squared growth relation.

In essence, the interest in the use of additives in MgO solid solution is a consequence of the following:

(a) The presence of Fe ions in solid solutions in this system inhibits non-viscous contributions (i.e., by dislocations) to creep and promotes viscous creep;
(b) The addition of Fe_2O_3 enhances viscous creep, either by grain boundaries or by lattice diffusion;
(c) The rate of grain growth is greatly depressed during MgO creep.

Thus, by adding the appropriate components, such as Fe_2O_3 (the representative example discussed in this section), to solid solutions of MgO, grain size may be controlled during creep, enabling the consequent design of better creep-resistant ceramics. Since MgO is considered to be one of the significant ceramics for high-temperature applications, it is very important to collect all the relevant information to that end.

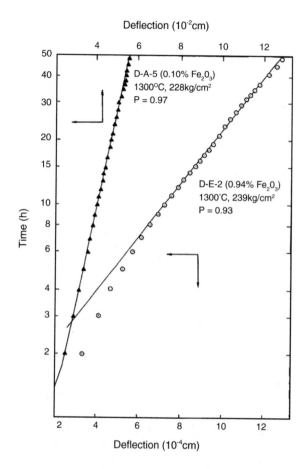

Fig. 13.19 Representative semilogarithmic plots of deflection versus time. Terwilliger et al. [20]. With kind permission of John Wiley and Sons

13.2   Creep in Single-Crystal MgO

Generally, creep tests may be performed by the application of stress via compression, tension, or deflection. This section will deal with the various types of tests that are available to do so.

13.2.1   Compression Creep in Single-Crystal Magnesia

Experiments were performed on single-crystal MgO by compressive deformation parallel to $\langle 100 \rangle$ up to 69% strain at temperatures between 1573 and 1773 K. The creep apparatus used for these experiments is shown in Fig. 13.21.

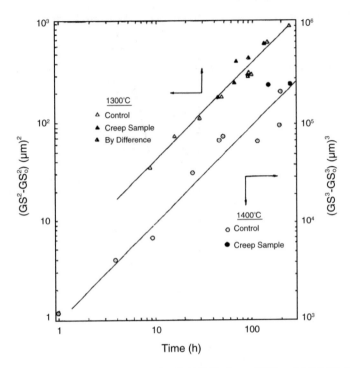

Fig. 13.20 Grain growth in magnesiowustite (0.10% Fe$_2$O$_3$ at 1300 and 1400 °C. Terwilliger et al. [20]. With kind permission of John Wiley and Sons

Table 13.5 lists pertinent data for the ten creep-tested specimens under uniaxial compression at constant force, F of 220–440 N. The deformation was calculated from the displacement data and corrected for apparatus compliance. The stress was calculated by dividing the applied force by the cross-sectional area. Although F was kept constant, the stress, σ, dropped as a function of strain. The results of these creep experiments appear in

Figure 13.22 as the log shear-strain rate $(\dot{\gamma})$ versus the log-normalized shear-stress $\left(\frac{\tau}{\mu}\right)$. The curves were fitted to $\dot{\gamma}$, σ and T using a semi-theoretical model for climb-controlled creep, given as:

$$\dot{\gamma} = \frac{\mu b}{kT} \left(\frac{\tau}{\mu}\right)^n A D_0 \exp\left(-\frac{Q}{RT}\right) \tag{13.25}$$

Strain and stress were converted to shear-strain and shear-stress in the above equation, following von Mises' equivalent strain rate and stress:

$$\dot{\gamma} = \frac{\dot{\varepsilon}}{3}$$

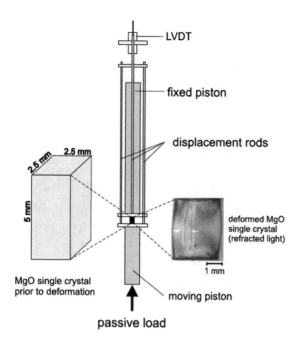

Fig. 13.21 Schematic of the creep apparatus loading column. The sample (2.5 × 2.5 mm diameter) is located between the SiC pistons (25 mm diameter). The displacement is measured by means of the strain cage consisting of two pairs of carbon rods that move independently and transfer the displacement to the LVDT rod and body respectively. A passive load is applied from the bottom and transferred to the load cell and then to the lower piston. A large furnace—not represented—surrounds the sample and SiC pistons. White arrows indicate deformation bands formed after 24% strain was imposed. Mariani et al. [14]. With kind permission of Elsevier. LVDT stands for linear variable differential transformer

and:

$$\tau = \frac{\sigma}{3} \tag{13.25a}$$

The shear modulus, μ, was the calculated using Eq. (13.26):

$$\mu = \mu_0 \left[1 + \frac{T - 300}{T_m} \left(\frac{T_m}{\mu_0} \frac{d\mu}{dT} \right) \right] \tag{13.26}$$

Here, μ_0 is the shear modulus at 300 K (taken as 125 GPa), $T_m = 3125$ and $\left(\frac{T_m}{\mu_0} \frac{d\mu}{dT} \right) = 0.68$, the temperature dependence of shear modulus. The Burgers vector for MgO $\frac{1}{2}\langle 110 \rangle$ is 2.98 Å at 300 K and atmospheric pressure.

It has been established that, above 0.5 T_m, the quasi-steady-state deformation in MgO single crystals occurs by dislocation glide and dislocation climb. Almost all

Table 13.5 Summary of experimental conditions and results. Mariani et al. [14]. With kind permission of Elsevier

Sample	Load (kg)	F (N)	T (K)	σ (MPa)	ε (%)	$\dot{\varepsilon}(s^{-1})$	t (h)	Comments
ps1	22	220	1673	26.8	24.4	2.2×10^{-7}	63	
ps2	22	220	1673	30.4	16.8	1.1×10^{-6}	14.4	
ps3	22	220	1673	27.3	34.8	1.3×10^{-7}	119.3	
ps4	33	330	1673	34.3	29.3	3.1×10^{-7}	46	Sticky LVDT
ps5	44	440	1673	50.1	23.4	5.3×10^{-6}	2.2	Slight buckling
ps6	44	440	1673	46.4	35.8	1.3×10^{-6}	16	Buckling
ps7	22	220	1773	24.9	23.2	5.4×10^{-7}	20.3	LVDT broke at end of test
ps8	44	440	1573	49.3	25.3	3.1×10^{-7}	67.4	
ps9	44	440	1773	31.1	68.7	2.4×10^{-6}	44.5	Thermocouple failed
ps10	44	440	1773	49.5	21	9.6×10^{-6}	1.4	

F is the force, T is the temperature, σ is the stress, ε is the natural strain, $\dot{\varepsilon}$ is the natural strain rate and t is time

Fig. 13.22 Diagram of $\log_{10}$ shear-strain rate versus $\log_{10}$ shear-stress/shear modulus curves obtained for temperatures of 1573, 1673 and 1773 K. Symbols represent experimental data, solid lines are best-fit linear regression curves and dashed lines are data from Yoo et al. 2002, plotted using the value of the stress exponent, $n = 4.5$, obtained in this study. The maximum error on the differential stress is ±2.5 MPa. Mariani et al. [14]. With kind permission of Elsevier

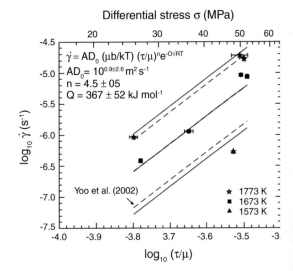

the strain is a consequence of dislocation glide, while the rate of deformation is controlled by climb. The stress exponent of 4.5 in these experiments, which is characteristic of power law creep, confirms that dislocation climb is the dominant creep deformation. Clearly, dislocation climb requires vacancy diffusion.

For detailed crystallographic information on slip systems, orientations and the dynamic evolution of microstructures, interested readers may consult the work of Mariani et al.

13.2.2 Tensile Creep in Single-Crystal Magnesia

The minimum creep rate was found to be:

$$\dot{\varepsilon} = A\sigma^n \exp\left(-\frac{4.1\,\text{eV}}{kT}\right) \tag{13.27}$$

where the activation energy for creep is 4.1 eV [8], which is basically Eq. (6.17) above. This work relates to single-crystal MgO with a $\langle 011 \rangle$ axial orientation, deformed by tensile creep at 1200–1500 °C and over a stress of 29.0–86.2 MN/m². The value of $n = 3.8$–4.5 and $A = 11 \times 10^{-2}$ (MN/m²)$^{-4}$ s^{-1}. Dislocation substructures developing during creep have been investigated by TEM and etch-pitting techniques. The tensile specimens had the dimensions and configuration shown in Fig. 13.23.

All three creep stages (primary, steady-state, and tertiary) were observed in these experiments. Figure 13.24 shows the tensile creep results, where the creep rates are plotted versus inverse temperature (Fig. 13.24a) and stress (Fig. 13.24b).

Specimens were tested at 1400 °C to 0.1 strain for the influence of the applied stress. The stress levels were 37.2, 44.8, 55.2, and 78.9 MN/m². The structures of the specimens tested at the three lowest stresses are shown in Fig. 13.25. At a stress of 37.2 MN/m², the dislocation density determined from the etch-pit was $\rho = 8.4 \times 10^{11}$ m^{-2}. At higher stresses (44.8 and 55.2 MN/m²), the etch-pit density increased and bands containing a very high density of pits were observed. These bands probably correspond to slip bands, especially at low creep strains. Figure 13.25 illustrates the etch pits and the slip bands. The wavy nature of the bands suggests that considerable cross-slip took place during creep. Only slip bands corresponding to one set of orthogonal $\{110\}\langle 1\bar{1}0 \rangle$ slip systems were observed (Fig. 13.26).

Slip traces for $\{010\}$ planes are at 45° to the tensile axis and cannot be differentiated from the $\{110\}$ slip traces. Slip traces for $\{111\}$ slip planes would be horizontal and vertical and were not observed. An increase in applied stress increases etch-pit density, as expected, since the dislocation density is a function of stress. Plots of the dislocation density versus stress and versus strain at 1400 °C are

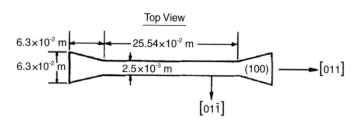

Fig. 13.23 Tensile creep specimen configuration. Clauer and Wilcox [8]. With kind permission of John Wiley and Sons

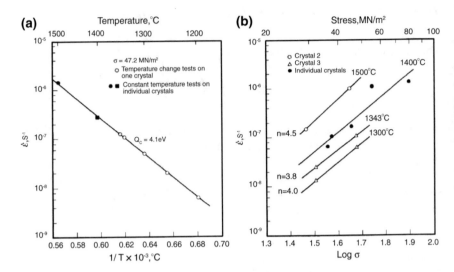

Fig. 13.24 Dependence of creep rate of $\langle 011 \rangle$-oriented MgO single crystals on **a** temperature at 47.2 MN/m^2 and **b** stress at temperatures indicated. Clauer and Wilcox [8]. With kind permission of John Wiley and Sons

seen in Fig. 13.27a, b, respectively. The density versus applied stress yields the relation $\rho \propto \sigma^{2.1}$ and, in subgrains, $\rho \propto \sigma^{1.4}$ for [100]-oriented MgO tested under compression. As seen in Fig. 13.27a, the dislocation density determined by etch-pit technique (optical microscope) is in good agreement with the density determined by the replica technique of electron microscopy. Usually, the etch-pit technique gives a lower dislocation density than TEM, since an etch-pit may originate from more than one dislocation.

The dislocation density increases with strain, as seen in Fig. 13.27b. At about 0.1 strain, the dislocation density appears to approach a constant value. TEM of thin foils parallel to one of the four {110} slip planes, having equal nonzero-resolved shear-stresses, are shown in Figs. 13.28 and 13.29. No sub-boundaries were observed and only an extensive, relatively uniform distribution of dislocations was present. A large number of dislocation loops are visible, which are the largest under great stress. With an assumed foil thickness of 5×10^{-7} m, the loop densities are 2.4×10^{17} m^{-3} and 2.6×10^{18} m^{-3} at 46.9 and 86.2 MN/m^2, respectively. The Burgers vector analysis sequence in Fig. 13.29 shows that most of the loops disappear (Fig. 11.39c), where $\mathbf{g} = 2\bar{2}\bar{2}$. If only $\langle 110 \rangle$ Burgers vectors are considered and it is assumed that slip systems with no resolved shear-stress do not operate, then these loops have $\mathbf{b} = a/2[101]$ and the slip plane was either $(\bar{1}01)$ or (010). It is suggested that the creep substructure is either mainly screw dislocations on $[\bar{1}01](101)$, $[\bar{1}01](010)$ or $[\bar{1}01](111)$ slip systems or edge dislocations on a $[\bar{1}01](010)$ slip system. Furthermore, if it is assumed that all oriented dislocations belong to the same slip system and that most of the oriented dislocations go out of

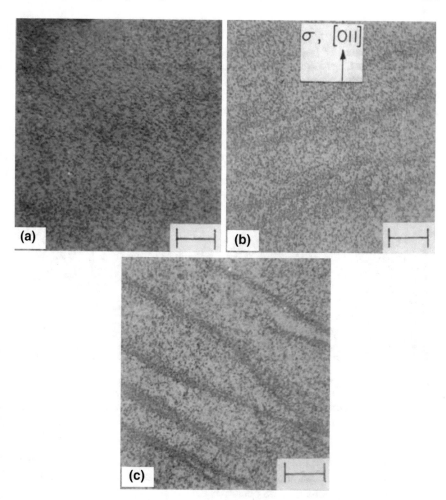

Fig. 13.25 Stress dependence of creep substructure after creep to 0.10 creep strain at 1400 °C; (100) surfaces are shown. Bar 30 μm. **a** $\sigma = 37.2$ MN/m^2, $\rho = 8.4 \times 10^{11}$ m^{-2}; **b** $\sigma = 44.8$ MN/m^2, $\rho = 9.2 \times 10^{11}$ m^{-2}; **c** $\sigma = 55.2$ MN/m^2, $\rho = 1.3 \times 10^{12}$ m^{-2}. Clauer and Wilcox [8]. With kind permission of John Wiley and Sons

contrast in Fig. 13.29c or d, then the most likely slip systems are either the [101] (010) or the [$\bar{1}$01](101) system. The trace of the [$\bar{1}$01](101) on the foil (110) plane is parallel to the [$\bar{1}$11] direction and, thus, any dislocation that tends to be in this direction would be nearly in the plane of the foil and, as such, long lengths would be observed. The long, relatively straight dislocation segments observed by TEM suggest the possibility of a glide-controlled creep mechanism. During creep, most dislocations are mobile and glide is controlled by a drag mechanism composed of charged atmospheres—perhaps acting as the rate-determining element. This concept originates from the possibility that dislocations in crystals have an ionic nature;

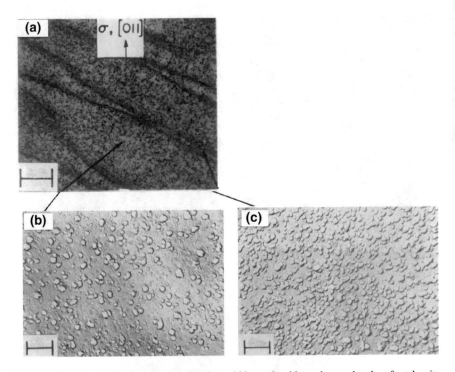

Fig. 13.26 Etch-pit distribution and density within and without heavy bands of etch pits. **a** $\rho(\text{avg}) = 1.3 \times 10^{12}$ m^{-2}, bar 30 μm; **b** $\rho = 9.2 \times 10^{11}$ m^{-2}, bar 5 μm; **c** $\rho = 1.5 \times 10^{12}$ m^{-2}, bar 5 μm. *Arrows* indicate types of regions represented by replica electron micrograph. Clauer and Wilcox [8]. With kind permission of John Wiley and Sons

as such, the line charge is compensated by the formation of a cloud of oppositely charged defects. It appears that a drag process controlled by the intrinsic diffusion of charged impurity ions may be operative.

13.2.3 Flexural (Bending) Creep in Single-Crystal Magnesia

Experiments on bending creep in single-crystal MgO were performed in a vacuum at the 800–1090 °C temperature range. In the vicinity of 1000 °C, creep is rate-controlled by the cross-slip of screw dislocations, having an activation energy of 1.5 ± 0.25 eV. Above 1300 °C, creep takes place at a constant rate, with an activation energy of 5.8 ± 0.73 eV and a stress exponent of 3. This creep is attributed to oxygen-ion-controlled edge-dislocation climb. Figure 13.30 shows a typical creep curve obtained at 1000 °C. This illustration is characteristic of transient creep. The derivative of two curves provides the data for the log–log plot shown in Fig. 13.31. The equation derived from the plot is an equation seen often in earlier chapters

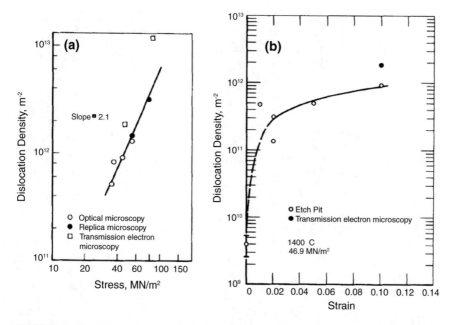

Fig. 13.27 a Stress dependence and **b** strain dependence of dislocation density determined at 1400 °C. Clauer and Wilcox [8]. With kind permission of John Wiley and Sons

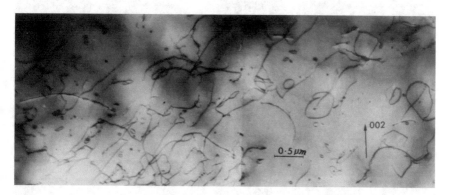

Fig. 13.28 Dislocation substructure in specimen after creep to 0.10 strain at 1400 °C and 86.2 MN/m^2. Foil plane is (110). Clauer and Wilcox [8]. With kind permission of John Wiley and Sons

$$\dot{\varepsilon} = At^{-1/2} \tag{13.28}$$

The above stress exponent is a result of 11 runs, and an expression for the creep rate near 1000 °C was given as:

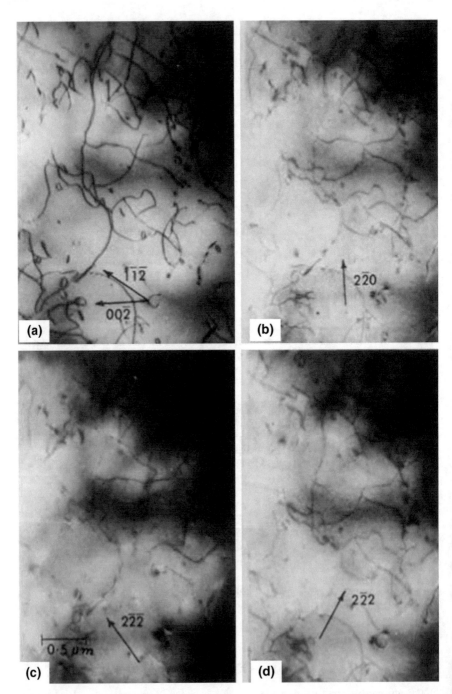

Fig. 13.29 Analysis of Burgers vectors in specimen after creep to 0.10 strain at 1400 °C and 86.2 MN/m². Foil plane is (110). Clauer and Wilcox [8]. With kind permission of John Wiley and Sons

Fig. 13.30 Representative creep time curve at 1000 °C for single-crystal MgO. Rothwell and Neiman [19]. With kind permission of AIP Publishing

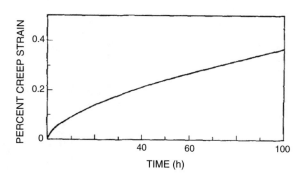

Fig. 13.31 Creep rate versus time for single-crystal MgO at 1000 °C. Data from two experiments. The *line* is drawn with a slope of $-\frac{1}{2}$. Rothwell and Neiman [19]. With kind permission of AIP Publishing

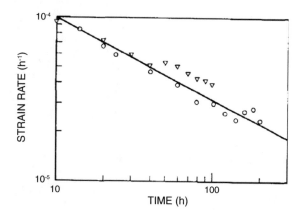

$$\dot{\varepsilon} = 20t^{-0.5}\left(\frac{\sigma}{\sigma_0}\right)^{5.2}\exp\left(-\frac{1.5}{kT}\right) \qquad (13.29)$$

The activation energy is the average of five determinations. A transition in creep behavior was observed between 1100 and 1300 °C (activation energy plots and time-dependent behavior). This seems to be associated with a diffusion-controlled mechanism occurring above 0.5 T_m (i.e., 1260 °C). This mechanism is supported by the appearance of polygonization after creep at temperatures above 1300 °C, as indicated by the etch-pit patterns shown in Fig. 13.32. This is further supported by the observation that the creep rate above 1300 °C was independent of time. Crystals crept at 1000 °C showed a discrete band pattern, as indicated in Fig. 13.33.

For the determination of the activation energy, plots of the strain rate versus the inverse temperature were constructed, like those found in Fig. 13.34. An activation energy of 5.85 ± 0.73 eV was the average of eleven determinations made. The obtained activation energy was independent of the stress applied in the 2000–8000 psi range.

Fig. 13.32 Etch pits in MgO showing polygonized structure after creep at 1600 °C. Edges of photograph are along the (100) direction (×500). Rothwell and Neiman [19]. With kind permission of AIP Publishing

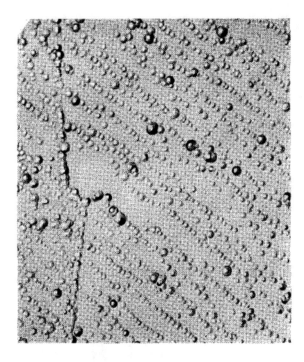

Fig. 13.33 Etch-pit pattern showing discrete slip bands formed by creep of MgO at 1000 °C. Edges of photograph are along ⟨100⟩ direction (×500). Rothwell and Neiman [19]. With kind permission of AIP Publishing

This activation energy value is in quite close agreement with the 5.22 eV of oxygen diffusion in MgO, perhaps suggesting that creep is associated with the oxygen-ion diffusion-controlled climb of edge dislocations.

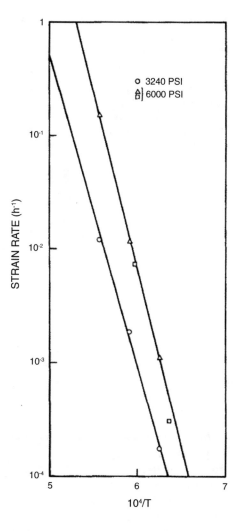

Fig. 13.34 Creep strain rate versus $10^4/T$ plot for determining activation energy. Two stress levels are represented here as indicated. Rothwell and Neiman [19]. With kind permission of AIP Publishing

The stress dependence on the strain rate was obtained by plotting the creep rate at the intersections of the different stress lines on the activation energy plots with a constant temperature ordinate. This is shown in Fig. 13.35 by the slope of 3, from which the stress exponent may be determined. The creep rate may be expressed as:

$$\dot{\varepsilon} = 8000\sigma^3 \exp\left(\frac{5.85}{kT}\right) \tag{13.30}$$

This equation suggests that creep can be described as the diffusion-controlled climb of edge dislocations over obstacles, such as impurities or sessile dislocations. Thus, it is assumed that this is the most likely mechanism for creep and that the diffusion of oxygen vacancies into jogs is the rate-controlling process.

Fig. 13.35 Stress
dependence of
high-temperature creep. The
slope of the line is 3. Data are
from six experiments
interpolated to a constant
temperature of 1400 °C.
Rothwell and Neiman [19].
With kind permission of AIP
Publishing

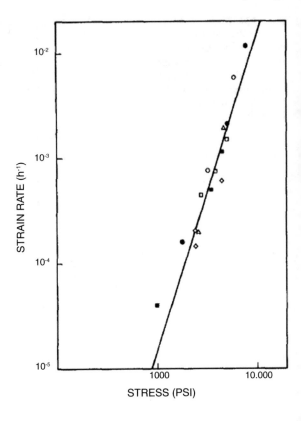

In an earlier work at higher temperatures, steady-state creep was measured in single-crystal MgO by means of the three-point bending deflection rate. The data from the 1450–1700 °C temperature range show that creep is a thermally activated process without a unique activation energy. The measured activation energy range is 3.5–7 eV. The activation energy indicated above in Eq. (13.3) was 5.85 eV, well within the range indicated here.

This was actually a measurement of the sample deflection rate, rather than of the creep strain rate. The loading of samples with successively greater weights was performed until a curve showing steady state deflection versus load was achieved. The sample was then unloaded and the temperature raised. This same procedure was repeated at the next temperature level and so on, obtaining the results displayed in Fig. 13.36. Details on the deflection rates and applied loads for particular slopes are listed in Table 13.6, which indicates that the creep rate depends on the load in accordance with a power law, and has an exponent in the range of 4–7.

The log of the deflection rate, as a function of the inverse temperature at constant load, is illustrated for two samples in Fig. 13.37. From the slopes of these curves, the activation energy was evaluated and found to be in the 3.5–7.0 eV range, as indicated earlier. This energy range may be explained in terms of self-diffusion

Fig. 13.36 Variation of deflection rate with load and temperature. Cummerow [10]. With kind permission of AIP Publishing

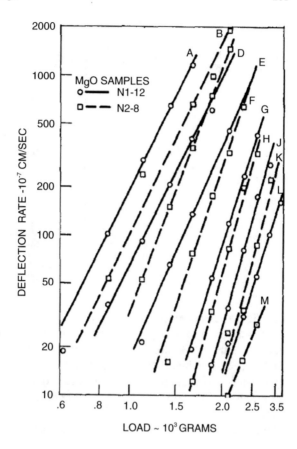

Table 13.6 Deflection rate versus load. Cummerow [10]. With kind permission of AIP Publishing

Curve	Slopes	Temp. °C
A	3.92	1704
B	4.02	1712
C	5.09	1656
D	3.94	1656
E	4.46	1605
F	5.84	1608
G	6.03	1551
H	7.08	1555
J	6.60	1503
K	7.30	1501
L	6.20	1454
M	4.73	1455

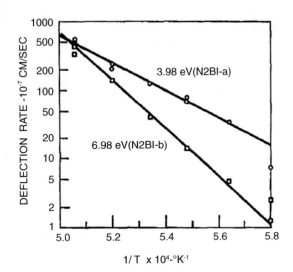

data, if it is assumed that the creep rate is limited by self-diffusion. The range of
exponents lends some support to this theory based on edge-dislocation climb over
barriers as the rate-limiting step. This was almost the same conclusion reached by
Rothwell and Neiman, considered above, although differences in the diffusing
specimens and in the diffusion details occurring must be noted. Meanwhile, there is
no electron microscopic evidence of barriers to climb. As stated by Cummerow,
some other climb mechanism, such as the dissolution of sessile edge loops or the
nonconservative motion of jogs, may be operative.

13.3   Creep Rupture in MgO

Only limited information is available on creep rupture per se, without diverging
to discussions on various factors involved in this process, such as GBS, etc.
When considering creep rupture, a fundamental concept involved is τ_r, time-
to-failure by rupture under a constant load during creep testing. The time-to-
rupture basically determines the service life of a material or product. Another
important factor is the strain at which rupture occurs; thus, strain and rupture, at
some load, must be interrelated. A known empirical relation has also been applied
to creep rupture in MgO [2], relating rupture time to stress, known as a 'power
law', given as:

$$\tau_r = A\sigma^n \tag{13.31}$$

Furthermore, an exponential relation has been suggested for τ_r, relating it to
temperature:

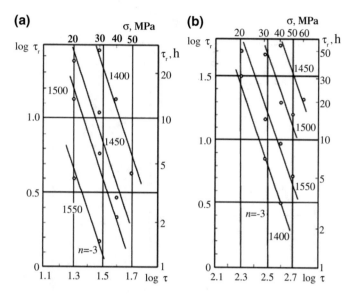

Fig. 13.38 Time to rupture τ_r versus the stress σ: **a** P-12; **b** P-25 (temperature in °C) n tangent of the slope angle of the straight lines. Bakunov et al. [2]. With kind permission of Springer

$$\tau_r = B\exp(bT) \qquad (13.32)$$

In the above equations, A, n, B, and b are all empirical coefficients. Recall the appearance of the creep curves in Chap. 1 for the σ constant (that depends on temperature) and for the T constant (when load/stress) determines the creep curves. This means that $\tau_r = f(\sigma, T)$. The rupture time is inversely proportional to the creep rate at the steady state, $\dot{\varepsilon}$, and τ_r = cons. The stress-dependent time-to-rupture of MgO, on a logarithmic scale, is presented in Fig. 13.38.

The designation of these specimens and other pertinent parameters are shown in Table 13.7.

The experiments were performed under constant stress in the 20–50 MPa range and in the 1450–1550 °C temperature range. Observe in Fig. 13.38 that τ_r decreases with increasing stress, as expected. Also note that the slopes of the lines

Table 13.7 Characteristics of the Experimental Samples. Bakunov et al. [2]. With kind permission of Springer

Batch index	Ceramic type	Predominant crystal size	Apparent density, g/cm³	True porosity, %	Bending strength, MPa, at temperature, °C				
					20	1400	1450	1500	1550
P-12	MgO	12	3.47	3.0	120	65	52	45	40
P-25	MgO	25	3.50	2.2	105	60	73	62	55

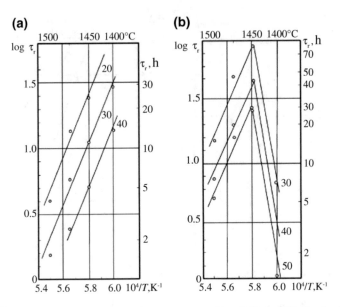

Fig. 13.39 Time-to-rupture, τ_r, versus the temperature, T: **a** P-12, **b** P-25 (the load in MPa is indicated). Bakunov et al. [2]. With kind permission of Springer

as expressed on the logarithmic scale remain constant. This indicates that the process mechanism does not change with temperature. The exponent, n, in Eq. (13.31) equals −3. The effect of temperature on rupture time is seen in Fig. 13.39.

Sample P-25 in Fig. 13.39b shows two branches of the curves. In the first branch, time-to-rupture increases with increasing $1/T$ (or with decreasing temperature). Beyond 1450 °C, the rupture time decreases with increasing $1/T$. The authors claim that the brittle-to-ductile transition temperature falls within the range of the experiments, since the strength reaches its maximum value at 1450 °C, which ordinarily occurs in the presence of plasticity.

Based on Fig. 13.39, the activation energy for rupture during creep testing was determined from the slopes, and the values are: 460 kJ mol^{-1} for the P-12 specimens and 440 kJ mol^{-1} for P-25, in the 1450–1550 °C temperature range, respectively. The time-to-rupture at all the tested temperatures decreased with the increase in steady-state creep, as illustrated in Fig. 13.40.

The relation between the creep before rupture, ε_r, and the steady-state rate is presented in Fig. 13.41.

A power law with exponent $n = -0.5$ is indicated for the type of MgO ceramics being tested. The relation between the time-to-rupture, τ_r, and the creep prior to rupture is expressed on a log–log scale in Fig. 13.42. A power law function is assumed with the exponent $n = 4$.

Fig. 13.40 Relation between the time-to-rupture, τ_r, and the rate of steady creep, $\dot{\varepsilon}$: *I* P-12 at 1400 °C; *II* general relation. Bakunov et al. [2]. With kind permission of Springer

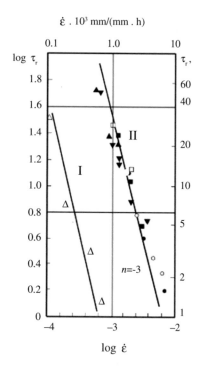

Fig. 13.41 Relation between the dip before rupture, ε_r, and the rate of steady creep, $\dot{\varepsilon}$: *I* P-12 at 1400 °C; *II* general relation. Bakunov et al. [2]. With kind permission of Springer

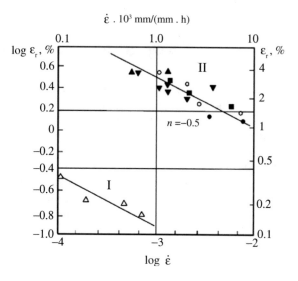

In conclusion, relations have been presented for the time-to-fracture in steady-state creep and applied stress; the time-to-failure by rupture and inverse temperature; the time-to-rupture and creep rate; and also the time-to-rupture and creep rupture, just prior to rupture.

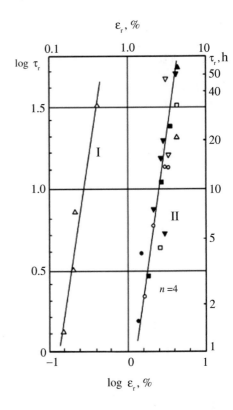

Fig. 13.42 Relation between the time-to-rupture, τ_r, and the dip prior to rupture, ε_r: *I* P-12 at 1400 °C; *II* general relation. Bakunov et al. [2]. With kind permission of Springer

13.4 Superplasticity in Magnesia Composites

Ceramics, especially those exhibiting fine grain sizes, have demonstrated their ability to form various superplastic components. The near-net-shape manufacturing of various components was one of the important reasons for expending a great deal of effort to discover more and more applicable systems (metallic and ceramic alike), that exhibit superplastic capabilities. Among the many systems, MgO plays an important role as a partner in certain superplastic ceramics. Some of these are superplastic spinels, based on $MgO \cdot xAl_2O_3$, where x is 1 or 2. The following example is a MgO-based composite—specifically, the $MgO \cdot Al_2O_3$ ceramic (known as spinel), which has been produced by HIP and has a fine grain size of 610 nm.

13.4.1 MgO·Al₂O₃

Recall that a creep rate was yielded by a number of aforementioned equations [Eqs. (4.1), (4.2), (8.3), and (12.1)], but now grain-boundary diffusion is added as:

$$\dot{\varepsilon} = A \frac{Gb}{kT} \left(\frac{b}{d}\right)^{P} \left(\frac{\sigma}{G}\right)^{n} \left(D_l + \frac{\pi\delta}{d} D_{GB}\right) \qquad (13.33)$$

in order to get the deformation associated with diffusion. A is a dimensionless constant, G is the shear modulus, d is the grain size, σ is the stress, d is the grain boundary, δ is the grain-boundary thickness, and p and n are the grain size and stress exponents, respectively. D_l and D_{GB} are the lattice and grain-boundary diffusion coefficients, respectively.

Before performing creep tests, the specimens were annealed at 1380 °C to relieve the stresses resulting from the HIP process. Then, compressive stress was applied to the specimens under constant stress in the 15–90 MPa range and at temperatures of 1350–1420 °C. The theoretical density of $MgO \cdot Al_2O_3$ is 3.58 g cm^{-3} and the obtained densities by HIP at the respective temperatures are listed in Table 13.8.

As may be seen in Table 13.8, at 1430 °C, the final density corresponds to the theoretical one with zero porosity and a grain size 0.8 μm. Often Norton's relation, given as:

$$\dot{\varepsilon}_s = A\sigma^n \qquad (13.34)$$

is used at high temperatures between the stationary strain rate and the stress, σ. The subscript, s, in Norton's equation indicates stationary strain. Here, A may be associated with grain size (and other microstructural features), while the stress exponent is an indication of the deformation process. In the case of a diffusion-controlled mechanism, factor A is expressed in terms of the grain size exponent, p, as:

$$A = A' \langle d \rangle^{-p} \qquad (13.35)$$

This same grain size exponent has already appeared in Eq. (13.33). Thus, A', in Eq. (13.35), takes into account any of the factors in Eq. (13.33) or any of the equations previously indicated, such as Eq. (4.1). As such, A' may be given as:

$$A' = \left(\frac{Gb}{kt}\right) b^p \frac{1}{G^n} D_l \qquad (13.35a)$$

Table 13.8 Density and grain size developments for different HIP conditions. Beclin et al. [4]. With kind permission of Elsevier

	HIPing temperature (°C)				
	1350	1380	1400	1410	1450
Green density (g cm³)	2.37	–	2.27	2.35	2.35
Final density (g cm³)	3.53	3.54	3.57	3.57	3.58
Porosity (%)	1.4	1.1	0.3	0.3	0
Grain size (μm)	0.48	0.51	0.51	0.57	0.8

By substituting the value of A' from (Eq. 13.35a) for A from Eq. (13.35), Eq. (4.1) is obtained. Equation (4.1) is cited, rather than Eq. (13.33), because the contribution of grain-boundary diffusion is also included in the latter. The grain size exponent $p = 2$–3. Express Eq. (13.34) in a logarithmic form, after replacing A in Eq. (13.35), as:

$$\ln \dot{\varepsilon}_s = \ln A' + n \ln \sigma - p \ln d \qquad (13.36)$$

A plot of $\ln \dot{\varepsilon}_s$ versus ε_s in a sample compressed at 60 MPa at 1400 °C is illustrated in Fig. 13.43.

There is a continuous decrease in the creep rate, which indicates that strain hardening is occurring in the sample, likely because strain-enhanced grain growth has occurred. The true strain indicated in Fig. 13.43 is 40%, which reflects the possibility of superplastic deformation in fine-grained spinel, in which the components are in a 1:1 ratio. Further support for the feasibility of superplastic deformation is the fact that no tertiary creep was observed in this ceramic (Fig. 13.44).

Using Eq. (13.36), the stress exponent, n, may be evaluated from the dependence of $\ln \dot{\varepsilon}_s$ on σ, while keeping A constant (Eq. 13.34). Values of n have been determined for materials with the same structure using the stress-jump method, as illustrated in Fig. 13.49. Table 13.9 lists the stress-dependent values of n determined at 1380 °C.

The values of n are lower than 2, which is usually observed in superplastic deformation, but similar values were observed in diffusion-controlled dislocation climb in $MgO \cdot 2Al_2O_3$ spinels. This subject will be discussed next.

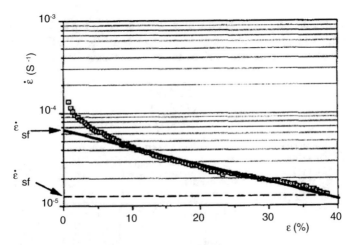

Fig. 13.43 Continuous linear decrease of the creep rate after a transient period ($T = 1400$ °C, $\sigma = 60$ MPa). Beclin et al. [4]. With kind permission of Elsevier

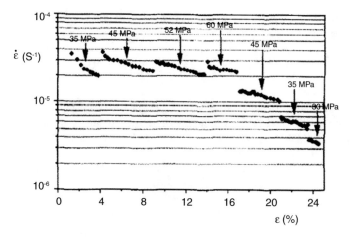

Fig. 13.44 Stress jumps from which the stress exponent, n, has been deduced ($T = 1380$ °C). Beclin et al. [4]. With kind permission of Elsevier

Table 13.9 Average values of the stress exponent obtained at different stresses ($T = 1380$ °C). Beclin et al. [4]. With kind permission of Elsevier

σ (MPa)	n	$\langle n \rangle$
15 → 20 → 30 → 45	1.97	
35 → 45	1.94	1.95
45 → 60	1.55	
50 → 60	1.47	1.49
45 → 52 → 60	1.46	
60 → 72 → 86.4	1.72	
60 → 75 → 90	1.85	1.78

Beclin et al. also investigated superplasticity in an earlier publication on MgO·Al$_2$O$_3$. Basically, the same experimental technique was used as above. In brief, the HIPed spinel had a grain size of 610 nm and uniaxial compression was applied in the superplastic (creep) experiments. Earlier studies [16] on MgO·2Al$_2$O$_3$ had served as guidelines for their investigations, indicating the importance of strain rate and temperature on fine-grained ceramics. In order to analyze creep at high temperatures, one of the equations already given [for instance Eqs. (1.9) or Eq. (13. 37)] might be used. Note that Eq. (1.9) uses B instead of A and Q instead of H.

$$\dot{\varepsilon}_s = A\sigma^n \exp\left(-\frac{\Delta H}{RT}\right) \qquad (13.37)$$

H, the activation energy, and n are characteristic of the strain rate (creep strain rate) and the deformation process. As mentioned above, A is associated with microstructure and its average grain size, $\langle d \rangle$. Grain size particularly effects diffusion-related processes. By means of Eqs. (13.35) and (13.37), one may write in logarithmic form:

$$\ln \dot\varepsilon = \ln A' + n \ln \sigma - p \ln\langle d\rangle - \frac{\Delta H}{RT} \tag{13.38}$$

The dependence of $\ln \dot\varepsilon$ on σ and T, respectively (Eq. 13.38), enables the determination of n and ΔH. In a stress- or ΔH-jump method, the values of n or ΔH may be evaluated by relating the change in $\ln \dot\varepsilon$ to $\ln \sigma$ or T^{-1}, so:

$$n = \left\{\frac{\Delta(\ln \dot\varepsilon_s)}{\Delta(\ln \sigma)}\right\}_T \tag{13.39}$$

and:

$$\Delta H = -R\left\{\frac{\Delta(\ln \dot\varepsilon_s)}{\Delta(\Delta T^{-1})}\right\}\sigma \tag{13.40}$$

In Fig. 13.45, creep curves are shown: (a) the creep strain versus time, and (b) the logarithm of the creep strain rate versus creep strain.

During this test, stress and temperature were kept constant. The applied stress was 60 MPa. The fact that the strain rate decreased (after transient creep) is an indication of strain hardening due to a microstructural change. Fine-grained spinel is, thus, a good candidate for superplasticity, since a true strain of 55% is observed in Fig. 13.45b. The original test material had fine-grained, equiaxed spinel, as shown in Fig. 13.46a, but some grain growth had occurred.

By increasing the stress to 150 MPa, cavitation is induced in the spinel and, therefore the decrease in density is associated with the decrease in the strain to 35%

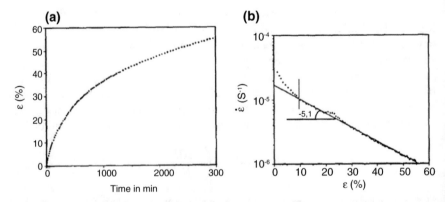

Fig. 13.45 Curves showing continuous linear decreasing of the creep rate after a transient ($\sigma = 60$ MPa, $T = 1723$ K): **a** ε versus t, and **b** $\ln \dot\varepsilon$ versus ε. Beclin et al. [3]. With kind permission of Elsevier

Fig. 13.46 TEM showing
the fine-grained
microstructure: **a** as-pressed
and then HIPed
polycrystalline MgO.Al$_2$O$_3$;
b deformed sample ($\varepsilon = 55\%$,
$T = 1723$ K, $\sigma = 60$ MPa).
Beclin et al. [3]. With kind
permission of Elsevier

at 1723 K. Cavity formation is a softening mechanism and, as such, works against
the strain hardening mentioned above. An important observation by SEM and
TEM was the lack of dislocations within the fine grains, and only a few rare,
individual dislocations were detected. Figure 13.46 is an illustration of fine-grained
microstructure.

The decrease in strain rate may be related to grain growth. This dynamic grain
growth is significant during superplastic deformation. Average grain growth may be
given as:

$$\langle \dot{d} \rangle = \dot{d}_a + \dot{d}_\varepsilon \tag{13.41}$$

where $\dot{d}_a$ is the stress-free grain growth rate after annealing, and $\dot{d}_e$ is the
GBS-induced grain growth rate, which is responsible for the superplastic strain rate,
$\dot{\varepsilon}$. One of the grain-growth-by-annealing models is based on the assumption
that grain-boundary velocity is driven by a mobility, M_b, by thermodynamic force
F, which is inversely proportional to the grain size, $\langle d \rangle$, to a power from 1 to 2,
given as:

$$\dot{d}_a = M_b \cdot F = M_b \cdot k\gamma_b \Omega \langle d \rangle^{1-m} \tag{13.42}$$

where the grain size exponent is $2 < m < 3$.

At intermediate strain rates, when the strain rate, $\dot{d}_\varepsilon$, is controlled by the rate of GBS, $\dot{d}_\varepsilon$ from Eq. (13.41) is given as:

$$\dot{d}_\varepsilon = \alpha b \langle d \rangle \dot{\varepsilon} \tag{13.43}$$

When the static grain growth effects are negligible, Eq. (13.41) may be integrated ($\langle \dot{d} \rangle = \dot{d}_\varepsilon$), resulting in:

$$\langle d \rangle = d_0 \exp(B\varepsilon) \tag{13.44}$$

Here, d_0 is the pretest grain size and $B = ab$. Using this value, Eq. (13.38) becomes:

$$\ln \dot{\varepsilon}_s = \ln A' + \ln \sigma - p \ln d_0 - pB\varepsilon - \frac{\Delta H}{RT} \tag{13.45}$$

Equation (13.45) has been obtained by expressing Eq. (13.44) in logarithmic form and substituting Eq. (13.44) for $\ln \langle d \rangle$ in Eq. (13.38). The $\ln (\langle d \rangle / d_0)$, as a function of ε for dynamic grain growth (according to Eq. 13.44) is a straight line and is plotted for $\sigma = 60$ MPa at 1723 K in Fig. 13.47. The slope of the curve provides B, resulting in $B = 1.87$.

As in their earlier work, the jump method was used to evaluate stationary-creep dependence on σ and T (Fig. 13.44). Compression creep experiments at constant stress and temperature are illustrated in Fig. 13.48. The jump time occurs when

Fig. 13.47 $\ln \langle d \rangle / d_0$ versus ε experimental curve showing a linear relationship. Beclin et al. [3]. With kind permission of Elsevier

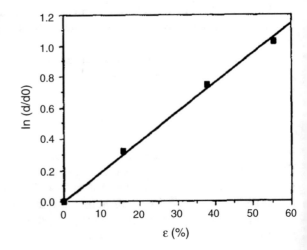

Fig. 13.48 General behavior observed in (ln $\dot{\varepsilon}/\varepsilon$) coordinates for a sample deformed during a change of σ (or T), which was carried out when the strain was ε_0. Beclin et al. [3]. With kind permission of Elsevier

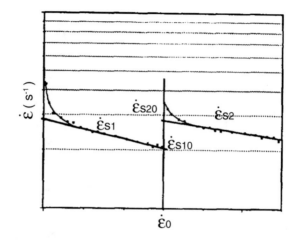

Table 13.10 Average values of the stress exponent obtained at different temperatures and stresses. Beclin et al. [3]. With kind permission of Elsevier

T(K)	σ (MPa)	$\langle n \rangle$
1623	90 → 120 → 150 → 120 → 90	2.1
1673	90 → 120 → 150 → 120 → 90	2.1
1723	60 → 90 → 120 → 150 → 180	2.0
1723	20 → 35 → 50 → 60 → 80	1.5
1723	10 → 15 → 20 → 30	1.7

$\varepsilon = \varepsilon_0$. Table 13.10 lists the stress exponents obtained at different stresses and temperatures.

The stress exponents listed in Table 13.10 were determined from the plots in Figs. 13.49 and 13.36, which basically reflects Eq. (13.39), which starts with the first jump:

$$n = \frac{(\ln \dot{\varepsilon}_{s20} - \ln \dot{\varepsilon}_{s10})}{(\ln \sigma_2 - \ln \sigma_1)} \qquad (13.46)$$

As seen in Table 13.10, the value of n in the 80–180 MPa stress range is somewhat above 2. For stresses below 80 MPa, it is about $n = \langle 1.5 \rangle$ in the stress range 60–35 MPa, but $n = \langle 1.7 \rangle$ in the stress range 30–10 MPa. The variation in the experimental $\ln \dot{\varepsilon}$ versus $\ln \sigma$ plot shown in Fig. 13.50 is an indication of stress exponent variation. Figure 13.49b and Eq. (13.40) serve to determine ΔH in the range of 1623–1723 K at a constant stress of 60 and 100 MPa (shown for 60 MPa in the plot) by temperature jumps. An analysis of Eq. (13.40), rewritten as Eq. (13.47), served to evaluate ΔH:

$$\Delta H = \frac{-R\left(\ln \dot{\varepsilon}_{s2_0} - \ln \dot{\varepsilon}_{s1_0}\right)}{\left(T_2^{-1} - T_1^{-1}\right)} \qquad (13.47)$$

Fig. 13.49 ln $\dot{\varepsilon}$ versus $\dot{\varepsilon}$
experimental curves showing:
a stress jumps: and
b temperature jumps. The
characteristic parameters,
n and ΔH, have been deduced
from the corresponding
jumps. Beclin et al. [3]. With
kind permission of Elsevier

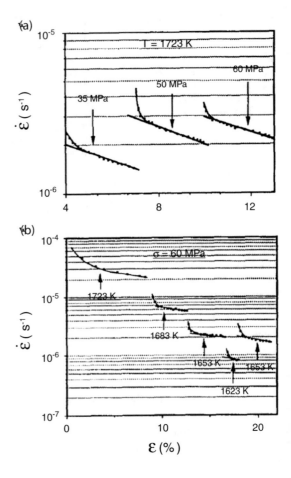

Fig. 13.50 ln $\dot{\varepsilon}$ versus ln σ
(experimental and predicted)
curves showing a sigmoidal
variation in the experimental
plot (T = 1723 K). Beclin
et al. [3]. With kind
permission of Elsevier

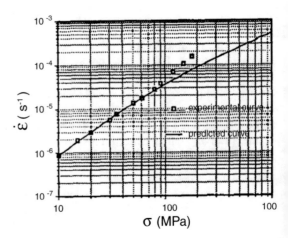

The values at 60 and 100 MPa are (647 ± 37) and (537 ± 53) kJ mol^{-1}, respectively. Moreover, the grain size exponent was determined from Eq. (13.45) and B (=1.87) from the plot of ln $\dot{\varepsilon}$ versus ε (Fig. 13.45b). At 1723 K and 60 MPa, $p = 2.72$. The values of n, p and ΔH determined above are characteristic of the creep mechanism.

It was seen from the creep experiments at 1623–1723 K under 20–180 PMa that MgO·Al$_2$O$_3$ can be deformed superplastically to a value of about 55% strain by compression. The measured activation energies are representative of the range for superplastic deformation. Superplasticity is accommodated via grain-boundary diffusion. The value of the grain size exponent obtained, being 2.72, is close to the value of $p = 3$, suggesting a significant contribution by grain-boundary diffusion.

13.4.2 MgO·2Al$_2$O$_3$

Unlike the case of MgO·Al$_2$O$_3$, a deformation of several hundred percent was obtained in fine-grained MgO·2Al$_2$O$_3$, deformed at high strain rates at temperatures of 1723–1885 K. Here, the mechanism of deformation was dislocation creep with a stress exponent of 2.1 ± 0.4. Despite this high deformation, the grains remained equiaxed and the evolution of the grain size was dependent only on temperature and strain rate. The initial microstructure determined whether the material would fracture or flow. An initial small grain size and supersaturated solid solution favored ductile flow. Such ductility is attributed to dynamic recrystallization, the effect of which is contrary to the onset of fracture. Factors promoting recrystallization also promote ductile flow, which is a prerequisite for superplasticity.

Deformation was performed by uniaxial compression under a vacuum above 0.01 Pa. The power law, seen in Eq. (13.35), was used to characterize flow. As mentioned above, superplastic deformation is possible if the power law exponent is $1 < n < 2.5$. The true stress and true strain at different initial strain rates at 1723 and 1880 K are shown in the plots in Figs. 13.51 and 13.52, respectively. In these figures, the strain is indicated by a negative sign, which is the usual way to indicate strain under compression. The strain rate has an effect on ductility and even on the transition from almost-brittle-to-ductile behavior. This is indicated in Fig. 13.51 (sample CT-4, for example) and Fig. 13.52 (e.g., sample CT-10).

As indicated above, the stress exponent may be determined from a plot of ln $\dot{\varepsilon}$ versus stress. The curves of the MgO·2Al$_2$O$_3$ are illustrated in Fig. 13.53, which also lists the n values for the various temperatures. The strain is $\varepsilon = -0.2$ and the equation for the strain rate, namely Eq. (13.37) is also shown. The activation energy may be calculated by this equation or by the one shown in the figure, resulting in $Q = 460 \pm 50$ kJ mol^{-1}. The plot of the strain rate (as a log strain rate) versus the inverse temperature (Eq. (13.37), but with Q instead of ΔH) is illustrated in Fig. 13.34.

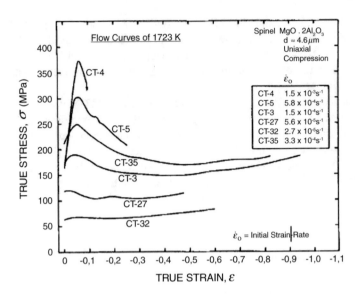

Fig. 13.51 Effect of deformation rate on true stress versus true strain curves at 1723 K. Panda et al. [16]. With kind permission of John Wiley and Sons

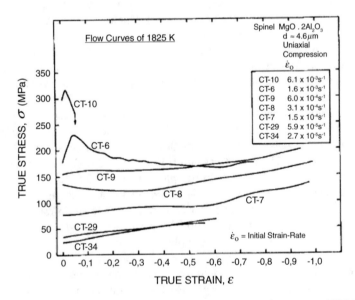

Fig. 13.52 Effect of deformation rate on true stress versus true strain curves at 1825 K. Panda et al. [16]. With kind permission of John Wiley and Sons

Again, grain size is an important factor in superplastic deformation. The grain size effect is shown in Fig. 13.55. The grain size influence on ductility is clearly seen in Fig. 13.55. Specimens with grain sizes 7.0 and 8.6 μm were ductile

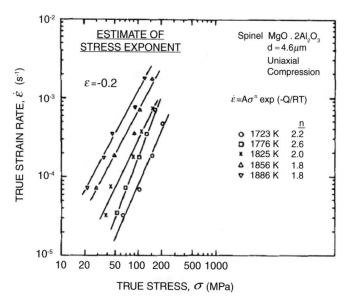

Fig. 13.53 Logarithmic plots to determine stress exponent ($\varepsilon = -0.2$). Panda et al. [16]. With kind permission of John Wiley and Sons

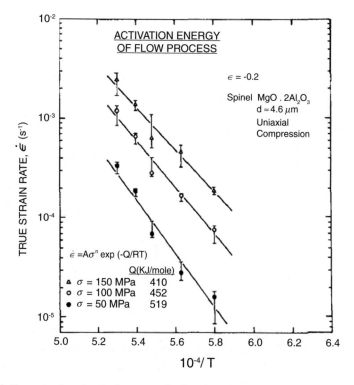

Fig. 13.54 Determination of activation energy for flow from Eq. (5) ($\varepsilon = -0.2$). Panda et al. [16]. With kind permission of John Wiley and Sons

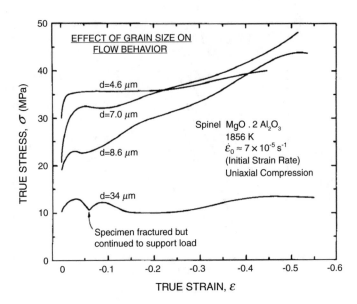

Fig. 13.55 Flow curves showing effect of grain size on flow behavior; 1856 K and $\dot{\varepsilon} = 7 \times 10^{-5}$ s^{-1}. Panda et al. [16]. With kind permission of John Wiley and Sons

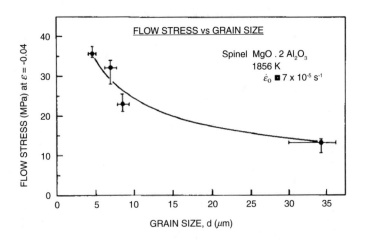

Fig. 13.56 Plot of flow stress versus grain size; $\varepsilon = -0.04$. Panda et al. [16]. With kind permission of John Wiley and Sons

with true strains at about 55%, whereas the large grain-sized specimen of $d = 34$ μm fractured at a strain of 6%. The variation of the flow stress with the grain size is shown in Fig. 13.56, which decreases with the increase in grain size (Fig. 13.54).

This suggests that diffusional creep is not the main mechanism operating. An increase in flow stress should be observed with grain size increase, if diffusional creep was the main creep mechanism. In fact, a dislocation mechanism is supported by a grain size-induced variation in the flow stress. The strain rate for such a dislocation mechanism is:

$$\dot{\varepsilon} = \rho_m \mathbf{b} v \qquad (13.48)$$

ρ_m is the mobile-dislocation density, $\mathbf{b}$ is the Burgers vector, and v is the average dislocation velocity. However, the dislocation mobility is likely to be limited by diffusion-controlled dislocation climb. Such a mechanism is supported by an obtained activation energy of 460 ± 50 kJ mol^{-1}, which is quite close the self-diffusion coefficient of the O ion, being 443 ± 50 kJ mol^{-1}, the slowest-diffusing species in the spinel.

13.5 Creep in Nano-MgO

MgO polycrystalline ceramics are brittle at room temperature, but at higher temperatures, in the $T \geq 0.5\ T_m$ range, they become ductile, dislocation slip occurs and they may strain-harden during deformation. However, nano-sized MgO is ductile and deforms under large strains without any work hardening (as in superplastic deformation). In Fig. 13.57, stress–strain curves at several temperatures are shown

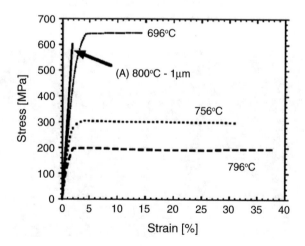

Fig. 13.57 Stress–strain curves recorded by compression of rectangular nc-MgO bars at constant cross-head speed and different temperatures. The curves exhibit elastic, perfectly plastic, behavior with no strain hardening. Specimen (A) was annealed to grow the grain size to 1 μm, and thus exhibited brittle behavior by compression at 800 °C (*arrowed solid line*), compared with the ductile behavior of its nanocrystalline counterpart specimen (*dashed curve*) at 800 °C. Domínguez-Rodríguez et al. [11]. With kind permission of Elsevier

Fig. 13.58 Strain rate versus
stress in nc-MgO subjected to
compression creep tests at
730 °C (*circles*) and 785 °C
(*triangles*). This dependence
was used to determine the
stress exponent $n = 2$.
Domínguez-Rodríguez et al.
[11]. With kind permission of
Elsevier

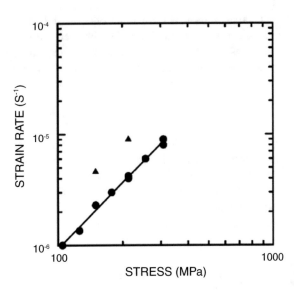

for nanocrystalline MgO; they indicate the possibility of making brittle MgO
ductile. The creep testing of these specimens was conducted at 100–300 MPa and at
the temperatures of 783 and 785 °C. The creep test results are shown in Fig. 13.58.
The stress exponent evaluated from the resulting curve was 2 ± 0.1. As usual, by
using the stress–strain curves (Fig. 13.57) and the value of the stress exponent, an
activation energy of (202 ± 9) kJ mol^{-1} was determined.

The microstructure of the equiaxed nanocrystalline grains is illustrated in
Fig. 13.59, before and after deformation. The equiaxed grains remained unchanged

Fig. 13.59 HRSEM images, using the secondary electrons, showing the surface microstructure of
the nc-MgO composed of equiaxed grains **a** prior to and **b** after the plastic deformation at 800 °C
(40% strain). No significant changes were visible after the deformation, except some
grain-boundary cavities and faceting of the surface grains which were in contact with the
compressing pad. *Arrows* indicate the applied load direction. Domínguez-Rodríguez et al. [11].
With kind permission of Elsevier

after the deformation. Compare Figs. 13.59a before and 13.59b after the deformation.

As stated above, for creep deformation related to point-defect diffusion in the lattice and grain boundaries, the strain rate was given as Eq. (13.33), which is reproduced here for convenience:

$$\dot{\varepsilon} = A \frac{Gb}{kT} \left(\frac{b}{d}\right)^{p} \left(\frac{\sigma}{G}\right)^{n} \left(D_l + \frac{\pi\delta}{d} D_{GB}\right) \tag{13.33}$$

Surprisingly, only limited information on creep deformation in nanocrystalline MgO is recorded in the literature. The importance of nano-sized ceramics is obvious, since they may attain plasticity, despite the fact that ceramics, with few exceptions, are inherently brittle or at best semi-brittle at room temperature.

13.6 Creep Recovery of MgO

Steady-state creep is a balance between work hardening and recovery. As such, steady-state creep is proportional to stress to an exponent n as:

$$\dot{\varepsilon}_s \propto \sigma^n \tag{13.34}$$

and stated as $\dot{\varepsilon} = A\sigma^n$. At low stresses, $n \sim 1$, and linear stress dependence may be interpreted as grain or grain-boundary diffusion-controlled creep. At higher stress levels, $n = 2.3$–4. It was believed that such high values of n are associated with dislocation movement. Two mechanisms have been suggested for dislocation-induced creep: (1) dislocation glide, associated with either the motion of jogged screw dislocations (nonconservative motion) or with dislocation dragging by a cloud of charged defects or (2) dynamic recovery by climb. Creep occurs because strain hardening, resulting from creep deformation, is annealed out by recovery. The existing balance of these two processes explains steady-state creep in ceramics.

An often-used technique for studying creep involves a sudden change in the applied stress and observation of the strain–time behavior. The aforementioned balance of an increment of strain, $d\varepsilon$, during a short time period, dt, by recovery, in order to maintain a constant flow stress, is given by:

$$d\sigma = \frac{\partial\sigma}{\partial\varepsilon} d\varepsilon + \frac{\partial\sigma}{\partial t} dt = 0 \tag{13.49}$$

Now designate $-r = -\frac{\partial\sigma}{\partial t}$, $h = \frac{\partial\sigma}{\partial\varepsilon}$ and recall that $\frac{d\varepsilon}{dt} = \dot{\varepsilon}_s$ (Eq. 13.49) may be clearly expressed as:

$$\dot{\varepsilon}_s = \frac{d\varepsilon}{dt} = \frac{r}{h} \qquad (13.50)$$

Compressive creep tests were performed up to 6% strain, to avoid barreling of the specimens. The height-to-diameter ratio of the cylindrical specimens is critical, so the experiments were carried out at $h/d = 1.5$. Creep strain–time curves are shown in Fig. 13.60; of the six runs, only three curves are shown for clarity.

The tests were carried out at 1596 K $(0.52\ T_m)$ and 62.6 MNm^{-2}. On reaching the steady state, the stress was reduced by a small amount, $\Delta\sigma \cong 0.05\sigma$ and, after this small stress decrease, the creep rate was zero (see Fig. 13.61). This stage continued until the creep rate reached a new steady state. When this new steady creep rate was established at the reduced stress, the stress was again increased by $\Delta\sigma$ to the original stress level and the instantaneous strain, $\Delta\varepsilon$, obtained on reloading was recorded (Fig. 13.61). This procedure was repeated. The strain–time behavior, following the series of stress decreases, was recorded and appears in Fig. 13.61. This procedure, of making small decreases to about $\sim 0.15\sigma$, throughout the incubation period, Δt, is linearly related to $\Delta\sigma$, so that $\Delta\sigma/\Delta t$ is a constant. Thus, this ratio is a measure of the recovery, $r\ (= -\partial\sigma/\partial t)$. At the same time, the specimen extension, $\Delta\varepsilon$, resulting from the increase in $\Delta\sigma$, provides the ratio $\Delta\sigma/\Delta\varepsilon$, previously designated as h. As such, Eq. (13.50) yields $\dot{\varepsilon}_s$.

See Fig. 13.62 for the relations between $\Delta\varepsilon$ and $\Delta\sigma$, and Δt and $\Delta\sigma$. Table 13.11 lists the predicted and experimentally observed steady-state creep rates, $\dot{\varepsilon}_s$. The coefficients of the work hardening rate and recovery rate, r, are related in Fig. 13.63 at a temperature of 1596 K. This temperature was below the sintering temperature and chosen so as to eliminate grain growth during creep.

In magnesia, it was found that the steady-state creep rate varies approximately as σ^3. This may be seen in Fig. 13.64 and that creep is controlled by recovery.

Fig. 13.60 Creep curves recorded at 1596 K and 84 MNm^{-2} for specimens of polycrystalline magnesia produced from two bars, labeled M and N, of nominally identical material. Birch and Wilshire [6]. With kind permission of Springer

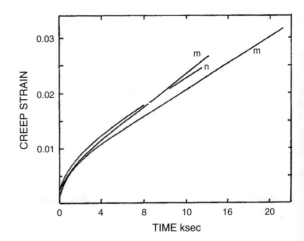

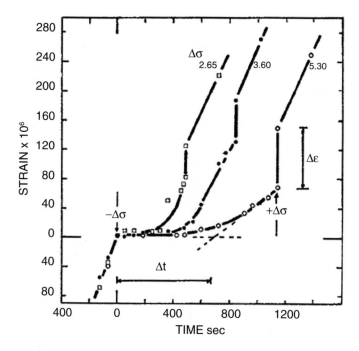

Fig. 13.61 Strain/time relationships following stress changes during steady-state creep of magnesia at 1596 K and 62.6 MNm^{-2}. The symbol □ denotes a stress decrease of 2.65 MN m^{-2} followed by a stress increase of 2.65 MN m^{-2} after a steady creep rate was established at the reduced stress. The symbols ● and ○ denote similar experiments with stress changes of 3.60 and 5.30 MN m^{-2}, respectively. Birch and Wilshire [6]. With kind permission of Springer

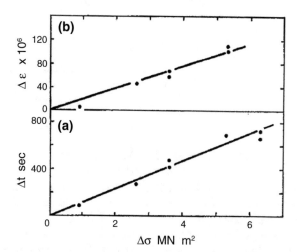

Fig. 13.62 **a** The relationship between the duration of the incubation period of zero creep rate, Δt, and the extent of the stress reduction, $\Delta \sigma$, during steady-state creep. **b** The dependence of the instantaneous strain, $\Delta \varepsilon$, observed on increasing the stress by $\Delta \sigma$ during steady-state creep of magnesia at 1596 K and 62.6 MN m^{-2}. Birch and Wilshire [6]. With kind permission of Springer

Table 13.11 A comparison of the predicted steady creep rate (calculated as r/h where r is the rate of recovery and h is the coefficient of strain hardening) and the measured steady creep rate, $\dot{\varepsilon}_s$, over a range of stresses at 1596 K. Birch and Wilshire [6]. With kind permission of Springer

(MN m^{-2} s^{-1} × 10^2)	h (MN m^{-a} × 10^{-4})	Predicted creep rate, r/h (set^{-1} × 10^7)	Observed creep rate, $\dot{\varepsilon}_s$ (set^{-1} × 10^7)
0.9	5.2	1.7	2.5
5.1	5.4	10.1	10.3
5.9	6.1	9.7	11.3
6.2	6.2	10.0	13.7
6.2	3.9	15.8	14.4
7.9	5.0	16.0	16.3
11.8	3.8	24.0	22.5
10.6	3.8	28.0	24.2
19.0	2.8	66.0	50.0
21.2	4.8	41.0	57.5

Fig. 13.63 The relationship between the rate of recovery and the steady creep rate for magnesia at 1596 K. Birch and Wilshire [6]. With kind permission of Springer

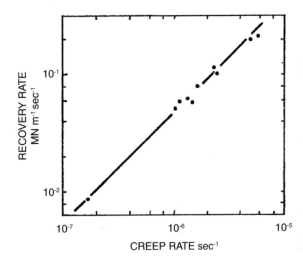

The results obtained from the steady-state creep measurements, in which n in Eq. (13.34) is $n \sim 3$ and recovery-controlled, suggest a model in which, during the rate-controlling creep, a 3-D dislocation network grows within the subgrains to form dislocation sources allowing slip to occur. It might also be interesting to compare such creep recovery data for single-crystal MgO.

Stress change experiments at temperatures above 1948 K with stresses lower than 4 MPa have been done. Specimens were deformed by compression along the $\langle 100 \rangle$ direction. As usual, the temperature was kept constant (± 2 K) and the applied stresses were in the 1.5–4 MPa range. The steady-state stress reductions (as done in the case of single-crystal MgO), were on the order of 0.05–1.70 σ_A,

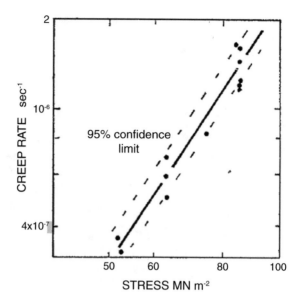

Fig. 13.64 The stress dependence of the steady creep rate of polycrystalline magnesia at 1596 K. Birch and Wilshire [6]. With kind permission of Springer

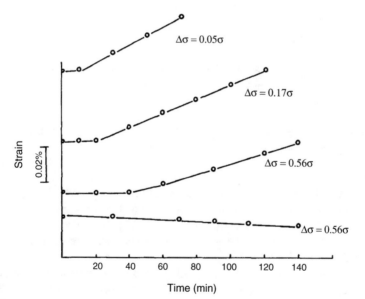

Fig. 13.65 Incubation periods following stress reductions during steady-state creep at 4 MPa and 1948 K. Ramesh et al.[17]. With kind permission of Springer

where σ_A is the applied stress. Each stress change was followed by a period of zero creep of duration Δt (incubation period), the length of which increased with increasing stress reduction (see Fig. 13.65). When the stress change was sufficiently large, negative deformation was observed (Fig. 13.65). The stress dependence of

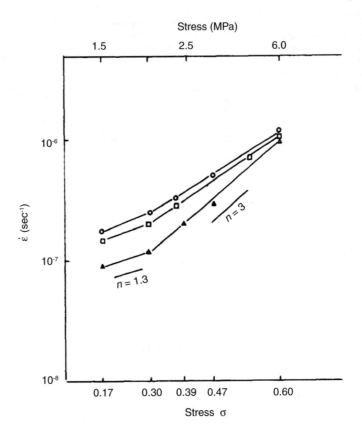

Fig. 13.66 Steady-state creep rates as a function of applied stress. ○ 2008 K; □ 1973 K; △ 1948 K. Ramesh et al.[17]. With kind permission of Springer

the steady-state creep rate under the conditions indicated above appear in Fig. 13.66. There is a transition in the stress exponent from 1, associated with low stresses, to 3, for high stresses (shown in Fig. 13.66). The recovery parameter, $\Delta\sigma/\Delta t$, was found to be linearly dependent on the creep rate prior to stress change, as illustrated in Fig. 13.67.

With the large stress reductions, negative creep was observed, implying that the stress had been reduced below the level of the internal stress—meaning that the creep process is governed by the effective stress ($\sigma_A - \sigma_i$). σ_i represents the internal stress developed during creep. In Fig. 13.65, this is indicated by negative strain at $\Delta\sigma = 0.56\sigma$.

A transition in the stress exponent, from $n = 1$ to 3, was observed with stress as the stress is increased. This transition is usually associated with a boundary mechanism in polycrystalline materials. Since the subgrain diameter is inversely proportional to the applied stress, any change in applied stress should change the mean subgrain diameter. This might induce partial sub-boundary migration. The

Fig. 13.67 Relationship between recovery parameter and steady-state creep rate before stress change. Ramesh et al.[17]. With kind permission of Springer

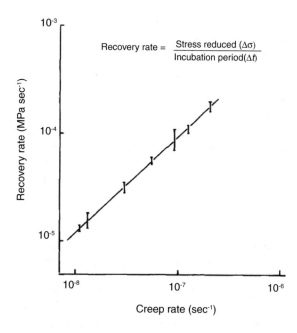

Fig. 13.68 Dislocation structure in the subgrain interior of MgO after a creep at 1973 K: **a** crept at 1.98 MPa and a strain of 0.04; **b** crept at 2.5 MPa and a strain of 0.038 followed by 30 min recovery after unloading to 1.98 MPa. Ramesh et al.[17]. With kind permission of Springer

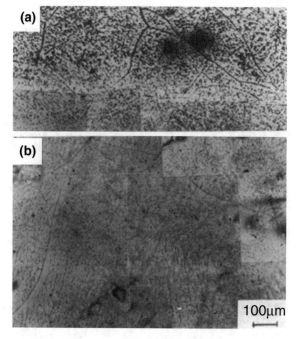

microstructure of a specimen crept to a strain of 0.038 at 1.98 MPa at 1973 K is seen in Fig. 13.68a. Inside the 300 μm subgrain, a 50 μm subgrain cell may be observed. However, in another specimen (see Fig. 13.68b), crept under nearly the same conditions and subsequently recovered, there is evidence of subgrain coarsening and the straightening of curved subgrain cells is visible. The microstructure shows dislocations blocked by subgrain boundaries and some which have been stopped when moving backwards. Back flow most likely involves the motion of dislocations from the interior to the boundaries. The negative creep shown in Fig. 13.65 may, thus, be associated with the effective stress of $(\sigma_A - \sigma_i)$. As such, the rate-controlling process is considered to be the rate of recovery and of the growth of the dislocation network. Following stress reduction, there are no sources capable of producing forward creep, and an incubation period of zero creep rate will result. Under these conditions, the direct relationship between creep rate and $\Delta\sigma/\Delta t$ (recovery rate), shown in Fig. 13.67, is obtained. The extent of the stress reduction plays a role in the relaxation of the bowed dislocations (i.e., straightening the bowing), permitting dislocations to move backwards after unloading.

Note after stress reduction, sub-boundaries that had been formed during creep move, indicating their growth (as revealed by the microstructure). Clearly, these sub-boundaries act to retard the forward movement of dislocations and to produce creep (the relaxation period). With sufficiently large stress reductions ($\Delta\sigma = 0.56\sigma$), the driving force provided by the internal stress, σ_i, causes the dislocations blocked at the sub-boundaries to move backwards after unloading. This observation is a possible explanation to the negative creep indicated in Fig. 13.65.

References

1. Addad A, Beclin F, Crampon J, Duclos R (2007) Ceram Int 33:1057
2. Bakunov VS, Lukin ES, Sysoev ÉP (2013) Glass Ceram 70:216
3. Beclin F, Duclos R, Crampon J, Valin F (1995) Acta Metall Mater 43:2753
4. Beclin F, Duclos R, Crampon J, Valin F (1997) J Europ Ceram Soc 17:439
5. Bilde-Sörensen JB (1972) J Am Ceram Soc 55:606
6. Birch JM, Wilshire B (1974) J Mater Sci 9:794
7. Boniecki M, Librant Z, Wajler A, Wesolowski W, Weglarz H (2012) Ceram Int 38:4517
8. Clauer AH, Wilcox BA (1976) J Am Ceram Soc 59:89
9. Coble RL (1963) J Appl Phys 34:1679
10. Cummerow RL (1963) J Appl Phys 34:1724
11. Domínguez-Rodríguez A, Gómez-García D, Zapata-Solvas E, Shen JZ, Chaim R (2007) Scr Mater 56:89
12. Herring C (1950) J Appl Phys 21:437
13. Langdon TG, Pask JA (1970) Acta Met 18:505
14. Mariani E, Mecklenburgh J, Wheeler J, Prior DJ, Heidelbach F (2009) Acta Mater 57:1886
15. Nabarro FRN (1948) Report of a conference on the strength of solids. Physical Society, London, pp 75–90
16. Panda PC, Raj R, Morgan PED (1985) J Am Ceram Soc 68:522
17. Ramesh KS, Yasuda E, Kimura S (1986) J Mater Sci 21:3147

18. Ratzker B, Sokol M, Kalabukhov S, Frage N (2016) Materials 9:493
19. Rothwell WS, Neiman AS (1965) J Appl Phys 36:2309
20. Terwilliger GR, Bowen HK, Gordon RS (1970) J Am Ceram Soc 53:241
21. Weertman J (1957) J Appl Phys 28:362
22. Yoo H-I, Wuensch BJ, Petuskey WT (2002) Solid State Ionics 150:207

Chapter 14
Creep in ZrO$_2$ Zirconia

Abstract ZrO$_2$ is a very refractory ceramic with excellent chemical inertness, corrosion resistance up to high temperatures and low thermal conductivity; it is also electrically conductive above ~ 600 °C. Due to all these properties, ZrO$_2$ has a very broad range of applications. Both polycrystalline and single-crystal zirconia were subject to creep tests. Alloying zirconia (composites) intends to enhance mechanical strength and to improve its physical properties. Zirconia has to be stabilized for technological applications. One of the most frequently used stabilizers is yttria. Stabilized zirconia might be partial or fully stabilized depending on the quantity of the stabilizer. Also in this ceramics superplasticity was observed, a section is devoted to nano-zirconia.

14.1 Introduction

A few introductory words are in place regarding ZrO$_2$, a very refractory ceramic with excellent chemical inertness, corrosion resistance up to high temperatures and low thermal conductivity; it is also electrically conductive above ~ 600 °C. Due to all these properties, ZrO$_2$ has a very broad range of applications. An important feature of zirconia is its polymorphism, which, as expected, is associated with volume changes. Pure zirconia has three crystal phases, depending on temperature: at very high temperatures, >2370 °C, it is cubic; at intermediate temperatures, 1170–2370 °C, its structure is tetragonal; and at low temperatures, below 1170 °C, it is monoclinic. Albeit pure zirconia is not stable, so a stress-induced metastable tetragonal phase can spontaneously transform into the more stable monoclinic zirconia.

A monoclinic transformation into the tetragonal phase is probably a Martensitic-type transformation, contributing to ceramic toughening. The largest volume change (increase) in the 3–5% range is associated with the tetragonal-to-monoclinic transformation, which is very rapid and induces extensive cracking, consequently destroying the mechanical properties of the fabricated components during cooling and rendering the pure zirconia useless for application. Several oxides, notably Y$_2$O$_3$, as well as MgO, CaO, etc., can slow down crystal structure changes or even completely eliminate them. The resulting, stabilized structure

© Springer International Publishing AG 2017
J. Pelleg, *Creep in Ceramics*, Solid Mechanics and Its Applications 241,
DOI 10.1007/978-3-319-50826-9_14

depends on the amount of the stabilizer and, with sufficient additives, even a high-temperature cubic structure can be preserved at room temperature. Clearly, a stabilized cubic structure does not undergo harmful transformations during heating and cooling. A controlled, stress-induced volume expansion of the tetragonal-to-monoclinic inversion is used to produce a very tough, hard, high-strength zirconia for mechanical and structural applications.

There are several different mechanisms that lead to strengthening and toughness in zirconias that contain tetragonal grains. Of particular interest are two strong, partially stabilized zirconias (PSZ) (available commercially): tetragonal zirconia polycrystal (TZP) and tetragonally stabilized zirconia (TSZ). The latter is an MgO-stabilized cubic zirconia, containing uniformly dispersed tetragonal precipitates. The stabilizing agents of TZP are rare earth oxides; the one most often used is yttria-stabilized zirconia. TZP exposed to water vapor degrades rapidly at 200–300 °C, thus requiring controlled conditions for use. Toughened zirconia is limited to use under 800 °C. The best thermal-barrier coatings are yttria-stabilized zirconia with yttria in the 6–8% range.

This chapter will now consider the creep properties of zirconia.

14.2   Creep in Polycrystals

Because of the instability of pure zirconia, the majority of the experimental data on creep in zirconia were gleaned from tests done on stabilized zirconia. Once again, various stabilizing agents are used, but most commonly yttria-stabilized zirconia is preferred. The usual, evaluated creep tests consist mainly of the application of compressive stress (or load), thus avoiding the opening of microcracks or the extension of pores and other flaws. There is little information available on the creep behavior of pure zirconia, apart from those doped with various additives. It is expected that 6 mol% (10.47 wt%) yttria-stabilized zirconia, which is on the verge of stabilization, will transform the tetragonal structure into a distorted cubic (fluorite structure) at ~ 700 °C. Thus, in the temperature range of these creep experiments (1100–1500 °C), the yttria-stabilized zirconia is expected to have a cubic fluorite structure. Most of the experiments were performed at an initial constant load of 4860 psi with a final stress, after 5% creep compression, of 4530 psi. This steady-state creep rate may be written as

$$\dot{\varepsilon} = \left(\frac{S\sigma^n}{d^m} \right) \exp\left(-\frac{Q}{RT} \right); \qquad (14.1)$$

where S is a structure factor. Assuming that the stress, grain size, and structural factor remain constant, Eq. (14.1) may be rewritten in a more familiar form, as

$$\dot{\varepsilon} = A \exp\left(-\frac{Q}{RT} \right) \qquad (14.1a)$$

Q, assumed to be constant in the tested temperature range, is obtained from a plot of log $\dot{\varepsilon}$ versus $1/T$. Figure 14.1a shows such a plot.

The best-fit line through the points, by least-square analysis, is given by $Q = 86 \pm 30$ kcal mol^{-1}. The stress exponent may be determined from the plots of creep rate versus stress. In Fig. 14.2, such a plot is shown on a logarithmic scale. Here, data on scandia stabilization is also included. The curve for the yttria-stabilized zirconia shows a break in the line. The best-fit lines for yttria-stabilized zirconia give $n = 1$ for the 1200–6000 psi range and $n \sim$ 6–7 for 6000–10,300 psi. The reliability of such a high value for n is questionable. The $n = 1$ exponent is usually associated with cation diffusion-controlled creep. The high value of n is probably associated with the cracks and likely a result of the growth and propagation of intercrystalline cracks.

Representative microstructures are shown in Fig. 14.3. Grain-boundary porosity may be seen in both micrographs (a) and (b).

14.3 Creep in Single-Crystal Zirconia

The single crystals used to investigate creep had higher yittria stabilizer contents than was found in the tested polycrystalline zirconias. In the case of the single crystals, 9.4 mol% Y_2O_3 stabilizer was measured. This creep investigation was conducted in the 1300–1550 °C range. This stabilization yielded a cubic zirconia (C-ZrO$_2$). The long axis of the specimen, which was the loading axis, was parallel to $[\bar{1}\bar{1}2]$, while the other two faces were parallel to $(100)[0\bar{1}\bar{1}]$ and $(010)[\bar{1}01]$. This

Fig. 14.1 a Log creep rate at 4840 psi YS inverse absolute temperature. (O) Yttria-stabilized zirconia, (▽) heat-treated scandia-stabilized zirconia, and (△) scandia-stabilized zirconia. Evans [9]. With kind permission of John Wiley and Sons

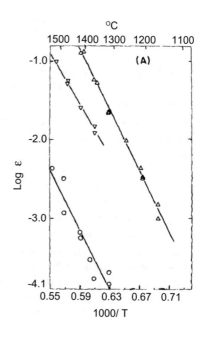

Fig. 14.2 Log creep rate versus log stress (psi). (O) Yttria-stabilized zirconia at 1485 °C and (Δ) scandia-stabilized zirconia at 1378 °C. Evans [9]. With kind permission of John Wiley and Sons

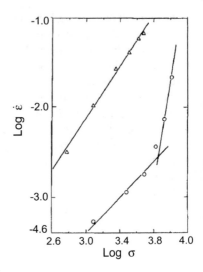

orientation provided a Schmidt factor of 0.47 for an easy $\{(001)[110]\}$slip system. The other two slip systems in this family, $(100)[0\bar{1}\bar{1}]$ and $(010)[\bar{1}01]$, both have Schmidt factors of 0.35, while the Schmidt factor for both $(\bar{1}11)[0\bar{1}1]$ and $(1\bar{1}1)[\bar{1}01]$, secondary slip systems, was 0.41. A constant creep load was applied under compression to the single-crystal specimens. These test results were analyzed by means of the steady-state creep relation [see Eq. (4.1)], given here as

$$\dot{\varepsilon} = A \frac{\mu b}{kT} \left(\frac{\sigma}{\mu}\right)^n \left(\frac{b}{d}\right)^p \left(\frac{p_{O_2}}{p_{O_2}^*}\right)^m \exp\left(-\frac{Q}{kT}\right) \qquad (14.2)$$

Although the relevant parameters were listed following Eq. (4.1), they are repeated here for the sake of convenience. A is a dimensionless constant, **b** is the Burgers vector of C-ZrO$_2$ (= 3.62 × 10^{-10} m), d is the grain size (only relevant for polycrystalline materials), $p_{O_2}^*$ is a reference oxygen partial pressure, and kT has its usual meaning. The creep parameters, n, p, m, and Q, depend on the details of the deformation and diffusion mechanisms. Their values may be obtained from a plot showing the variation of the creep rate, $\dot{\varepsilon}$, after changes in σ, T, and the partial pressure of oxygen, $p_{O_2}^*$. For single crystals, the exponent $p = 0$. Figure 14.4 is a plot of the creep rate variation with strain at the indicated temperatures. The evaluated value of n and the activation energies are also shown. These dislocation structures were characterized by TEM (Figs. 4.2 and 4.3) and discussed. The results indicate that deformation occurs principally by the glide of $a/2\langle110\rangle$ dislocations on the primary slip plane. Also, the creep parameters and dislocation structure suggest that the creep mechanism changes in the temperature range investigated.

The creep parameters evaluated by the incremental changes during deformation (Fig. 14.4) are also listed in Table 14.1.

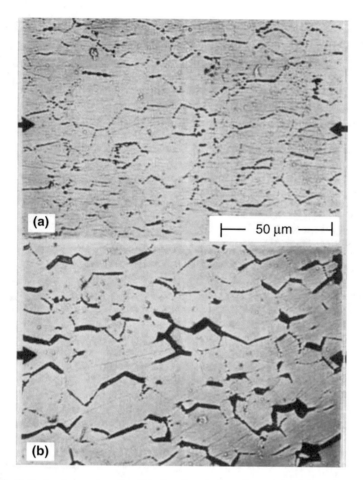

Fig. 14.3 Optical micrographs of polished and etched sections of yttria-stabilized zirconia. (**a**) Creep strain of 0.67% with stress of 4860 psi at 1382 °C and (**b**) creep strain of 2.3% with stress of 4860 psi at 1535 °C. *Arrows* show compression axis

Fig. 14.4 Typical creep experiment in Y_2O_3-stabilized ZrO_2 at stresses of 84 and 66 MPa and three temperatures, 1466, 1516–1518, and 1558 °C. Fernandez et al. [11]. With kind permission of John Wiley and Sons

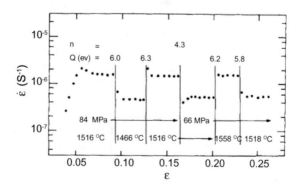

Table 14.1 Creep-Law Parameters (Eq. (14.2)) Determined by Incremental Changes During Deformation[a]. Fernandez et al. [11]. With kind permission of John Wiley and Sons

Deformation temperature range (°C)	n[b]	Q(eV)[c]	m[d]
1300–1400	7.3 ± 0.5(7)	7.4 ± 0.3(5)	0(4)
1450–1550	4.5 ± 0.4(4)	6.1 ± 0.3(7)	0(4)

[a]The number of determinations is indicated in parentheses
[b]Stress exponent
[c]Activation energy
[d]P_{O_2} exponent

Fig. 14.5 Literature data for creep of polycrystalline ZrO$_2$, normalized by temperature and grain size, as a function of stress, and compared with the single-crystal data of this study. Fernandez et al. [11]. With kind permission of John Wiley and Sons

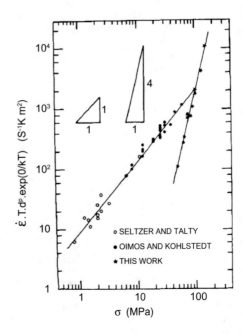

The steady-state creep rate was shown in Fig. 4.1 and also extensively discussed. In brief, it may be seen that there is a break in the $\dot{\varepsilon}$ versus 1/T line, suggesting two mechanisms for creep at high and low temperatures with a transition between the two regimes during the temperature interval 1400–1450 °C. The least-squares fit of these lines yields for the activation energies $Q_1 = 6.2 \pm 0.4$ eV at $T \geq 1450$ °C and $Q_2 = 7.7 \pm 0.4$ eV at $T \leq 1400$ °C.

The high-temperature creep of the single-crystal ZrO$_2$ is also plotted in Fig. 14.5 (with $p = 0$), and is compared with data on polycrystals from the literature. The slope determined was $n = 4.1 \pm 0.3$. Figure 14.5 shows that, as expected, the creep resistance of single crystals is better than that of polycrystalline ZrO$_2$ with a similar composition at all the stresses <100 MPa. This fact is clearly related to the effect of grain boundaries. The rate-controlling mechanism at

Fig. 14.6 Survey of high-temperature diffusivities in Y_2O_3-stabilized ZrO_2 from creep, self-diffusion, and dislocation loop annealing experiments. Treating these data as a single set yields D_0 of 10^{-3} and Q of 5 eV [see Eq. (4.3)]. Fernandez et al. [11]. With kind permission of John Wiley and Sons

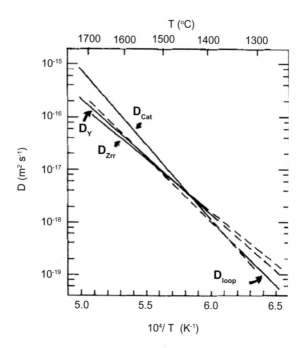

the higher temperatures is dislocation climb, which is supported by the stress exponent, 4.5, and by the TEM observations. The high activation energy of ~ 6 eV (see Table 14.1) should correspond to the activation energy for the diffusion of Zr or Y in c-ZrO_2, but it does not agree with measured interdiffusion data of 4.05 and 4.38 for Zr and Y, respectively. In Fig. 14.6, the diffusion of cations, D_{cat}, from interdiffusion measurements is shown together with annealing data from dislocation prismatic loops, D_{loop}. This diffusion data is expressed as a single data set. After applying a least-squares fit to this set, it yields the following equation for the diffusion coefficient in Y_2O_3-stabilized ZrO_2 at high temperatures:

$$D = 1 \times 10^{-3} \left(^{+0.6\times10^{-3}}_{-0.2\times10^{-3}} \right) \exp\left(-\frac{5 \pm 0.5\,\text{eV}}{kT} \right) \text{m}^2\,\text{s}^{-1} \qquad (14.3)$$

At low temperatures, the rate-controlling mechanism is dislocation cross-slip, which may be as important in controlling creep as is recovery via dislocation climb at elevated temperatures. This is supported by the higher values of n and Q. At the lower temperatures, n is ~ 7.5 and $Q \sim 7.5$ eV (see Table 14.1). A high density of dislocations and significant cross-slip are characteristic of such cross-slip-controlled creep, along with a stress-dependent activation energy. In this case, according to Poirier, when cross-slip and dislocation climb operate at the same time, the creep rate is expressed as

$$\dot{\varepsilon} = \dot{\varepsilon}_{\text{cross-slip}} + \dot{\varepsilon}_{\text{climb}} = \dot{\varepsilon}_{O_1}\left(\frac{\sigma}{\mu}\right)^{n_1}\exp\left(\frac{Q_1}{kT}\right)$$
$$+ \dot{\varepsilon}_{O_2}\left(\frac{\sigma}{\mu}\right)^{n_2}\exp\left(-\frac{Q_2}{kT}\right)$$

(14.4)

In Eq. 14.4, subscripts 1 and 2 refer to cross-slip and dislocation climb, respectively. The effects of the Y$_2$O$_3$ content on the stabilization of ZrO$_2$ with increased temperature have previously been presented. The amount of yttria is 21 mol% and the temperature range is 1400–1800 °C. Again, two regimes of creep were observed in single crystals both below 1500 °C and between 1500 and 1800 °C. A transformation in the creep-controlling mechanism is indicated around 1500 °C—from a glide-controlled to a recovery-controlled creep mechanism. These creep tests were performed similarly to those done on the 9% yttria-stabilized zirconia. The stress, σ, and the temperature, T, were changed incrementally to determine the values of n and Q, as illustrated in Fig. 14.7.

When the controlling mechanism is diffusion, the data analysis may be accomplished by an equation similar to Eq. (14.2):

$$\dot{\varepsilon} = A\frac{\mu b}{kT}\left(\frac{\sigma}{\mu}\right)^n D_0\exp\left(-\frac{Q}{kT}\right)$$

(14.5)

Note that $D_0\exp(-Q/kT)$ is D, the diffusion coefficient. The meanings of the symbols are the same as in Eq. (14.2). Equation 14.5 is a consequence of p, the exponent being 0 for single crystals (the term = 1 when exponent $p = 0$), and the use of $m = 0$, the exponent of the last term for creep when not a function of p_{O_2} at

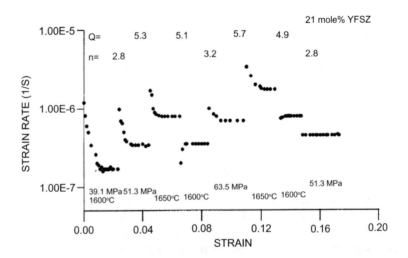

Fig. 14.7 Typical creep curve of 21 mol% stabilized zirconia plotted as strain rate ($\dot{\varepsilon}$) versus strain (ε). Garcia et al. [12]. With kind permission of Elsevier

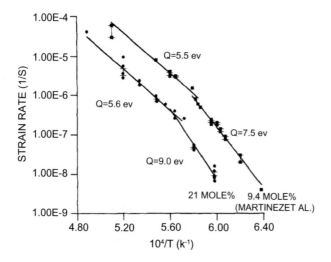

Fig. 14.8 Steady-state creep rate, normalized to 100 MPa, plotted versus the reciprocal of the temperature. The *solid line* represents the results obtained by Martinez et al. in the case of 9.4 mol %-Y-FSZ [yttria fully stabilized zirconia]. Garcia et al. [12]. With kind permission of Elsevier

pressures of 10^{-5} atm or higher. (This was also found for c-ZrO$_2$ polycrystals by Dimos and Kohlstedt when $\dot{\varepsilon}$ is between 10^{-9} and 1 atm.) Q and n may be determined by these tests. When keeping the load constant and changing the temperature, Q may be determined, as well:

$$Q = \frac{kT_1T_2}{\Delta T} \ln\left(\frac{T_2\dot{\varepsilon}_2}{T_1\dot{\varepsilon}_1}\right) \tag{14.6}$$

The stress exponent may be obtained by maintaining the temperature constant and changing the load. Thus

$$n = \frac{\ln\left(\frac{\dot{\varepsilon}_2}{\dot{\varepsilon}_1}\right)}{\ln\left(\frac{\sigma_2}{\sigma_1}\right)} \tag{14.7}$$

Suffice it to say that the activation energy may also be determined with Eq. (14.5) by plotting $\ln \dot{\varepsilon}$ versus $1/T$, bearing in mind that AD_0 is constant. (One may normalize the strain rate to 100 MPa and use the value of n obtained from Eq. (11.7)). Such a plot is shown in Fig. 14.8.

On the basis of Fig. 14.7, Q and n were evaluated. Their values are listed in Table 14.2. This table shows the two temperature regions, above and below 1500 °C, with the respective values of Q and n. At all the temperatures investigated, the material showed good creep resistance with a strain rate about 10 times lower than in the 9.4 mol% zircona tested under the same conditions of T and σ.

Table 14.2 Creep-law parameters determined by incremental changes during deformation. In parenthesis, we have written the number of determinations considered to calculate the mean value of the parameters shown. We have observed no difference in n and Q for positive or negative steps of Q or T. Garcia et al. [12]. With kind permission of Elsevier

Interval of temperature (°C)	Stress exponent (n)	Activation energy (Q) in eV
1400–1500 °C	5.4 ± 0.4(5)	8 + 1(6)
1500–1800 °C	2.9 + 0.2(13)	5.8 + 0.5(13)

TEM foils were prepared to characterize the dislocation microstructure. Specimens of the two regions were inspected at the representative temperatures of 1400 and 1600 °C. Figure 14.9 shows the TEM microstructure of the 1400 °C sample (the TEM microstructure at 1600 °C is not shown). A high dislocation density was observed, causing the forming nodes to react with each other, often pinned. Extinction studies indicate that most of the dislocations have a $\frac{1}{2}\langle 1\bar{1}0\rangle$-type Burgers vector and lie on the primary (001) slip plane. A stereopair analysis of the (001) zone axis, using $\mathbf{g} = (2\bar{2}0)$, showed the existence of some cross-slip. The (111) cross-slip is seen in Fig. 14.10.

In the samples deformed at 1600 °C, two planes, (001) and (110), have been observed by TEM, indicating: (i) the dislocation density is lower than the annealing effect during the cooling and under the low stresses used while deforming the samples; (ii) in both cases, very long dislocations with Burgers vector of type 1/2 (110) were observed. Dislocation climb is an active mechanism in this range of temperatures.

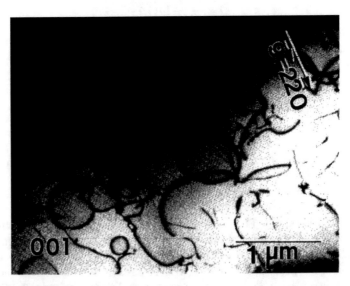

Fig. 14.9 TEM micrograph of the primary slip plane (001), showing the microstructure at 1400 °C. Garcia et al. [12]. With kind permission of Elsevier

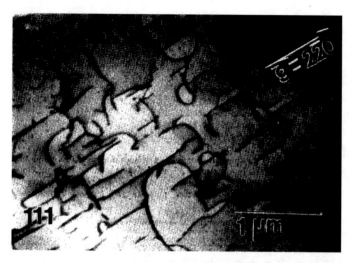

Fig. 14.10 TEM micrograph showing the dislocation microstructure in the case of a slice parallel to the cross-slip plane (111), at 1400 °C. Rodriguez et al. [23]. With kind permission of Elsevier

The suggested creep model for the low-temperature region (<1500 °C) and the elevated temperature interval (>1500 °C) is based on the concept that steady-state creep, or rather the strain rate $\dot{\varepsilon}$ is a consequence of the dynamic equilibrium between the rates of dislocation creation and annihilation. The rate of dislocation creation is controlled by the mean dislocation-glide velocity from the sources. However, the annihilation process is controlled by diffusion, which is associated with dislocation climb. The following is an expression for these two opposing processes:

$$\dot{\varepsilon} = \frac{\mu b}{kT} \frac{\alpha^3}{(1+\alpha)^2} \left(\frac{\sigma}{\mu}\right)^2 \left(\frac{\sigma - \sigma_i}{\mu}\right) D \tag{14.8}$$

In the above relation α is the ratio of the dislocation mobilities for glide and climb (the other symbols are as previously defined). As such

$$\alpha = \frac{\mu_{glide}}{\mu_{climb}} \tag{14.9}$$

For dislocation glide, $\alpha < 1$, and for $\alpha > 1$, climb becomes more favorable than glide. σ_i was defined earlier as "internal stress." This is the average stress exerted on one dislocation by the others, related to σ by

$$\sigma_i = \left(\frac{\alpha}{1 - \alpha}\right)\sigma \tag{14.10}$$

By combining Eqs. (14.8) and (14.10), one obtains

$$\dot{\varepsilon} = \frac{\mu b}{kT} \left(\frac{\alpha}{1-\alpha}\right)^3 \left(\frac{\sigma}{\mu}\right)^3 D_{\text{diff.}} \tag{14.11}$$

and (see, for example, Philibert on diffusion) $D_{\text{diff.}}$ is

$$D_{\text{diff}} = \frac{D_{Zr} \cdot D_Y}{x_{Zr} D_Y + x_Y D_{Zr}} \tag{14.12}$$

Based on the experimental results and D_{diff}, A (from the creep law, Eq. (14.5)) may be evaluated by means of

$$A = \frac{\dot{\varepsilon} kT}{D_{\text{diff}} \mu b} \left(\frac{\mu}{\sigma}\right)^3 \tag{14.13}$$

Using Eqs. (14.5) and (14.11), and noting that $D_{\text{diff}} = D_0(\exp\text{-}Q/T)$, there is an additional relation for A in terms of the mobilities' ratio:

$$A = \left(\frac{\alpha}{1+\alpha}\right)^3 \tag{14.14}$$

A plot of α versus $1/T$ is shown in Fig. 14.11. Despite the great scatter, it may be inferred from the plot that, at T $> \sim 1500$ °C, $\alpha > 1$ and about ≤ 1 at temperatures below 1500 °C. From the definition of α (Eq. (14.9)), this means that the glide mobility of dislocations is controlling the deformation when $\alpha < 1$; and for values >1, it means that dislocation glide becomes easier than climb, which is the limiting step. In a mathematical analysis using Burton's theoretical consideration, the creep mechanism at temperatures below and above T $= 1500$ °C is corroborated as being solute-drag-controlled below 1500 °C and controlled by climb above it—being the main mechanisms controlling creep in 21 mol% Y$_2$O$_3$ fully stabilized ZrO$_2$ single

Fig. 14.11 α, ratio of mobilities versus $1/T$ for the 21 mol%-Y-FSZ. Rodriguez et al. [23]. With kind permission of Elsevier

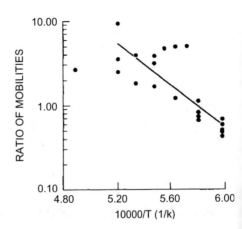

crystals. In conclusion, note that the 21 mol% zirconia is more creep resistant than the ~ 9 mol% stabilized zirconia.

14.4 Zirconia-Based Composite

The main reason for alloying ceramics (or materials in general) is to enhance mechanical strength or to improve other physical properties of materials. The additives used to do so are either in solution or as a separate phase. In most cases, the quantities of the additives used dictate if they are going to be applied in solution or as a second-phase precipitate. Clearly, the outcome of such additives to a material, whether in solution or precipitated, depends on the phase relations between the components of each specific ceramic. Virtually unlimited possibilities exist for the addition and combination of these additives. Therefore, numerous experimental investigations have been and are being performed in order to achieve this practical end, as well for the sake of pure research and knowledge.

In the following section, a good example of a superplastic composite is presented and discussed—the zirconia alumina system, which is very popular and widely investigated.

14.4.1 Yttria-Stabilized Zirconia Alumina

Tensile creep was applied to a 3 mol% yttria-stabilized zirconia-20 wt% alumina, which also exhibits the characteristic of superplasticity. As in all the prior cases, the steady-state creep rate may be expressed by Eq. (4.2), reproduced here for convenience:

$$\dot{\varepsilon} = \frac{AD\,Gb}{kT} \left(\frac{b}{d}\right)^p \left(\frac{\sigma}{G}\right)^n \tag{4.2}$$

Constant stress in the 4–100 MPa range was applied in 1600–1750 K temperature range and the test was performed in air. The experimental technique used at a constant temperature was either the application of a single stress or of changing stresses. The specimens were two-phase materials, of zirconia and alumina. The equiaxed grains of the as-received specimens were preserved after annealing and had grain sizes dependent on temperature and annealing time. An example of equiaxed grain sizes in the microstructure appears in Fig. 14.12. The variation of strain rate with strain is shown for some specimens in Fig. 14.13. In this figure, L stands for the grain size and the subscripts, z and a stand for zirconia and alumina, respectively. Here, up to $\sim 10\%$ strain represents the primary (transient) creep, which is followed by the steady state, after the decrease in the strain.

Fig. 14.12 Scanning electron micrograph of the zirconia-20% alumina composite with a grain size of 0.7 μm. Owen and Chokshi [20]. With kind permission of Springer

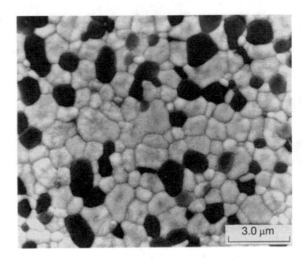

Fig. 14.13 Variation in strain rate with strain for the ZRO$_2$-20% AI$_2$O$_3$ composite tested at 1665 $\bar{L}_z = \bar{L}_a = 0.7$μm. σ (MPa): (□) 60, (◇) 30, (△) 14, (○) 4.5. Owen and Chokshi [20]. With kind permission of Springer

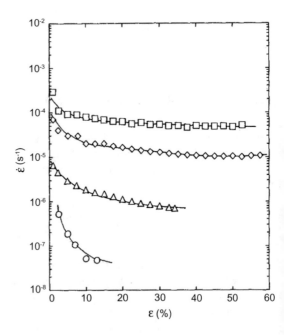

The variation in the steady-state creep under stress is illustrated in Fig. 14.14. Note that the single-stress test results and the stress-change experiments fall on the same line, indicating a good agreement between the two types of experiments with regard to the stress application. The obtained stress exponent in Fig. 14.14 is $n = 2.8 \pm 0.2$ and it does not change over the strain-rate range $\sim 10^{-8}$–10^{-4} s^{-1}. The effect of temperature may be obtained, as usual, from an Arrhenius-type

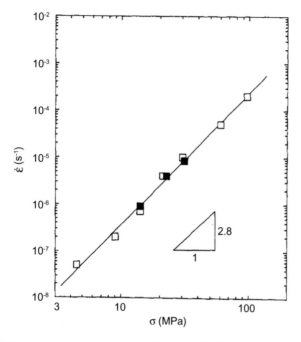

Fig. 14.14 Variation in steady-state creep rate with stress for the composite tested at 1665 K. $\bar{L}_z = \bar{L}_a = 0.7\mu m$. (□) single σ, (■) σ change. Owen and Chokshi [20]. With kind permission of Springer

Fig. 14.15 Arrhenius plot of the variation in strain with inverse temperature for the ZrO$_2$-20% Al$_2$O$_3$ composite. (Δ) $\sigma = 96$ MPa, $Q_{app} = 570 \pm 10$ kJ mol^{-1}; (□) $\sigma = 21$ MPa, $Q_{app} = 550 \pm 25$ kJ mol^{-1}. Owen and Chokshi [20]. With kind permission of Springer

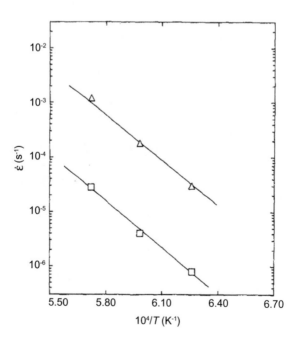

Fig. 14.16 Variation in
steady-state strain rate with
grain size for the composite
tested at 1665 K: the grain
size may be defined in terms
of the zirconia or alumina
phases or their volume
average, $\sigma = 21$ MPa. p: ($\square$)
ZrO₂, 2.0 ± 0.1; (Δ) Al₂O₃,
2.6 ± 0.6; (O) Av. 2.1 + 0.1.
Owen and Chokshi [20]. With
kind permission of Springer

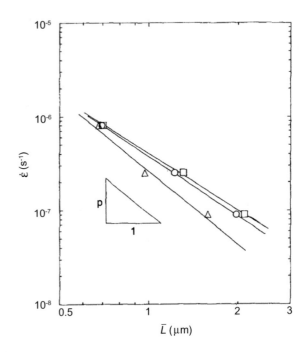

relation, shown in Fig. 14.15. Constant stresses of 21 or 896 MPa were applied at temperatures of 1598, 1665, and 1748 K, while the grain size was at a constant level of $\bar{L}_z = \bar{L}_a = 0.7$ μm. Notice that increasing the temperature increases the creep rate, as expected. The activation energies calculated for 21 and 96 MPa are 550 ± 25 and 570 ± 10 kJ mol^{-1}, respectively.

From the variation of the strain rate with grain size, the grain-size exponent may be evaluated at constant stress. A plot of strain rate versus grain size on a logarithmic scale (Eq. 4.2) results in Fig. 14.16. The average grain-size exponent, p, is 2.1, as indicated in Fig. 14.16. Additional experiments at 96 MPa revealed that $p = 2$ (Fig. 14.17).

Normalizing the data by means of the shear modulus taken from literature, a plot can be obtained. For this purpose, the experimental values are: slope = 2.8, $p = 2$, and $Q = 285$ (averages of a single temperature and temperature cycling are not shown) and the use of Eq. (2.4) yield the plot shown in Fig. 14.18. In the above, the volume averages for zirconia and alumina were used for $\bar{L}_x$. It is possible to write a constitutive equation for creep in the zirconia 20 wt% alumina composite as

$$\dot{\varepsilon} = (3.3 \pm 0.08) \times 10^8 \left(\frac{Gb}{kT}\right)\left(\frac{b}{\bar{L}}\right)^{2.\pm0.1} \times \left(\frac{\sigma}{G}\right)^{2.8\pm0.2} \exp\left[\left(-\frac{585 \pm 45}{RT}\right)\right]$$

$$(14.15)$$

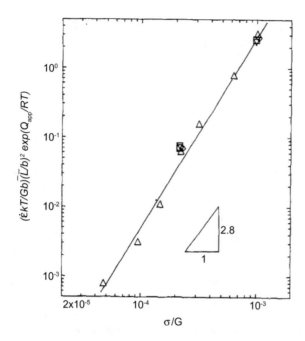

Fig. 14.17 Variation in the temperature and grain-size-compensated strain rate with the stress normalized by the shear modulus. T (k): ($\square$, ∇, $\bigcirc$) 1598, (Δ) 1665, ($\Diamond$) 1748. $\bar{L}_x$ = (μm): ($\square$) 0.7, (∇) 1.3, ($\bigcirc$) 2.1, (Δ, $\Diamond$) 0.7

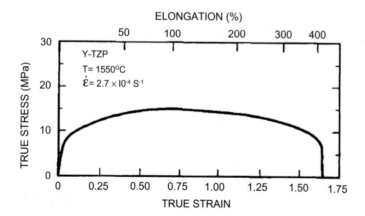

Fig. 14.18 True stress-true strain curve for Y-TZP tested at 1550 °C and a true strain rate of 2.7×10^{-4} s^{-1}. The flow stress exhibits a plateau over an extended range of strain. Nieh et al. [19]. With kind permission of Elsevier

By eliminating various possible creep mechanisms (Owen and Chokshi) and based on the experimental data, such as n, p, activation energy magnitude, and by preserving the equiaxed grains following creep without grain growth, and by not observing significant dislocation activity, the conclusion, based on a detailed analysis of the possible mechanisms, is that creep, in this case, occurred due to a GBS/grain rearrangement process.

14.5   Superplasticity in Zirconia

Among the superplastic ceramics, ZrO$_2$ with Y$_2$O$_3$ additive also exhibits superplastic behavior. The principal requirement is a very fine grain-sized material on the order of <10–20 μm. The growth of the grains must be retarded as much as possible during the forming temperatures. One of the means for preventing grain growth is by including fine, second-phase particles and distributing them uniformly, so that they pin the grain boundaries. Another structural requirement is grain-boundary mobility, to accommodate strain changes without grain-boundary separation under tension. Despite the many ceramics that fulfill the microstructural requirements, only certain ceramics classify as "superplastic materials." Usually, the temperatures required for deformation are above the homologous temperature, and not only compressive but also tensile deformation may be involved. An additional requirement is having proper strain-rate sensitivity, m—namely, $m > 0.3$. Recall that m appears as an exponent in equation

$$\sigma = k\dot{\varepsilon}^m \tag{14.16}$$

This equation is related to Eq. (8.7), given as

$$\dot{\varepsilon} = A\sigma^n \tag{8.7}$$

From this expression

$$\sigma^{\frac{1}{n}} = A^{-\frac{1}{n}}\dot{\varepsilon}^{\frac{1}{n}} = k\dot{\varepsilon}^m \tag{8.7a}$$

Here, $k = A^{-1/n}$ and m is the reciprocal of n.

Another feature of superplasticity is that the deformed material gets thinner in a very uniform manner, without forming a neck during tensile deformation (see Fig. 14.19). The first to report superplasticity in Y-TZP were Wakai et al. [30] who observed superplastic behavior in fine-grained Y-TZP by both tensile and compressive deformation. They indicated a 200% elongation at 1450 °C at a strain rate of 2.8 × 10^{-4} s^{-1}. Nieh shows an elongation >160% in his plot, which is seen in Fig. 14.18. A comparison between the undeformed and superplastically deformed specimens is illustrated in Fig. 14.19. The strain rate versus the flow stress is shown in Fig. 14.20. Recall that the strain-rate sensitivity, m, is the reciprocal of the stress exponent, n, which is defined at constant ε as

Fig. 14.19 Undeformed and superplastically deformed Y-TZP specimens. An elongation of over 350% is noted. Nieh et al. [19]. With kind permission of Elsevier

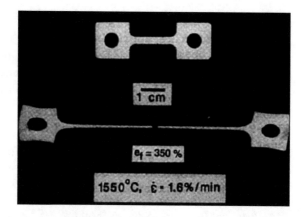

Fig. 14.20 True strain rate as a function of true flow stress (at ε—0.4) for Y-TZP at 1450 °C. The stress exponent is calculated to be approximately 3. For a direct comparison, the data from Wakai et al. are included. Nieh et al. [19]. With kind permission of Elsevier

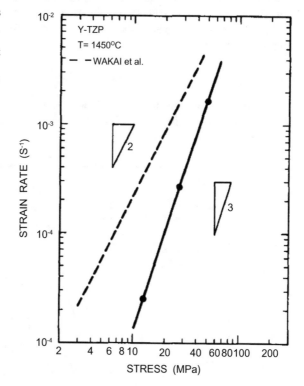

$$n = \frac{d \ln \dot{\varepsilon}}{d \ln \sigma} \qquad (14.17)$$

A stress exponent of 3 was evaluated, compared to the value of 2 obtained by Wakai et al. [30]

14.5.1   Superplasticity in Pure ZrO$_2$

Very limited information, if any, exists on superplasticity in pure ZrO$_2$. This is likely because of its instability under certain conditions. However, one could mention an early work by Hart and Chaklader, who did explore superplasticity in pure zirconia. The total strain that may be produced was found to depend on the density of the ZrO$_2$. In order to evaluate superplasticity, creep deformation was carried out with a three-point bending device. Due to its very fast transformation, monoclinic-to-tetragonal (between 1160 and 1205 °C), this creep deformation was carried out both below and above the transformation temperatures under isothermal conditions. Several typical deflection-temperature plots are shown in Fig. 14.21 for zirconia specimens having relative densities between 0.840 and 0.915 (higher densities over 95% could not be prepared due to specimen cracking). It was assumed that deformation in zirconia during the phase transformation is associated with GBS and that porosity combined with enhanced grain ductility produces greater deformation.

The total strain depends on the relative density, which may be described by the empirical relation

$$\varepsilon_{\text{total}} = A \exp(BP) \tag{14.18}$$

A and B are constants and P is the porosity. This is shown in Fig. 14.22, where the log of the total deflection versus the porosity is shown.

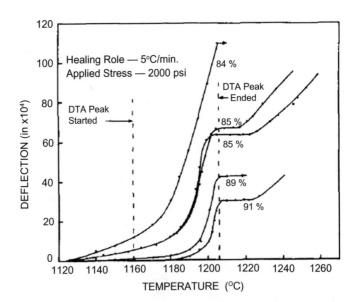

Fig. 14.21 Creep deformation of zirconia of different relative densities at a constant heating rate (5 °C/min.). Note that no creep occurred during 1205–1225 °C (just above the transformation temperature range 1160–1205 °C). Hart and Chaklader [14]. With kind permission of Elsevier

Fig. 14.22 Semi-log plot of the total deflection (d_T) during the monoclinic-to-tetragonal transformation versus the porosity of zirconia. Data were corrected for the dimensional change associated with the phase transformation ($\Delta L \sim 3\%$). Hart and Chaklader [14]. With kind permission of Elsevier. Materials Research Bulletin

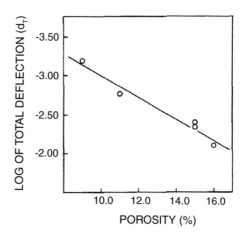

14.5.2 Stabilized Zirconia

As indicated earlier, there are three known phases: <1170 °C, 1170–2370 °C, and cubic >2370 °C. The addition of some other oxides, such as magnesium oxide (MgO, magnesia), yttrium oxide (Y_2O_3, yttria), calcium oxide (CaO, calcia), etc., stabilizes zirconia. Depending on the amount of the stabilizer, one may obtain PSZ, stabilized tetragonal zirconia or cubic zirconia. The main aspect of pure zirconia instability is the cracking on cooling due to volume change. The volume expansion caused by the cubic-to-tetragonal-to-monoclinic transformation induces large stresses, and these stresses cause ZrO_2 to crack upon cooling from high temperatures. By adding small percentages of yttria (the most popular stabilizer), these phase changes are eliminated and the resulting zirconia has superior thermal, mechanical, and electrical properties. Yttria-cubic stabilized zirconia (Y-CSZ) is formed in solid solution with zirconia at higher concentrations of yttria than the Y-TZP, which only has a metastable, tetragonal phase.

14.5.2.1 Partially Stabilized Zirconia (PSZ)

The addition of about 3 mol% (~ 5 wt%) yttria partially stabilizes polycrystalline zirconia. Compression of the fine-grained PSZ in the 1220–1330 °C temperature range—where the stability range is of the tetragonal phase—provided almost 100% strain. Plots of stress versus strain at an initial strain rate of 2.4×10^{-5} s^{-1} at several temperatures are illustrated in Fig. 14.23. At higher temperatures of 1297 and 1327 °C, however, at the strain rates indicated, strains approaching 100% are observed in Fig. 14.24.

For the activation energy, one of the relations presented earlier, is rewritten somewhat differently here as:

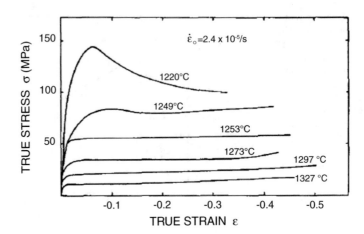

Fig. 14.23 True stress versus true strain curves. Temperature influence for an initial strain rate, $\dot{\varepsilon}_0 = 2.4 \times 10^{-5}\,\text{s}^{-1}$. Duclos et al. [8]. With kind permission of Elsevier. Materials Research Bulletin

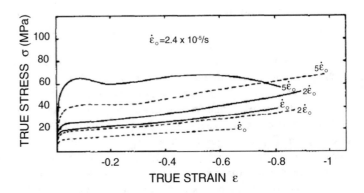

Fig. 14.24 True stress versus true strain curves at 1297 and 1327 °C: strain rate influence. *Full lines*, $T = 1297$ °C. *Dotted lines*, $T = 1327$ °C. Duclos et al. [8]. With kind permission of Elsevier. Materials Research Bulletin

$$\dot{\varepsilon} = \dot{\varepsilon}_0 \left(\frac{\sigma}{\mu}\right)^n \exp\left(-\frac{Q}{kT}\right) \tag{14.19}$$

and then rearranged as

$$\frac{\sigma}{\mu} = \left(\frac{\dot{\varepsilon}}{\dot{\varepsilon}_0}\right)^{1/n} exp\left(\frac{Q}{nkT}\right) \tag{14.20}$$

From the slope of the stress versus the inverse temperature, at an initial strain of $2.4 \times 10^{-5}\,\text{s}^{-1}$, the activation energy may be evaluated. The slope of $Q/n = 510$

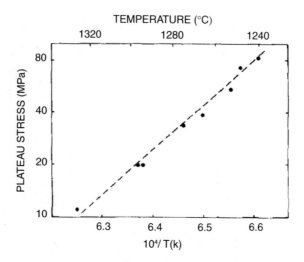

Fig. 14.25 Temperature dependence of the plateau stress for an initial strain rate $\dot{\varepsilon}_0 = 2.4 \times 10^{-5}\,\mathrm{s}^{-1}$. Duclos et al. [8]. With kind permission of Elsevier

kJ mol^{-1}. n was determined to be 1.1 ± 0.1 from the strain-rate changes at constant temperature in the 1240–1327 °C range. Derived from the values of Q/n and n, the activation energy is 570 ± 50 kJ mol^{-1}. The grain size was stable and the equiaxed shape was retained even under strains of $\sim 100\%$. No elongation of the grains occurred. In Fig. 14.26, a TEM microstructural micrograph indicates grain-size stability and shape.

The deformed specimens showed cavitation at triple-point junctions, but it was not a general phenomenon and has taken place only at low temperatures or at high strain rates. The conditions of deformation are listed in Table 14.3, including the densities. Cavitation may be observed in Fig. 14.27 at the external surface in a specimen strained to 40% at 1223 °C having an initial strain rate of $2.4 \times 10^{-5}\,\mathrm{s}^{-1}$.

Inside the specimen (from Fig. 14.27), the TEM illustration in Fig. 14.28 shows cavitation at triple points, likely initiated in the glassy phase. Cavitation gradually disappeared as the temperature increased. At the highest experimental temperatures, 1297 and 1327 °C, only weak cavitation was observed with the fastest strain rate ($1.2 \times 10^{-4}\,\mathrm{s}^{-1}$), as indicated in Fig. 14.28. Furthermore, note that no dislocations are observed in the grains, which are, therefore, dislocation free (Fig. 14.29).

In PSZ, the large strain is a consequence of GBS controlled by diffusion, while the dislocation contribution is claimed to be negligible. Despite the cavity formation, cavities were not involved in the fracture, not even at 100% strain, since diffusion was acting to relax the stress concentration and making the cavity shape less sharp (i.e., blunting (rounding)) it. Grain size and shape stability were maintained during and after deformation.

Since only limited research on superplasticity in PSZ is available (to the author's best knowledge), no confirmation of the claimed mechanism may be made at this time.

Fig. 14.26 TEM of a sample strained to 97% at 1327 °C. $l_0 = 1.2 \times 10^{-4}$/s. Scale bar = 0.5 μm. Duclos et al. [8]. With kind permission of Elsevier

Table 14.3 Deformation condition influence on the density. Duclos et al. [8]. With kind permission of Elsevier

Test temperature (°C)	Final strain (%)	Initial strain rate (10^{-5}/s)	Density
As sintered	–	–	5.98
1223	40	2.4	5.76
1240	42	2.4	5.92
1275	45	2.4	5.98
1297	45	2.4	6.00
1297	80	2.4	5.98
1297	85	4.8	5.94
1297	82	12	5.83
1327	63	2.4	5.97
1327	86	4.8	5.98
1327	99	12	5.90

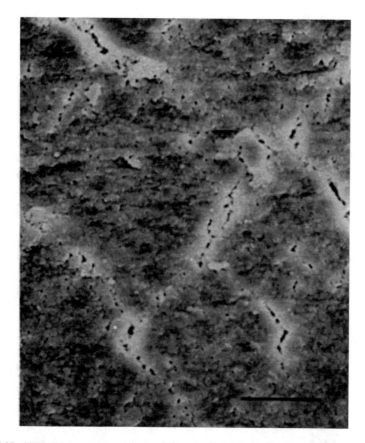

Fig. 14.27 SEM of the external surface of a sample strained to 40% at 1223 °C. $\dot{\varepsilon}_0 = 2.4 \times 10^{-5}\,\text{s}^{-1}$. Scale bar = 10 μm. Duclos et al. [8]. With kind permission of Elsevier

14.5.2.2 Tetragonally Stabilized Zirconia

In this section, the subject is superplasticity in tetragonal zirconia either without any dopants or glass-forming additives or only in small percentages. The objective is to consider superplasticity per se in yttria-stabilized materials without other influences on the superplastic behavior resulting from the addition of other components. Fine-grained Y-TZP is claimed to exhibit 700% elongation before fracture (Schissler et al.). The largest elongation-to-failure exhibited by a superplastic yttria-stabilized zirconia reported (Nieh et al.) is $\sim 800\%$ containing 3 mol% (~ 5 wt%) yttria. Those experiments conducted were in the 1623–1923 K temperature range with strain rates in the $\sim 10^{-5}$–10^{-3} range. The variation in the local true strain for specimens tested at 1823 K at different strain rates is illustrated in Fig. 14.30. On the plot, the local true strain versus normalized distance from the fracture tip is indicated. The stresses increased continuously until large true strains

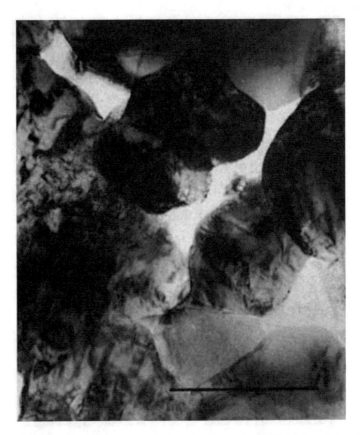

Fig. 14.28 TEM showing cavities in a sample strained to 40% at 1223 °C. $\dot{\varepsilon}_0 = 2.4 \times 10^{-5}\,\text{s}^{-1}$. Scale bar = 0.5 pm. Duclos et al. [8]. With kind permission of Elsevier

of ~ 1. At the same time, grain growth occurred, since the stress in superplastic ceramics is proportional to the grain size, according to

$$\sigma \propto d^a \tag{14.21}$$

In some cases, the grain size is symbolized by L. In Fig. 14.30, the variation in local true strain with normalized distance from the fracture tip is shown for 1823 K, as are the strain rates.

The local true strain is defined as $\ln(A_0/A_f)$, where A_0 and A_f are the initial and final areas).

The elongation-to-failure is ~ 320, 700, 350, and 175% at the strain rates indicated in Fig. 14.31, respectively $(2.7 \times 10^{-5}$, 8.3×10^{-5}, 2.7×10^{-4} and $1.7 \times 10^{-3})$. The variations of the grain aspect ratio with the local true strain for specimens tested at 1823 K and the strain rates of 2.7×10^{-5}–1.7×10^{-3} are presented in Fig. 14.31. All the points are quite close to the fitted line, with no

Fig. 14.29 TEM of a sample strained to 82% at 1297 °C. Cavities are indicated by *arrows*. $\dot{\varepsilon}_0 = 1.2 \times 10^{-4}\,\mathrm{s}^{-1}$. Scale bar = 1 μm. Duclos et al. [8]. With kind permission of Elsevier

definite indication that the aspect ratio varies with the true strain, and the value of the largest true strain is about 1.4. The variations (growth) in the average grain sizes at constant temperatures and various strain rates, and at CSR and various temperatures, are seen in Fig. 14.32.

Cavity formation during superplastic deformation in Y-TZP was observed in the microstructures. A SEM micrograph of a specimen tested to an elongation-to-failure of $\sim 115\%$ is illustrated in Fig. 14.33.

Note the cavities at the triple points. In Fig. 14.34, cavity stringers may be seen in a specimen pulled to an elongation of $\sim 700\%$ at 1823 K and at a strain rate of $8.3 \times 10^{-5}\,\mathrm{s}^{-1}$. In all the other specimens, where the elongation was <400%, extensive cavity interlinkage was observed perpendicular to the tensile axis. An illustrative example is found in Fig. 14.35. Some consequences of cavity interlinkage are crack formation and extension. In addition, grain growth was observed. Deformation-enhanced grain growth appears in Fig. 14.36.

Cavitation to about 30% may be observed in superplastic deformation at the optimum strain rate of $8.3 \times 10^{-5}\,\mathrm{s}^{-1}$, with a corresponding elongation of 30%.

Fig. 14.30 Variation in local true strain with normalized distance from the fracture tips. Schissler et al. [24]. With kind permission of Elsevier. Materials Research Bulletin

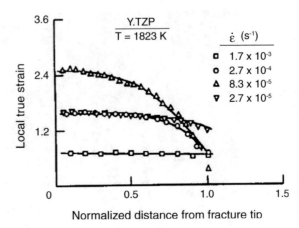

Fig. 14.31 Variation in grain aspect ratio with local true strain. Schissler et al. [24]. With kind permission of Elsevier

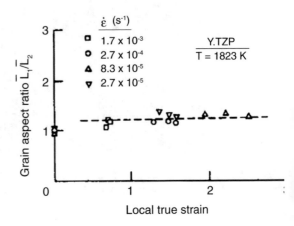

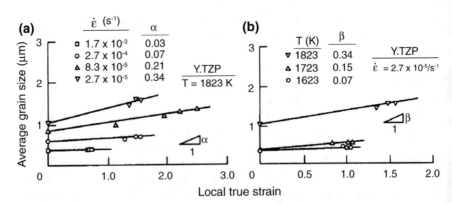

Fig. 14.32 Variation in average grain size with local true strain for (**a**) a fixed temperature of 1823 K and (**b**) a fixed strain rate of 2.7×10^{-5} s^{-1}. Schissler et al. [24]. With kind permission of Elsevier

Fig. 14.33 Scanning electron micrograph of a polished and thermally etched section of a specimen tested to an elongation-to-failure of ~115% at 1623 K and a strain rate of 2.7×10^{-5} s^{-1}. Schissler et al. [24]. With kind permission of Elsevier

Fig. 14.34 Optical micrograph illustrating the formation of cavity stringers in a specimen tested to an elongation-to-failure of ~700% at 1823 K and a strain rate of 8.3×10^{-5} s^{-1}. The tensile axis is horizontal. Schissler et al. [24]. With kind permission of Elsevier

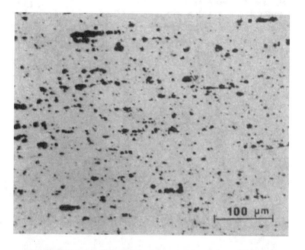

Fig. 14.35 Optical micrograph illustrating extensive cavity interlinkage perpendicular to the tensile axis in a specimen tested to an elongation-to-failure of ~150% at 1723 K and a strain rate of 2.7×10^{-5} s^{-1}. The tensile axis is horizontal. Schissler et al. [24]. With kind permission of Elsevier

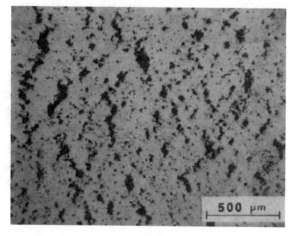

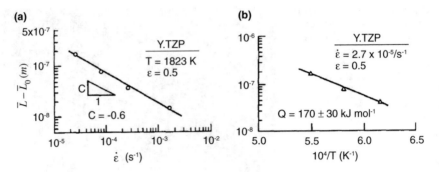

Fig. 14.36 Variation in deformation-enhanced grain growth with (**a**) strain rate and (**b**) temperature. Schissler et al. [24]. With kind permission of Elsevier

Fig. 14.37 Variation in cavitation area fraction with local true strain for specimens tested at 1823 K. Schissler et al. [24]. With kind permission of Elsevier

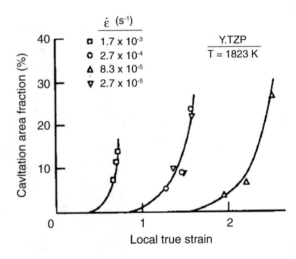

The variation in cavity formation with true strain is illustrated in Fig. 14.37. A summary of the experimental observations regarding Y-TZP follows:

(a) An optimum elongation may be achieved in fine-grained Y-TZP at 1823 K and at a strain rate of 8.3×10^{-5} s^{-1} with a grain size of ~ 0.3 μm;
(b) Superplastic deformation is accompanied by grain growth;
(c) Cavity formation to $\sim 30\%$ occurs mostly at triple points; and
(d) Cavity interlinkage occurs in a direction perpendicular to the tensile axis.

   Thus, superplastic elongation is enhanced by: (i) high rate sensitivity; (ii) limited simultaneous grain growth; (iii) reduced cavity formation; and (iv) hindered cavity interlinkage transverse to the tensile axis. Based on Fig. 14.36, the grain growth due to deformation in a superplastic ceramic may be expressed for:

$$\text{strain - dependent grain growth:} \quad \bar{L} - \bar{L}_0 \propto \dot{\varepsilon}^{-0.6} \tag{14.22}$$

$$\text{temperature - dependent grain growth:} \quad \bar{L} - \bar{L}_0 \propto \left(-\frac{170000}{RT} \right) \tag{14.23}$$

Using the experimental results from Figs. 14.32 and 14.36, the kinetics of deformation-enhanced simultaneous grain growth may be given as

$$\bar{L} - \bar{L}_0 = \varLambda \varepsilon \dot{\varepsilon}^{-0.6} \exp\left(-\frac{170000}{RT} \right) \text{kJ mol}^{-1}, \tag{14.24}$$

where $\varLambda$ is a constant.

14.5.2.3 Cubic Stabilized Zirconia (CSZ)

As early as 1998, Evans et al. claimed that superplasticity had not been observed in 8-mol%-yttria-stabilized cubic zirconia (8Y-CSZ), probably due to its larger grain size and high grain-growth rates. Doping with glassy phases or other additives can induce superplasticity in CSZ (which is a technologically important material for fuel cell applications). For example, pure and 5 wt% colloidal silica additives were used to produce superplasticity in cubic zirconia. In this manner, 180% true strain (505% engineering strain) could be achieved within 1 h at 1500 °C. For the purpose of comparison, several specimens were prepared in addition to the 5 wt% colloidal silica: pure 8Y-CSZ; 1 wt% colloidal silica; and 1 wt% borosilicate glass (composition: 83.3 mol% SiO_2, 1.5 mol% Al_2O_3, 11.2 mol% B_2O_3, 3.6 mol% Na_2O_3, and 0.4 mol% K_2O). Tests were performed in the 1300–1500 °C temperature range under compression. The objective of these additives was to limit grain growth during the deformation while enhancing GBS. The addition of appropriate intergranularly located phases (additives) can modify grain growth not only by affecting grain-boundary mobility, but also the interfacial energy. Grain growth may be expressed as

$$d^n - d_0^n = 2M\gamma\Omega t \tag{14.25}$$

Here, d is the instantaneous grain size at time t, d_0 is the initial grain size at $t = 0$, n is the grain-growth exponent, M is mobility, γ is grain-boundary energy, and Ω is the atomic volume. The intergranular second phase may enhance superplasticity by improving the resistance against cavity nucleation and by inducing GBS and rotation. The silicate added in these experiments is amorphous and its purpose is to refine the grain size and promote GBS in the Y-CSZ. A comparison of the influences of the above additives on grain size is found in Fig. 14.38. It is obvious that the 5 and 1 wt% SiO_2s are the most effective in grain-size reduction. The microstructures of the compared additives appear in Fig. 14.39. These microstructures clearly indicate that the most effective grain-size reduction is

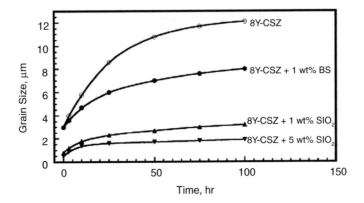

Fig. 14.38 Comparison of the grain growth of pure 8Y-CSZ to those containing 1 wt% borosilicate glass, 1 wt% silica, and 5 wt% silica at 1400 °C. Sharif and Mecartney [26]. With kind permission of Elsevier

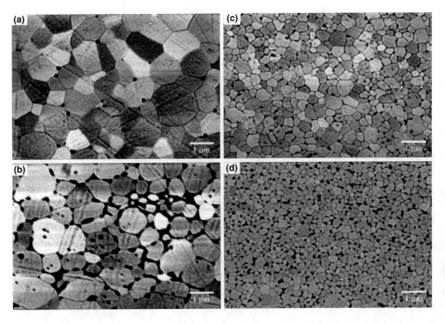

Fig. 14.39 Comparison of the initial microstructures (**a**) pure 8Y-CSZ, (**b**) 8Y-CSZ with 1 wt% borosilicate glass, (**c**) 8Y-CSZ with 1 wt% colloidal silica, and (**d**) 8Y-CSZ with 5 wt% colloidal silica. Sharif and Mecartney [26]. With kind permission of Elsevier

achieved by the SiO₂ additives. The greatest amount of pure silica was the most effective in limiting grain growth. High-temperature (1400 °C) deformation is also compared between the specimens with the same components indicated above and pure Y-CSZ by means of plots of strain rate versus stress, as shown in Fig. 14.40.

Fig. 14.40 Comparison of the steady-state strain rates of various samples at 1400 °C. Sharif and Mecartney [27]. With kind permission of Elsevier

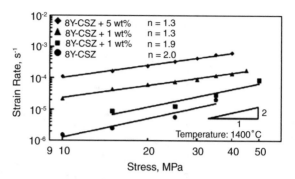

The strain rate of borosilicate glass (■) is only slightly enhanced, compared to the strain rate of pure 8Y-CSZ. Strain rates for the 5 wt% silica samples were about two orders of magnitude greater than those in pure 8Y-CSZ. As indicated in earlier equations, the steady-state strain rate (or creep rate) is reproduced here as

$$\dot{\varepsilon} = A d^{-p} \sigma^n \exp\left(-\frac{Q}{RT}\right) \tag{14.26}$$

One immediately recognizes that Eq. (14.26) is equivalent to Eq. (8.3), when A from Eq. (14.26) is made equal to

$$\frac{A' D_0 G b}{kT} \frac{b^p}{G^n} \tag{14.26a}$$

To eliminate confusion, the A taken from Eq. (8.4) is indicated in Eq. (14.26a) as A'. Here, p is the inverse grain-size exponent, n is the stress exponent, and G is the shear modulus. A good example of plastic deformation is that of a cylinder of Y-CSZ with 5% SiO_2. The stress exponent may be determined from the slope of the plot of the strain rate versus stress, as shown in Fig. 14.42. The stress exponents are included in Fig. 14.42 for the indicated temperatures. The activation energy was calculated from the slope of the strain-rate plot versus the inverse temperature (Fig. 14.42). It was observed that the activation energy increases with increasing stress from 341 kJ mol^{-1} at 10 MPa to 411 kJ mol^{-1} at 45 MPa.

Fig. 14.41 Sample of 8Y-CSZ + 5 wt% silica before deformation and after deformation at 1450 °C. Sharif and Mecartney [27]. With kind permission of Elsevier

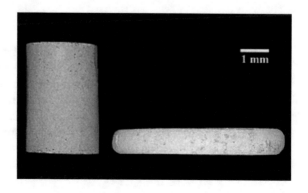

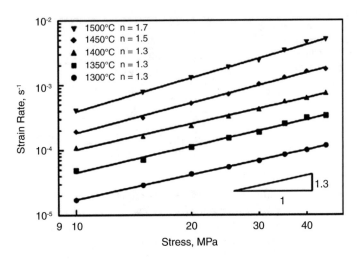

Fig. 14.42 High-temperature deformation of 8Y-CSZ + 5 wt% silica at 1300–1500 °C and calculated stress exponents. Sharif and Mecartney [26]. With kind permission of Elsevier

The plots presented above provide the following observations: (i) there was much less grain growth in 5 wt% silica containing 8Y-CSZ than in pure 8Y-CSZ; (ii) high strain rates could be applied, such as 5 × 10^{-3} s^{-1}; (iii) the total true strains of 180%, namely the 505% engineering strain could be obtained during a 1 h compression with a 5 wt% SiO$_2$ additive; (iv) no significant strain hardening occurred, so flow was not hindered; and (v) the presence of the glassy phase promotes GBS, while minimizing grain growth. These essential features provide the methods for achieving superplasticity in ceramics.

In a recent publication (Shirooyeh et al.), the basics of the above results were confirmed. A maximum superplastic elongation of more than 500% was recorded at a testing temperature of 1703 K. Also, the presence of an amorphous second phase (5% colloidal SiO$_2$) is effective in limiting grain growth. However, the activation energy reported was 600–670 kJ mol^{-1}, unlike the values reported in the work of Sharif and Mecartney.

14.5.2.4   Composite Stabilized Zirconia

One of the most familiar additives to induce superplasticity in Y-CSZ is a stable single-phase material formed with high Y$_2$O$_3$ in solid solution in Zr$_2$O$_3$—alumina. The addition of finely dispersed alumina to zirconia meets some of the essential requirements for superplasticity, namely, it suppresses grain growth during sintering and high-temperature deformation and prevents cavity nucleation. The role of alumina is to pin the grain boundaries, thus limiting grain growth, while promoting a high deformation rate. Rapid grain growth after annealing is seen in Y-CSZ in Fig. 14.44. Compare Fig. 14.44 with Fig. 14.45 in regard to the limiting effect of

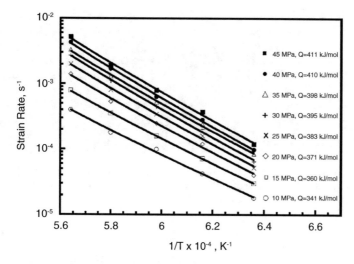

Fig. 14.43 Arrhenius-type plot for calculating activation energies for high-temperature deformation of 8Y-CSZ+5 wt% silica under 10–45 MPa compressive stress. Sharif and Mecartney [26]. With kind permission of Elsevier

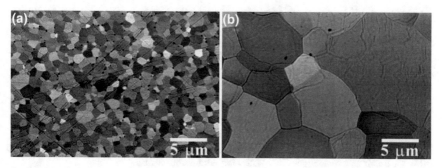

Fig. 14.44 Rapid static grain growth in 8Y-CSZ: (**a**) as hot isostatically pressed and (**b**) after annealing statically for 75 h at 1400 °C. Sharif and Mecartney [28]. With kind permission of Elsevier

the alumina on static grain growth after processing and annealing under the same conditions in both cases. The initial grain size of the samples containing 10 wt% alumina was 1.1 μm and they grew to only 2.2 μm after annealing at 1400 °C for 75 h (five times smaller than in the samples without alumina), as illustrated in Fig. 14.46. The deformation of the samples containing the alumina below a 35 MPa stress at temperatures above 1300 °C was quite uniform, and a specimen deformed at 10 MPa and 1450 °C is shown in Fig. 14.47. Dynamic grain growth after 115% strain at the high temperature of 1400 °C at 25 MPa is another example of high-temperature deformation seen in Fig. 14.48. Using Eq. (14.26), the activation energy for the rate-controlling mechanism may be calculated.

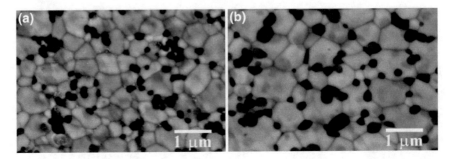

Fig. 14.45 Effect of addition of 10 wt% alumina on static grain growth of 8Y-CSZ (**a**) as hot isostatically pressed and (**b**) after annealing statically for 75 h at 1400 °C. Sharif and Mecartney [28]. With kind permission of Elsevier

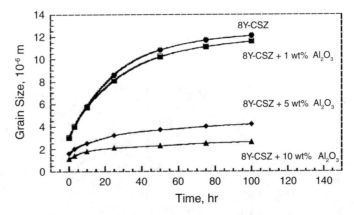

Fig. 14.46 Effect of addition of various amounts of alumina on static grain growth of 8Y-CSZ at 1400 °C. Sharif and Mecartney [28]. With kind permission of Elsevier

Fig. 14.47 8Y-CSZ+10 wt% alumina samples before and after creep at 1450 °C, 10 MPa. Sharif and Mecartney [28]. With kind permission of Elsevier

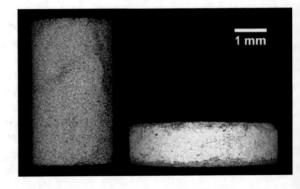

Fig. 14.48 Grain growth for
8Y-CSZ+10 wt% alumina
after total true strain of 115%
at 1400 °C, 25 MPa. Sharif
and Mecartney [28]. With
kind permission of Elsevier

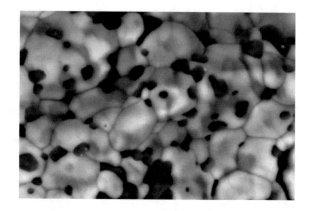

The values of p and n are required for the activation energy calculation according
to Eq. (14.26). The inverse grain-size exponent, p, may be calculated from the
slope of the line obtained by plotting the strain rate, $\dot{\varepsilon}$, versus d on a logarithmic scale.
Such a plot is shown in Fig. 14.49. A value of 2.2 was obtained for p. The stress
exponent, n, was calculated from the slope of the straight line obtained by plotting $\dot{\varepsilon}$
versus stress, as illustrated in Fig. 14.50. The stress exponent decreased with
increasing temperature from a value of 2.1 at 1300 °C to 1.7 at 1450 °C. The values
of n at each temperature are also indicated in Fig. 14.50. A plot of the strain rate, $\dot{\varepsilon}$
versus the inverse temperature (Fig. 14.51) enables the determination of the activation energy from the slope of the straight line. The values derived from the
Arrhenius plots at the indicated stresses vary from 683 kJ mol^{-1} at 10 MPa to 597
kJ mol^{-1} at 35 MPa. Thus, the activation energy decreases with increasing stress.

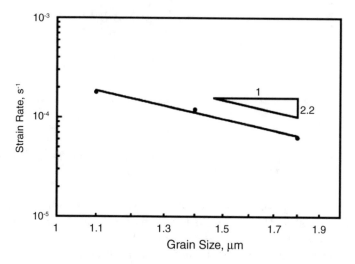

Fig. 14.49 Effect of grain size on steady-state strain rate of 8Y-CSZ+10 wt% alumina at 1400 °C,
20 MPa. Sharif and Mecartney [28]. With kind permission of Elsevier

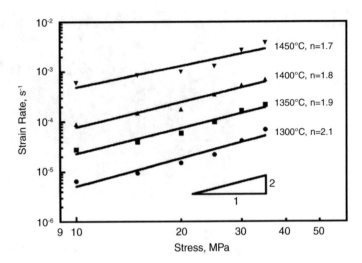

Fig. 14.50 Calculation of stress exponent for creep of 8Y-CSZ+10 wt% alumina at 1300–1450 °C. Sharif and Mecartney [28]. With kind permission of Elsevier

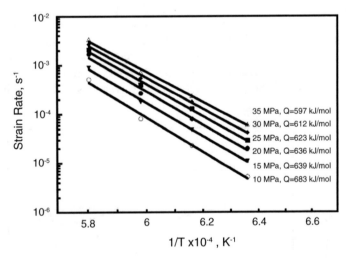

Fig. 14.51 Arrhenius-type plot for creep of 8Y-CSZ+10 wt% alumina at 10–35 MPa. Sharif and Mecartney [28]. With kind permission of Elsevier

Knowing n, one may calculate its inverse value to obtain m, which varies from 0.48 at 1350 °C to 0.59 at 1450 °C. See Eq. (14.16) for the relation between stress and m.

An increase in the value of m was observed with increasing temperature. Hence, it may be concluded that superplasticity in this ceramic is enhanced at higher temperatures. This indicates the diffusion dependence of the rate-controlling deformation mechanism. In fine-grained ceramics, it has been proposed that superplasticity occurs with GBS, accommodated by a diffusional mechanism. The total true strain of 130% true strain (260% engineering strain) was obtained at 1450 °C/10 MPa. The addition of finely dispersed inert phases may be an effective way to induce superplasticity in oxide ceramics. A similar mechanism was mentioned earlier for 3Y-TZP.

14.6 Creep in Nano-Zirconia

GBS remains a dominant deformation process during creep in both nanocrystalline and submicron-grained zirconia. It is believed that significant segregation at the grain boundaries reduces grain growth, the segregant being Y. It was observed that the grain-size-compensated diffusion-creep rate was essentially the same in cubic, tetragonal, and monoclinic zirconia. The commonly applied Eq. (4.2) is reproduced here for the reader's convenience and applied, in this case, to nanosized yttria-stabilized zirconia:

$$\dot{\varepsilon} = \frac{ADGb}{kT} \left(\frac{b}{d}\right)^p \left(\frac{\sigma}{G}\right)^n \tag{4.2}$$

Clearly, $D = D_0 \exp(-Q/RT)$. Deformation often occurs by diffusion creep, particularly when dislocation activity is limited. The diffusion of matter may occur via lattice (Nabarro-Herring creep) or along grain boundaries (Coble). In Fig. 14.52, the variations in strain rate with grain size for tetragonal and cubic zirconia are shown in the grain-size range of 0.5–10 μm. Also shown in this figure are data from nanocrystalline monoclinic zirconia. In Fig. 14.53, theoretical predictions for Coble creep are compared with tetragonal and monoclinic zirconia experimental data. The plot is of $\dot{\varepsilon}\bar{L}_0^3$ versus stress at 1273 K. $\bar{L}_0$ is the initial grain size in nanoscale. No difference in the creep behavior of nanocrystalline, tetragonal, and monoclinic zirconia is observed. The experimental data are consistent with the Coble creep model, using the grain-boundary diffusion coefficient for tetragonal zirconia (close to the tetragonal line), although there is trend toward nonlinearity.

Recall that grain-boundary diffusion is associated with Coble creep, according to Eq. (3.42); now below, it is expressed somewhat differently as

$$\dot{\varepsilon}_{\text{Coble}} = \frac{33\delta D_{\text{GB}}}{kT} \left(\frac{b}{d}\right)^3 \sigma \tag{14.27}$$

During high-temperature processes, grain growth may also occur. Grain growth may be expressed as

$$\bar{L}_f^N - \bar{L}_0^N = K_g t, \tag{14.28}$$

where L_f represents the grain sizes at time t, L_0 is the initial grain size, N is the grain-growth exponent, and K_g is a temperature-dependent constant. This grain growth is thermally activated and, thus

Fig. 14.52 The influence of
grain size on creep in cubic
(8Y), tetragonal (3Y) and
monoclinic (0Y) zirconia.
Chokshi [5]. With kind
permission of Elsevier

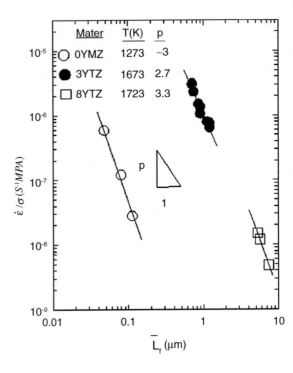

Fig. 14.53 Comparison of
theoretical models with data
for tetragonal (2.5Y) and
monoclinic (0Y) ZrO$_2$.
Chokshi [5]. With kind
permission of Elsevier

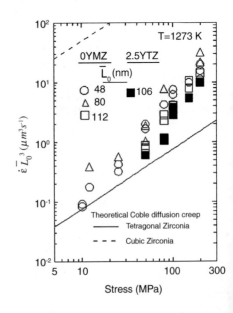

$$K_g = K_{0g} \exp\left(-\frac{Q_g}{RT}\right) \tag{14.29}$$

Q_g is the activation energy for grain growth. $N = 2$–3; both values give similar fits to the experimental data. The variation of K_g may be expressed by an Arrhenius plot, as in Fig. 14.54.

The creep rates and diffusion coefficients are very similar for cubic, tetragonal, and monoclinic zirconia; however, there is a very significant difference in the rates of grain growth. The tetragonal zirconia exhibits very slow grain growth compared with the cubic zirconia. This is due to the significant segregation of Y at the grain boundaries in tetragonal zirconia and the lack of such segregation in monoclinic and cubic zirconia. Furthermore, GBS is the creep mechanism in nanocrystalline and submicron-sized grains in zirconia. The level of segregation of Y at the grain boundaries is much reduced in nanocrystals, by about a factor of 2. Creep deformation may be expressed as indicated earlier by Eq. (4.2). In order to evaluate the transport in grain boundaries along the Coble lines, one must use the stress and grain-size exponents, $n = 1$ and $p = 3$ and $Q = Q_{GB}$, respectively.

The microstructures of nanosized specimens are seen in Fig. 14.55. The variations in the strain rate during stress testing performed at 1423 K for grain sizes in the 65–380 nm range are shown in Fig. 14.56. In addition, the theoretical Coble

Fig. 14.54 Variation in kinetic grain-growth constants versus inverse temperature for cubic (8Y), tetragonal (3Y), and monoclinic (0Y) zirconia. Chokshi [5]. With kind permission of Elsevier

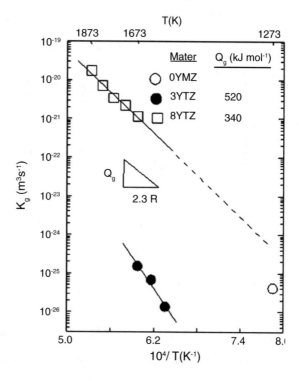

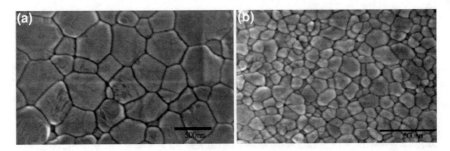

Fig. 14.55 Microstructures of the specimens used in this study with grain sizes of (**a**) 310 nm and (**b**) 83 nm. Ghosh and Chokshi [13]. With kind permission of Elsevier

curve for diffusion creep is indicated (*a*). The experimental data for all the grain sizes fall on a straight line with a slope of $n \approx 3$.

Since there have been very few studies thus far on creep in nanocrystalline zirconia, valuable supplementary experimental data are provided here, taken from experiments performed by various researchers listed in Fig. 14.57.

The variation of strain rate with grain size, if all data for grain sizes from 65 to 380 nm are taken as one set, may be seen in Fig. 14.58. If the data are divided by grain size, <120 nm and >120, then $p = 2 \pm 0.2$ and $p = 2.2 \pm 0.4$, respectively. According to the activation energy measurements, a value of ~ 550 kJ mol^{-1} was obtained for the nanocrystalline samples. Figure 14.59 is a plot of the strain rate versus the inverse temperature.

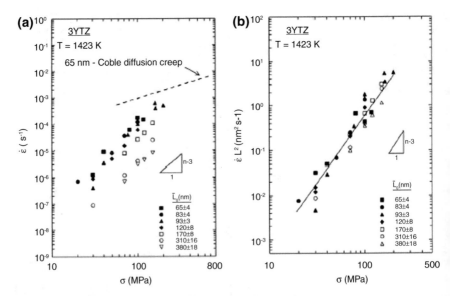

Fig. 14.56 **a** Variation in strain rate with stress for specimens with grain sizes ranging from 65 to 380 nm. **b** Grain-size-compensated creep rate versus stress for n- and c-3YTZ [yttria tetragonal zirconia] samples. Ghosh and Chokshi [13]. With kind permission of Elsevier

Fig. 14.57 Previous reports
on variation in creep rate with
stress in nano-zirconia.
Chokshi [6]. With kind
permission of Elsevier

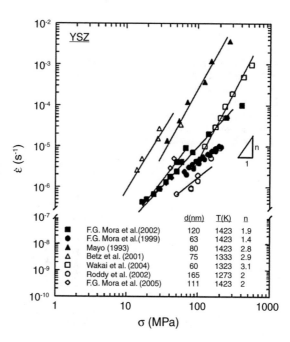

Fig. 14.58 Variation in
strain rate with grain size for
n- and c-3YTZ, giving an
inverse grain-size exponent of
$p \approx 2$ over the entire range of
grain sizes. Chokshi [6]. With
kind permission of Elsevier.
Note the data with symbol (■)
are from Gutierrez-Mora et al.
[17] and (●) also from
Gutierrez-Mora et al. [18]

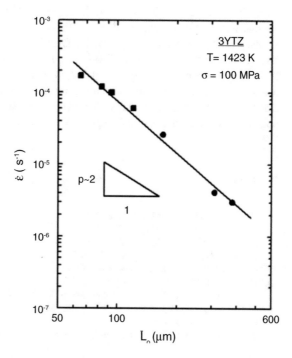

Fig. 14.59 Variation in
strain rate with inverse
temperature, giving an
activation energy of
540 kJ mol^{-1}. Chokshi [6].
With kind permission of
Elsevier Note that the
symbols (●) and (▲) are
from Gutierrez-Mora et al.
[17] and Mayo, respectively

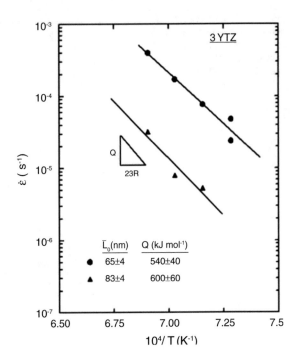

Grain shape was retained after creep, remaining equiaxed and the measurements indicate the large contribution of GBS to creep. This finding is similar to observations made of coarser grained 3Y-TZP, suggesting similar deformation mechanisms. The experimental creep data for samples with grain sizes between 65 and 380 nm appear in Fig. 14.56b, in terms of the variations in the grain-size-compensated creep rate (p = 2) with stress. There have been reports that nanocrystalline materials are more creep resistant than the common (macroscopic) structures, but this claim is not well proven, since it seems that nanomaterials are not more resistant to creep deformation Mora et al. [16].

References

1. Burton B (1982) Phil Mag A 45:657
2. Burton B (1982) Phil Mag A 46:607
3. Burton B (1990) Phil Mag Lett 62:383
4. Burton B (1989) Mater Sci Technol 5:1005
5. Chokshi AH (2003) Scripta Mater 48:791
6. Chokshi AH (2014) Scripta Mater 86:13
7. Dimos D, Kohlstedt DL (1978) J Am Ceram Soc 70:531
8. Duclos R, Crampon J, Amana B (1989) Acta Metal 37:877
9. Evans PE (1970) J Am Ceram Soc 53:365
10. Evans ND, Imamura PH, Mecartney ML, Bentley J (1998) ORNL/CP-97149, CONF-980713

11. Fernandez JM, Melendo MJ, Rodriguez AD, Heuer AH (1990) J Am Ceram Soc 73:2452
12. Garcia DG, Fernandez JM, Rodriguez AD, Eveno P, Castaing J (1996) Acta Mater 44:991
13. Ghosh S, Chokshi AH (2014) Scripta Mater 86:13
14. Hart JL, Chaklader ACD (1967) Mater Res Bull 2:521
15. Mayo MJ (1997) Nanostruct Mater 9:717
16. Gutierrez-Mora F, Jimenez-Melendo M, Dominguez-Rodriguez A (2005) J Am Ceram Soc 88 (6):1529
17. Gutierrez-Mora F, Jimenez-Melendo M, Dominguez-Rodriguez A (2002) J Eur Ceram Soc 22:2615
18. Gutierrez-Mora F, Jimenez-Melendo M, Dominguez-Rodriguez A, Chaim R, Heft M (1999) Nanostruct Mater 11:531
19. Nieh TG, McNally CM, Wadsworth J (1989) J Metals 41:31
20. Owen DM, Chokshi AH (1994) J Mater Sci 29:5467
21. Philibert J (1991) Atom movements: diffusion and mass transport in solids. Editions de physique, Les Ulis, p 504
22. Poirier JP (1985) Creep in crystals. Cambridge University Press, New York
23. Rodriguez AD, Eveno P, Castaing J (1996) Acta Mater 44:991
24. Schissler DJ, Chokshi AH, Nieh TG, Wadsworth J (1991) Acta Metal Mater 39:3227
25. Seltzer MS, Talty PK (1975) J Am Ceram Soc 58:124
26. Sharif AA, Mecartney ML (2003) Acta Mater 51:1633
27. Sharif AA, Mecartney ML (2003) Acta Materialia 51:1633
28. Sharif AA, Mecartney ML (2004) J Eur Ceram Soc 24:2041
29. Shirooyeh M, Dillon RP, Sosa SS, Imamura PH, Mecartney ML, Langdon TG (2015) J Mater Sci 50:3716
30. Wakai F, Sakaguchi S, Matsuno Y (1986) Adv Ceram Mater 1:259

Chapter 15
Creep in Silicon Carbide (SiC)

Abstract Silicon carbide (SiC) is a technologically important ceramic, due to its high hardness, optical properties, and thermal conductivity. The high strength of SiC is a consequence of the strong covalent bonds (similar to diamonds) providing resistance to high pressures. The properties of SiC, which are similar to those of diamonds, have opened the gem industry to this material for use as a possible diamond substitute. However, a very important application of SiC is in microelectromechanical systems (MEMS), such as in wide-band gap semiconductors and power semiconductors, due its inherent strength and durability. Creep in polycrystalline and single-crystal SiC is the subject of this chapter. SiC is reinforced with fibers among them SiC fibers. Creep rupture is evaluated in a section devoted to this subject. Superplasticity observed in SiC is also discussed.

15.1 Introduction

Silicon carbide (SiC) is a technologically important ceramic, due to its high hardness, optical properties, and thermal conductivity. The high strength of SiC is a consequence of the strong covalent bonds (similar to diamonds) providing resistance to high pressures. The properties of SiC, which are similar to those of diamonds, have opened the gem industry to this material for use as a possible diamond substitute. However, a very important application of SiC is in microelectromechanical systems (MEMS), such as in wide-band gap semiconductors and power semiconductors, due to its inherent strength and durability.

Among the strength properties, creep is an important gradient for high-temperature applications (the subject of this chapter), but the major use and many applications of SiC are as fiber, so consideration will also be given to its creep property as fiber.

© Springer International Publishing AG 2017   357
J. Pelleg, *Creep in Ceramics*, Solid Mechanics and Its Applications 241,
DOI 10.1007/978-3-319-50826-9_15

15.2   Creep in Polycrystals

It has been suggested (Krishnamachari and Notis) that creep in polycrystalline silicon carbide is a Coble creep, which, as previously stated, is a grain-boundary-diffusion creep mechanism. Creep tests in the 1573–1673 K temperature range and at stress levels of 34.47–86.19 MPa were performed by means of four-point bending tests. Creep deformation at two stresses is illustrated in Fig. 15.1.

In a four-point bending test, the stress-strain relation is usually evaluated from the elasticity, giving the outer-fiber stress by:

$$\sigma = 1.5 \frac{(L - a)P}{bh^2} \tag{15.1}$$

This formula clearly takes into account the specimen's dimensions; thus b is the sample width 0.683 cm (0.27 in.) and the total height h is 0.253 (0.1 in.). P is the applied load, L the distance between supporting points, 3.795 cm (1.5 in.), and a is the distance between load points, 1.265 (0.5 in.). The strain is given as:

$$\varepsilon = \frac{6hx}{(L - a)(L + 2a)} \tag{15.2}$$

The vertical point displacement is x. The strain rate is given (derivative with respect to time) from Eq. (15.2) as:

$$\dot{\varepsilon} = \frac{6h\dot{x}}{(L - a)(L + 2a)} \tag{15.3}$$

Fig. 15.1 Creep deformation of SiC. Krishnamachari and Notis [7]. With kind permission of Elsevier

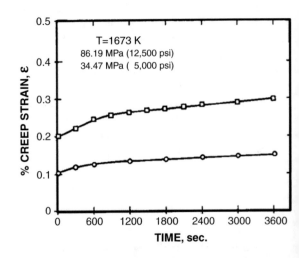

A plot of strain rate versus stress allows for the determination of the stress exponent, giving $n = 0.9 \pm 0.18$ (Fig. 15.2).

The temperature dependence of the creep rate in the usual Arrhenius-type plot is presented in Fig. 15.3 for four stresses. The apparent activation energy is $Q = (146.51 \pm 25.14)$ kJ mol^{-1}. Under the test conditions at the various temperature and stress levels, the creep in SiC may be described by the

Fig. 15.2 Stress dependence of creep rate in SiC. Krishnamachari and Notis [7]. With kind permission of Elsevier

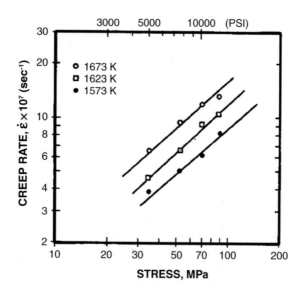

Fig. 15.3 Temperature dependence of creep rate in SiC. Krishnamachari and Notis [7]. With kind permission of Elsevier

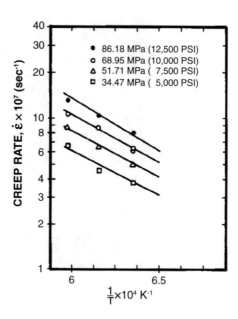

well-known mechanisms discussed earlier and reproduced here for the sake of convenience:

(1) Nabarro-Herring creep, which is lattice-diffusion controlled:

$$\dot{\varepsilon}_{NH} = \frac{13.3\Omega\sigma}{G^2 kT} D_{0l} \exp\left(-\frac{Q_l}{RT}\right) \tag{15.4}$$

(2) Coble-creep controlled by grain-boundary diffusion:

$$\dot{\varepsilon}_{\text{Coble}} = \frac{47.7W\Omega\sigma}{G^3 kT} D_{0gb} \exp\left(-\frac{Q_{gb}}{RT}\right) \tag{15.5}$$

(3) Dislocation creep:

$$\dot{\varepsilon}_d = \left(\frac{A\mu b}{kT}\right)\left(\frac{\sigma}{\mu}\right)^n D_{0d} \exp\left(-\frac{Q_d}{RT}\right) \tag{15.6}$$

In the familiar relations above, some of the symbols must be redefined in line with their manner of notation. Thus, here, G refers to grain size, W is grain-boundary width, A is Dorn's parameter, and μ is the shear modulus. The subscripts refer to lattice (l) and grain boundary (gb). Clearly, the product of $D_0 \exp\left(-\frac{Q}{RT}\right)$ refers to the relevant diffusion coefficient.

A fractured surface is illustrated in the SEM micrograph in Fig. 15.4. Some included particles are seen but, in general, the fractured surface is free of inclusions.

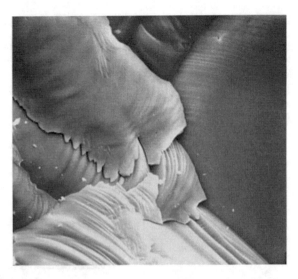

Fig. 15.4 SEM micrograph of fractured surface (×2000). Krishnamachari and Notis [7]. With kind permission of Elsevier

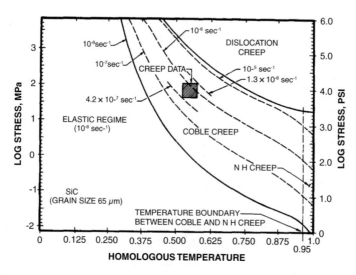

Fig. 15.5 Deformation map for SiC (grain size 65 μm). Krishnamachari and Notis [7]. With kind permission of Elsevier

There is no indication of a second phase at the grain boundaries; thus, grain-boundary diffusion is likely to be the creep-controlling mechanism. Deformation maps for SiC (originally suggested by Ashby) are presented in Fig. 15.5. CSR contours may be superimposed on the same stress-temperature space plot. In this plot, regions of the mentioned creep mechanisms are displayed. Solving Eqs. (15.4)–(15.6) for stress as a function of temperature is required in order to outline the boundaries of those mechanisms. Table 15.1 shows the values necessary to plot the SiC deformation map.

The values of $D_{gb}W$ for Coble diffusion, calculated from Eq. (15.5) and by a least-squares analysis, are plotted in Fig. 15.6 and provide the following relation:

$$D_{gb}W = 3.3x10^{-9}x \exp\left(-\frac{146.51 \pm 25.14\,\text{kJ}\quad\text{mol}^{-1}}{RT}\right) \quad (15.7)$$

One can see that the value of $D_{gb}W$ for C diffusion (shown in Farnsworth and Coble) is not in good agreement with the present data. They are, however, in reasonably good agreement with the extrapolated values for silicon or carbon self-diffusion coefficients in SiC. The activation energy for creep is about half that of Si self-diffusion in SiC, suggesting that Si-controlled grain-boundary diffusion is the most likely creep process in SiC. The above suggestion is based on the concepts: (1) the value of Coble creep is about half that of lattice diffusion (i.e., grain-boundary diffusion is about half that of lattice diffusion, namely, $Q_{gb}/Q_l \sim 0.5$; Atkinson and Monty [1]); and (2) the agreement between $D_{gb}W$ and the extrapolated values of Si self-diffusion shown in Fig. 15.6.

Due to the other controversies found in earlier publications (excluded in this book), it felt worthwhile to add (below) a more recent work on creep in SiC, so that

Table 15.1 Materials constants for SiC required for plotting a deformation map. Krishnamachari and Notis [7], Lane et al. [11]. With kind permission of Elsevier

Constant	Value
Vacancy volume, $\times 10^{23}$ (cm^{23})	2.08[a]
Grain houndary Width, $W \times 10^8$ (cm)	10 or 100[b]
Pre-exponential grain-houndary diffusion constant, D_{gb}° (Silicon) (cm^2/s)	3.24×10^{-2} or 3.24×10^{-3c}
Pre-exponential dislocation creep constant, $Ab \times 10^4$ (cm^2/s)	4.6[d]
Pre-exponential lattic diffusion constant for carbon diffusion, $D_L^{\circ} \times 10^2$ (cm^2/s)	3.0[e]
Limit between creep end elastic region, $\times 10^8$ (s^{-1})	1.0
Activation energy for grain boundary diffusion (Silicon), Q_{gb}. (kJ/mol)	161[f] (38.5 kcal/mol)
Activation energy for lattice diffusion of carbon, Q_L, (kJ/mol)	590.2[g] (141 kcal/mol)
Activation energy for dislocation creep (carbon diffusion), Q_d (kJ/mol)	590.2[h] (141 kcal/mol)
Melting paint, $T_m(^{\circ}$K)	2.973
Stress exponent for dislocation creep, n	4.5
Grain size G (μm)	65
Shear modulus, $\mu \times 10^5$ (MPa)	1.52 (2.2 $\times 10^7$ psi)

$a = A/\rho\, N_a$ (A: atomic weight, ρ: density, $N_a = 6.02 \times 10^{23}$)

b, c: 1. $W = 10$ Å (similar to most metallic systems [18]) and D_{gb}°(silicon) = 10 D_L° (silicon), where D_L° (silicon) =|3.24 $\times 10^{-3}$ cm^2/s obtained from the diffusion data of silicon reported by Goshtagore [9], or

2. $W = 100$ Å (typical for ceramics such as Al$_2$O$_3$, [18]) and D_{gb}° (silicon) = D_L° (silicon) = 3.24 $\times 10^{-3}$ (cm^2/s)

Note In both cases, D_{gb}° (silicon) W = 3.24 $\times 10^{-9}$ (cm^2/s) remains the same

d [17], e, g, h; Reference [1], f: $Q_{gb} = 1/2\, Q_L$ ([14. 15])

it might resolve the conflicting reports from the literature, shedding new light on creep in SiC.

15.3 Creep in Single-Crystal SiC

Of the six polymorphs of SiC, the creep properties of 6H α-SiC is considered in this section. Dislocation motion is the only mechanism available for macroscopic plastic deformations in SiC. At about 800 °C, Niihara suggests the onset of plasticity by basal slip on the (0001) <1120> plane as being the dominant system causing dislocation motion. Substantial dislocation mobility within the 850–900 °C range was observed around scratches in hot stage TEM. Very few compressive-creep data exist for SiC, but those that do indicate that single-crystal SiC is very creep resistant when stressed, so as not to activate basal slip; however, that basal-slip deformation can occur at relatively low temperatures when basal slip has been activated.

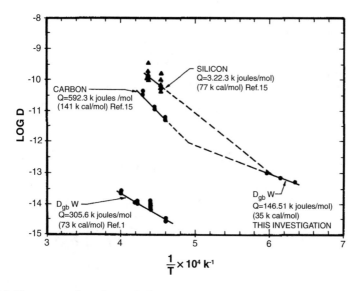

Fig. 15.6 Temperature dependence of effective diffusion coefficients. Krishnamachari and Notis [7]. With kind permission of Elsevier

The electronics-grade SiC specimens were arranged so that their long axes were parallel to the loading direction, aligned either parallel to [0001], or at 45° from it toward the $[11\bar{2}0]$ direction. (The Schmidt factors are 0 and 0.5 for $(0001)\langle 11\bar{2}0\rangle$ slip, respectively). The strain versus time plot is presented in Fig. 15.7. Whenever primary creep was present, its extension was not large, but the steady-state creep was readily observed. From the linear portion of the strain versus time plots, the steady-state creep rate was determined by means of a linear-regression analysis. The steady-state creep analysis of the experimental data was done according to Eq. (13.4). A multiple-regression routine was used to fit Eq. (13.4) to the creep data. (For those interested in multiple regression, the work of Newton and Spurrell may be consulted.) The measured creep rates and regression fit of the data (in Fig. 15.8) indicate that Eq. (13.4) describes the data quite well. The analysis of the flow-stress behavior for this alignment (45° from [0001]) gave a stress exponent of 3.1 ± 0.41 and an activation energy of 277 ± 24 kJ mol^{-1}. This activation-energy value is much lower than that of the self-diffusion of C or Si in SiC, being 714 and 697 kJ mol^{-1}, respectively. These values indicate that the mechanism controlling deformation is thermally assisted activation over the Peierl's-barrier stress. Recall that the Peierl's barrier is important for dislocation motion; here, it specifically refers to covalently bonded crystals, such as ceramics. This barrier is a result of the requirement to break and restore bonds at the core of dislocations every time a dislocation moves one atomic distance. Dislocations are thermally assisted to overcome barrier stresses.

The creep rates for the c-axis in SiC are shown in Fig. 15.9 and listed in Table 15.2. A regression analysis of this orientation yielded a stress exponent,

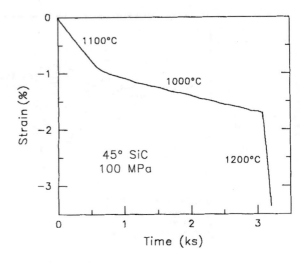

Fig. 15.7 Typical strain versus time behavior for the 45° single-crystal α-SiC creep specimens. Corman [8]. With kind permission of John Wiley and Sons

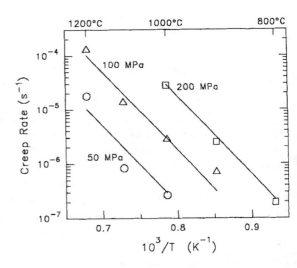

Fig. 15.8 Steady-state creep rate data for single-crystal α-SiC measured at 45° to [0001]. Corman [8]. With kind permission of John Wiley and Sons

$n = 4.93 \pm 0.16$, and an activation energy of 180 ± 27 kJ mol^{-1}, which is again smaller than the self-diffusion values of Si and C in SiC, but in agreement with creep data measured in SiC (Carter et al. [2]), obtained by the chemical vapor deposition (CVD) method (175 kJ mol^{-1}).

Despite the orientation of the specimen in the c direction [0001], where basal slip is not expected to occur, the expectation of good creep resistance was not met–creep resistance was poor. Creep rates versus inverse temperatures for several forms of SiC are shown in Fig. 15.10. The creep specimens were aligned 45° from [0001], where the creep resistance was expected to be low and it was thought that creep would occur by slip along the basal plane, as seen

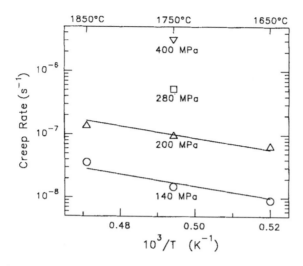

Fig. 15.9 Steady-state creep rate data for [0001] aligned single-crystal α-SiC. Corman [8]. With kind permission of John Wiley and Sons

Table 15.2 Steady-State Creep Rates for Single-Crystal α-SiC. Corman [8], Farnsworth and Coble [12]. With kind permission of John Wiley and Sons

Specimen	Orientation	Stress (MPa)	Temperature (°C)	Creep rate (s⁻¹)	Cumulative strain (%)
1	45°	100	1100	1.42×10^{-5}	0.90
			1000	2.91×10^{-6}	1.71
			1200	1.36×10^{-4}	3.36
2	45°	200	1000	2.80×10^{-5}	7.19
			900	2.50×10^{-6}	7.90
			800	1.95×10^{-7}	8.02
		100	900	7.29×10^{-7}	8.22
3	45°	50	1100	8.32×10^{-7}	0.41
			1000	2.65×10^{-7}	0.72
			1200	1.78×10^{-5}	2.69
4	[0001]	200	1750	9.90×10^{-8}	0.65
			1850	1.42×10^{-7}	1.37
			1650	6.70×10^{-8}	1.90
5	[0001]	400	1750	3.08×10^{-6}	1.46
6	[0001]	280	1750	5.25×10^{-7}	0.65
7	[0001]	140	1750	1.50×10^{-8}	0.28
			1650	9.15×10^{-9}	0.35
			1850	3.36×10^{-8}	0.62

in Fig. 15.11 (side surfaces). Slip traces and steps may also be seen parallel to the basal plane. In the case of specimens tested in the c direction (skewed), slip steps parallel to the basal planes were observed (see Fig. 15.12).

Fig. 15.10 Creep-rate data
for several forms of SiC,
interpolated or extrapolated to
200 MPa: single crystal is
[0001] alignment; other data
from Corman [8]. With kind
permission of John Wiley and
Sons

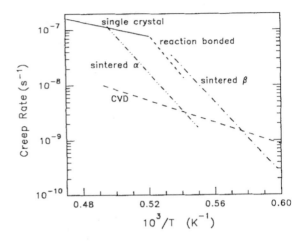

Fig. 15.11 Reflected-light
micrographs of the surfaces of
a 45° creep specimen: *top*
$\{10\bar{1}0\}$ type face, *bottom* face
45° between (0001) and
$\{11\bar{2}0\}$. The compression
axis is shown by the *arrows*.
Corman [8]. With kind
permission of John Wiley and
Sons

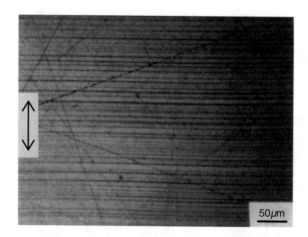

Fig. 15.12 Reflected-light micrograph of the side surface of a [0001] α-SiC creep specimen showing slip steps parallel to the basal plane. The *oblique lines* are polishing scratches. The [0001] compression axis is shown by the *arrows*. Corman [8]. With kind permission of John Wiley and Sons

From the experimental data and according to Eq. (13.4) one may write the following:

for specimens oriented 45° between [0001] and $[11\bar{2}0]$ directions, at 800–1200 °C under 50–200 MPa stresses:

$$\dot{\varepsilon} = 0.155\sigma^{3.32} \exp\left(-\frac{277\,\text{kJ} \quad \text{mol}^{-1}}{RT}\right) \tag{15.8}$$

and for stresses parallel to [0001] at temperatures of 1650–1850 °C under stresses of 140–400 MPa, the steady-state creep rate is:

$$\dot{\varepsilon} = 2.01 \times 10^{-14}\sigma^{4.91} \exp\left(-\frac{180\,\text{kJ} \quad \text{mol}^{-1}}{RT}\right) \tag{15.9}$$

The creep resistance is lower in single crystals than in both CVD SiC measured under the same conditions and in polycrystalline SiC. It is not exactly known why (under the same experimental conditions) there should be a difference in these results. In this author's opinion, grain boundaries, although they may be sites for sliding, they have a dual effect: (a) GBS, which does not favor good creep resistance, and (b) strengthening by blocking the easy passage of dislocations through the boundaries. The CVD SiC, although showing a preferred orientation, still has grains (and, consequently, grain boundaries), like polycrystalline SiC, so it is reasonable to assume that the strengthening effect may override the GBS, due to the greater creep resistance of the CVD together with the polycrystalline SiC (as opposed to single crystals).

15.4   Creep Results in SiC

15.4.1   Introduction

In a further attempt to resolve reported controversies regarding creep and creep mechanisms in SiC (a very useful structural material), the following text will present significant landmarks in the relevant experimental research, assuming that the differences in the reported results have been caused by the different shapes and sizes of the specimens, the various techniques involved, impurities, etc. Below, some additional creep results are considered regarding: CVD SiC, SiC fibers, and composite SiC (with additives to improve creep-resistant properties).

15.4.2   CVD SiC

To obtain measurable creep, the required stress and temperature must be chosen. Specimens should be aligned, so that the deposition direction is $45°$ to the applied stress axis. At this orientation, the highest resolved-shear stress on these planes provides measurable deformations at the stresses applied at the proper temperatures. The compressive-creep tests were performed in a protective nitrogen atmosphere, the purpose of which was to eliminate $\beta \rightarrow \alpha$ transformation. A typical creep curve is shown in Fig. 15.13. The steady-state creep curves versus $\log(\sigma/G)$ are shown in Fig. 15.14a, b at the temperatures indicated. The activation energy for steady-state creep is illustrated in Fig. 15.15a, b at the stresses indicated. The results indicate that Eq. (15.10) describes creep properly, resulting in 174 ± 5 kJ mol^{-1}:

$$\dot{\varepsilon} = A\left(\frac{\sigma}{G}\right)^n \exp\left(-\frac{Q}{kT}\right) \tag{15.10}$$

Fig. 15.13 Creep curve of CVD SiC (1923 K). Carter et al. [3]. With kind permission of John Wiley and Sons

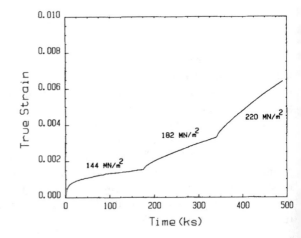

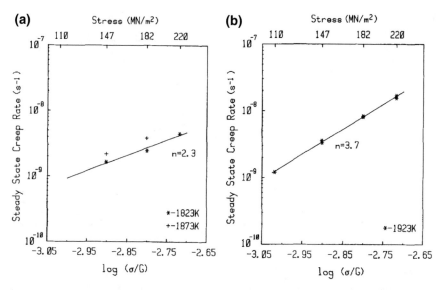

Fig. 15.14 a Steady-state creep rate versus log (σ/G) for CVD SiC samples crept at temperatures <1923 K; **b** SiC samples crept at 1923 K. Carter et al. [3]. With kind permission of John Wiley and Sons

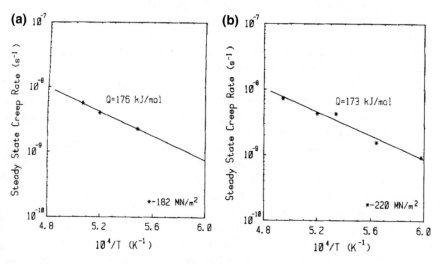

Fig. 15.15 Steady-state creep rate versus 10^4/T (K^{-1}) for CVD SiC samples crept at **a** 182 MN/m^2; **b** at 220 MN/m^2. Carter et al. [3]. With kind permission of Professor Davis and John Wiley and Sons

Note that A in Eq. (15.10) may be determined from a plot of $\log \dot{\varepsilon} + \frac{Q}{2.303kT}$ versus $\log\left(\frac{\sigma}{G}\right)$, as seen in Fig. 15.16. The values of G and n are found in the literature (Carter et al. [3]). TEM micrographs were intended to show the grain

Fig. 15.16 $\log \dot{\varepsilon} +$
$Q/2.303\,kT$ versus $\log(\sigma/G)$
for CVD SiC. Carter et al. [3].
With kind permission of
Professor Davis and John
Wiley and Sons

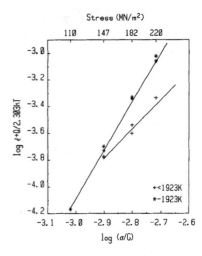

size, shape, and distribution. A set of TEM micrographs appears in Fig. 15.17. The sectioning was such that the normal to the foil was parallel to the applied stress direction, however, a few were perpendicular. Grain growth during deformation occurred in grains oriented along the preferred orientation direction of low-angle grain boundaries, caused by gliding dislocations (that play an important role in the deformation process). Grain growth is a direct consequence of the creep process.

From lattice images appearing in Fig. 15.18 and from diffraction patterns; it was determined that the primary α polytype is 6H SiC. The ratio of β to α is about 60:40, which existed in the as-deposited material remaining after creep. The dislocations in the large grains of the as-received material were analyzed to determine their Burgers vectors, as shown in Fig. 15.19. Additional dislocation structures often observed in CVD materials are seen in Figs. 15.20 and 15.21.

The dislocation tangles seen in Fig. 15.20a are also seen in the as-received and crept material and are believed to be significantly altered during the deformation process. Many of these dislocations were dipoles, whose formation resulted from dislocation-dislocation interaction. The series of loops seen in Fig. 15.20b was caused by pinching off of the dipoles. The progressive annihilation of dislocation loops, as indicated by the decreasing diameter of the series of loops, may be observed at the bottom left of Fig. 15.20b. In the crept SiC, slip bands were the dominant feature at all the stresses and temperatures (see Fig. 15.21), but they were never seen in the as-received material. A Burgers vector analysis of the dislocations in the slip bands appears in Fig. 15.22. These dislocations are glide-dissociated, $a/2 <110>$. The specific Burgers vectors seen in the slip bands are $a/6[2\bar{1}\bar{1}]$ and $a/6[1\bar{2}1]$. Dislocations within a slip band are seen forming a cavity, as indicated in Fig. 15.21b. Cavity formation occurs at grain boundaries having high angles between the basal planes in specimens that have undergone large deformation. A low density of cavities was observed in a deformed specimen; these cavities were

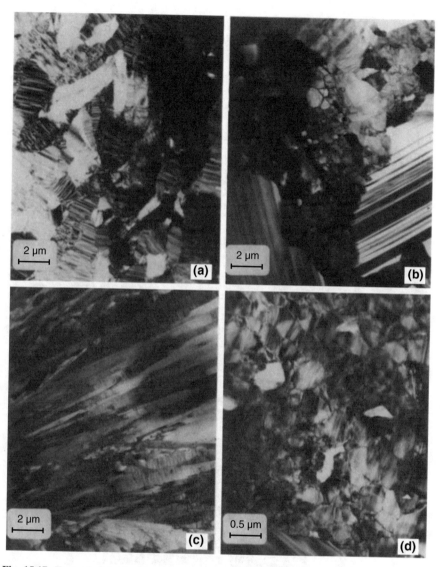

Fig. 15.17 Transmission electron micrographs of CVD SiC, showing **a** heavy faulting of as-deposited material and **b–d** tremendous range of grain sizes and shapes. These micrographs show as-crept material, but are typical of both as-received and crept material. Carter et al. [3]. With kind permission of Professor Davis and John Wiley and Sons

formed during the creep process and appear in Fig. 15.23. One may compare the activation energy of creep in CVD SiC (175 kJ mol^{-1}) with the diffusions of Si or C in α and β-SiC (which are very high). C, for example, shows about 563 kJ mol^{-1} for grain-boundary diffusion in β-SiC, meaning that, in this case, the creep mechanism is unlikely to be controlled by diffusion.

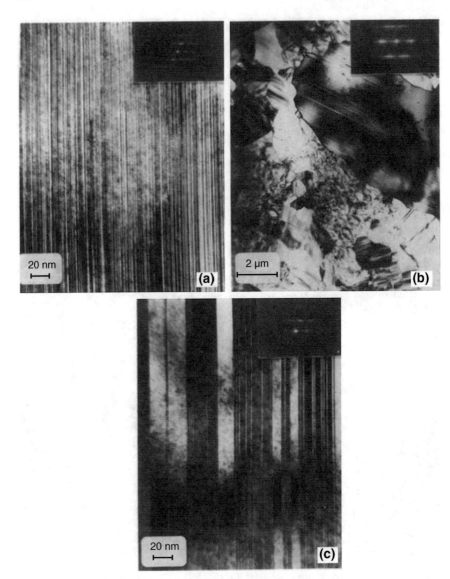

Fig. 15.18 Transmission electron micrograph and associated electron diffraction patterns for CVD SiC. **a** Lattice images of as-deposited material, which has an almost random stacking sequence. **b** Micrograph of grain which grew during argon anneal; comparison of its diffraction pattern to that of **a** shows that this structure is much less random. **c** Lattice image of material deformed at 1973 K and 182 MN/m^2; *dark areas* are α SiC and *light areas* β SiC. Carter et al. [3]. With kind permission of Professor Davis and John Wiley and Sons

Here, the values of the stress exponents suggest that either a dislocation or GBS is the creep-controlling mechanism. The above TEM micrographs suggest a dislocation-glide-controlled mechanism, since they show no indication of grain-boundary activity

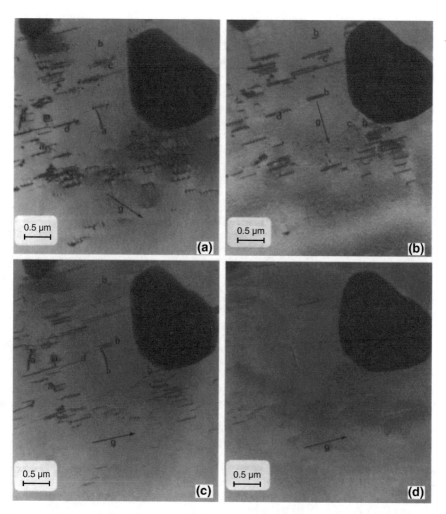

Fig. 15.19 Partial series of set of transmission electron micrographs taken for Burgers vector analysis of as-received CVD SiC. Burgers vector of labeled dislocations are as follows: **a** **b** = $a/6[11\bar{2}]$, **b** **b** = $a/6[114]$, **c** **b** = $a/3[\bar{1}\bar{1}\bar{1}]$, **d** **b**[110], (a) **g** = $[13\bar{1}]$, **z** $\cong [\bar{2}11]$; (b) **g** = $[111]$, **z** = $\cong [\bar{2}11]$; (c) **g** = $[0\bar{2}2]$, **z** $\cong [\bar{2}11]$, (d) **g** = $[\bar{1}\bar{1}2]$. Carter et al. [3]. With kind permission of Professor Davis and John Wiley and Sons

(separation or slide). The presence of slip bands seen in Figs. 15.21, 15.22 and 15.23 above prove that dislocation glide is occurring in the CVD SiC subjected to creep. Following Weertman's proposal [4], Carter et al [3]. suggest that, in materials where the Peierls stress is high, SiC among them, dislocation glide may be controlled by overcoming the Peierls stress. Dislocation glide is believed to be the creep mechanism at low temperatures (1673–1873 K) and the rate-controlling mechanism is the overcoming of the Peierls stress. At higher temperatures (1923–2023 K), climb

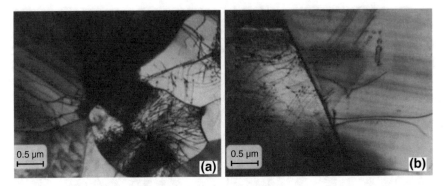

Fig. 15.20 Transmission electron micrographs of CVD SiC after creep, showing **a** dislocation tangle (deformed at 1923 K and 220 MN/m^2 and **b** trails of dislocation loops that have pinched off from dipole in material determined at 1823 K and 182 MN/m^2. Carter et al. [3]. With kind permission of Professor Davis and John Wiley and Sons

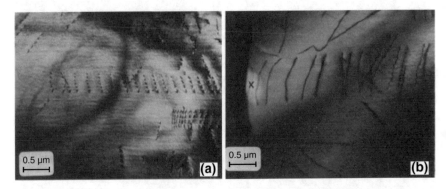

Fig. 15.21 Transmission electron micrograph of CVD SiC deformed at 2023 K and 220 MN/m^2, showing, **a** dislocation slip bands (dislocation density of 5.3×10^{12} m/m^3) and **b** dislocation in slip band forming a cavity (*X*) at grain boundary. Carter et al. [3]. With kind permission of Professor Davis and John Wiley and Sons

can occur, and the stress exponent indicates that dislocation glide/climb controlled by climb may become the steady-state creep-controlling mechanism.

15.4.3   Creep in SiC Fibers

SiC fibers, usually in graded forms, are known as 'Nicalon.' The continuous length SiC fibers are used to reinforce ceramic matrix composites (CMC) for high-temperature structural applications. Standard grade Nicalon has optimum mechanical properties and high-temperature performance for most applications. With little or no oxygen in its structure, the fiber displays high stiffness, high thermal stability, and high room temperature strength. In order to understand the

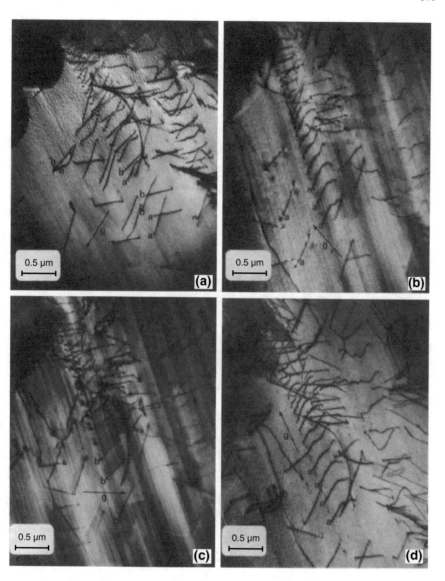

Fig. 15.22 Partial series of transmission electron micrographs taken for Burgers analysis of slip bands in sample shown in Fig. 15.21. Burgers vector analysis of labeled dislocations are as follows: **a** $\mathbf{b} = a/6[2\bar{1}\bar{1}]$, **b** $\mathbf{b} = a/6[1\bar{2}1]$. **a** $\mathbf{g} = [\bar{2}20]$, $z \cong [112]$; **b** $\mathbf{g} = [31\bar{1}]$, $z \cong [114]$; **c** $\mathbf{g} = [13\bar{1}]$, $z \cong [114]$; **d** $\mathbf{g} = [20\bar{2}]$, $z \cong [111]$. Carter et al. [3]. With kind permission of Professor Davis and John Wiley and Sons

performance of SiC fiber in CMC, it is important to know the tensile strength and creep-rupture properties of the multifilament tows (consisting of SiC fibers) and to compare the multifilament behavior with single-fiber behavior under the same conditions.

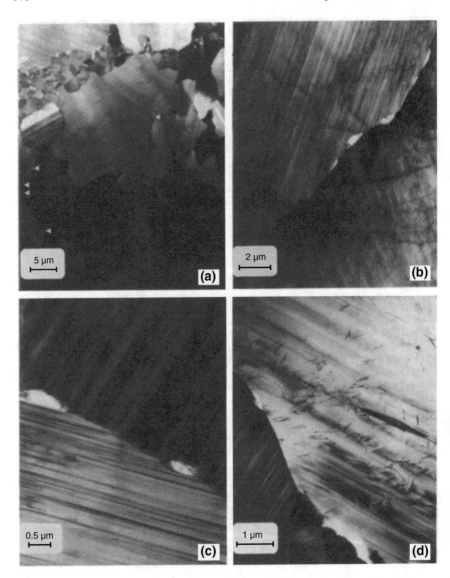

Fig. 15.23 Transmission electron micrographs of crept CVD SiC, showing cavities which were formed on grain boundaries during creep. **a** Material deformed at 2023 K and 220 MN/m² showing low volume fraction of cavities (*arrowed*). **b, c** Same material at higher magnification; specimens were oriented such that cavities were more pronounced, which put dislocations out of contrast. **d** Material deformed at 1923 K and 220 MN/m², showing dislocations which form cavities partially in contrast. Carter et al. [3]. With kind permission of Professor Davis and John Wiley and Sons

The creep-rupture measurements were made at 1200 and 1400 °C on as-produced tows and single fibers under inert (argon) and oxidizing (air) conditions. Low oxygen content SiC (0.5 wt%)-based fiber materials were tested for creep. Typical creep curves for different temperatures and applied stresses tested in air are shown in the usual manner of creep strain versus time in Fig. 15.24 for experimental Hi-Nicalon fibers and Nicalon fiber (12 wt% oxygen), referred to as CGN (Nippon Carbon Inc., Japan). The Hi-Nicalon was found to be more creep-resistant than the CGN throughout its entire lifetime.

In the case of the Hi-Nicalon, long steady-state creep was observed, while in the CGN, only primary creep existed in most cases. Additional creep curves are shown in Fig. 15.25 at various temperatures and stresses. These creep tests were performed in an argon atmosphere, either at varying temperatures and constant stress or at various stresses and a constant temperature of 1400 °C; the results may be seen in Fig. 15.26. Fibers, preheated at 1600 °C, were creep tested at 1200–1550 °C and at a constant stress of 0.45 GPa or at 1400 °C and at various stresses. The creep results are shown in Fig. 15.27. The average time-to-failure versus temperature is shown on a logarithmic scale in Fig. 15.28. The apparent activation energies in air for the as-received fibers and for the preheated Hi-Nicalon were determined from the steady-state creep versus the inverse temperature; the resulting values appear in Fig. 15.29. The activation energies in the argon atmosphere of the as-received fibers are lower than those in air, derived from similar plots found in Fig. 15.29, giving 193 kJ mol^{-1} for a stress of 0.7 GPa, and 292 kJ mol^{-1} for a stress of 0.45 GPa, respectively. This analysis was once again performed using Eq. (13.4), known as "Dorn's equation," reproduced here as:

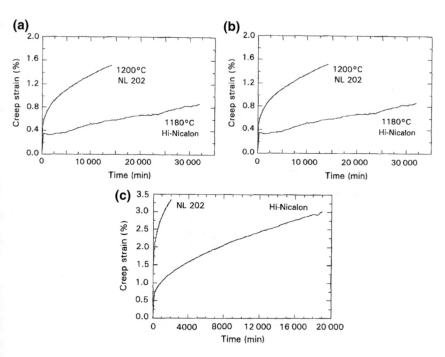

Fig. 15.24 Creep curves for CGN and Hi-Nicalon fibers tested in air at **a** 1180–1200 °C and 0.45 GPa, **b** 1180 °C and 0.70 GPa, and **c** 1300 °C and 0.45 GPa. Bodet et al. [5]. With kind permission of Springer Publishing Company

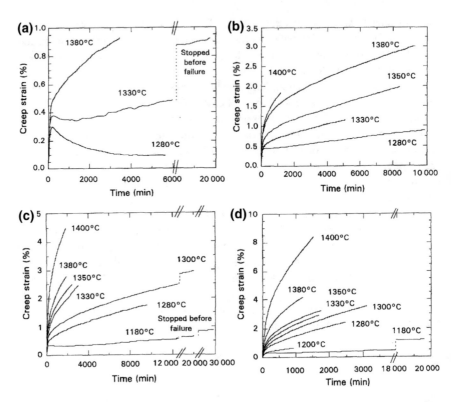

Fig. 15.25 Creep curves for the Hi-Nicalon fibers tested in air for an applied stress of **a** 0.15 GPa, **b** 0.30 GPa, **c** 0.45 GPa and **d** 0.70 GPa. Bodet et al. [5]. With kind permission of Springer Publishing Company

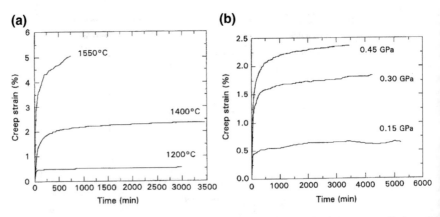

Fig. 15.26 Creep curves for the Hi-Nicalon fibers tested in argon **a** for an applied stress of 0.45 GPa, **b** at 1400 °C. Bodet et al. [5]. With kind permission of Springer Publishing Company

Fig. 15.27 Creep curves for the Hi-Nicalon fibers preheat treated at 1600 °C for 1 h and tested at 1300 °C in argon. Bodet et al. [5]. With kind permission of Springer Publishing Company

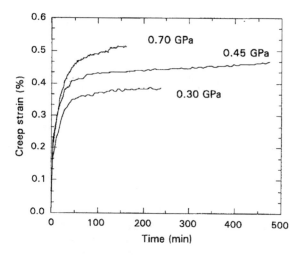

Fig. 15.28 Effect of temperature on the times-to-failure of as-received Hi-Nicalon fibers on a logarithmic scale. The single-sided error bars represent one standard deviation. Applied stress: (Δ) 0.15 GPa, (□) 0.30 GPa, (♦) 0.45 GPa, (▲) 0.70 GPa. Bodet et al. [5]. With kind permission of Springer Publishing Company

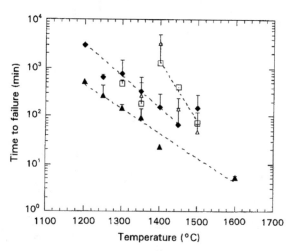

$$\dot{\varepsilon}_{ss} = A\sigma_{app}^{n} \exp\left(-\frac{Q}{RT}\right) \qquad (15.10)$$

The stress exponents in air were determined by linear-regression analysis and were found to be in the range of 1.96–3.04; the various values in the investigated range are indicated in Fig. 19.29b. For the heat-treated fibers tested in *Ar*, it was found that the stress exponents are 1.9 and 2.0 at temperatures of 1400 and 1300 °C, respectively. Figure 15.30 presents the Dorn plots for creep tested in an argon atmosphere. Table 15.3 summarizes the creep experiments done on SiC fibers.

One can conclude that the experimental Hi-Nicalon (SiC fibers), which are low in oxygen content (O wt% <0.4), have greater creep resistance than the Si-C-O commercial Nicalon fibers tested under the same conditions in the same creep

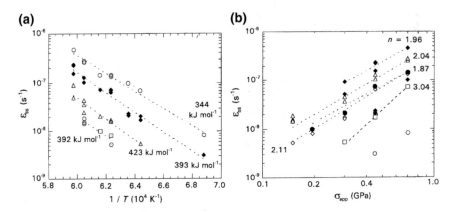

Fig. 15.29 Dorn plots for Hi-Nicalon fibers tested in air. **a** Steady-state creep rate versus reciprocal temperature at (○) 0.15 GPa, (□) 0.194 GPa, (Δ) 0.30 GPa, (♦) 0.45 GPa, (○) 0.7 GPa. **b** Steady-state creep rate versus applied stress at (○) 1180 °C, (□) 1280 °C, (♦) 1300 °C, (◊) 1330 °C, (●) 1350 °C, (A) 1380 °C, (♦) 1400 °C. Bodet et al. [5]. With kind permission of Springer Publishing Company

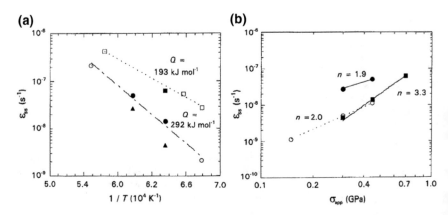

Fig. 15.30 Dorn plots for Hi-Nicalon fibers tested in argon. **a** Steady-state creep rate versus reciprocal temperature at (○) 0.45 GPa and (□) 0.7 GPa for as-received fibers, and at (▲) 0.3 GPa, (●) 0.45 GPa and (■) 0.7 GPa for preheat-treated fibers. **b** Steady-state creep versus applied stress at (○) 1400 °C for as-received fibers, and at (■) 1300 °C and (●) 1400 °C for fibers preheat-treated in argon for 1 h at 1600 °C. Bodet et al. [5]. With kind permission of Springer Publishing Company

experiments. Hi-Nicalon also has better creep-rupture properties. However, these fibers, when subjected to creep tests at 1400 °C in air, show ultra-plastic, localized deformation at high stress levels. Additionally, great deformation (about 14%) is obtained at 1500 °C in argon. Based on the stress exponents and activation energies for creep, Bodet et al [5]. propose that the active creep mechanism functioning in this case is a process involving the de-wrinkling of initially wrinkled carbon layer

Table 15.3 Creep parameters for the Hi-Nicalon and the CGN fibers tested in air and argon. Bodet et al. [5]. With kind permission of Springer Publishing Company

Atmosphere	T(°C)	σ_a (GPa)	$\dot{\varepsilon}_m$ (s⁻¹) Hi-Nicalon	$\dot{\varepsilon}_{min}$ (s⁻¹)CGN	l_f (min) Hi-Nicalon	l_f (min) CGN	ε_f (%) Hi-Nicalon	ε_f (%) CGN
Air	1180	0.45	3.10×10^{-9}	4.31×10^{-9}	31821*	593	0.86	0.80
		0.70	8.25×10^{-9}	1.82×10^{-9}	19298	2351	1 21	1.3
	1200	0.16		1.40×10^{-9}		15.312		-0.14
		0.45		6.77×10^{-9}		14181		1.52
		0.70	1.47×10^{-9}	8.33×10^{-9}	916	1461	0.65	1.27
	1280	0.15	0	-4.81×10^{-9}	5626	773	0.09	0.02
		0.30	5.39×10^{-9}		9833		0.89	
		0.45	1.73×10^{-8}	2.65×10^{-7}	9612	204	1.76	0.84
		0.70	7.08×10^{-8}		2477		2.44	
	1300	0.45	2.09×10^{-8}	8.23×10^{-8}	19010	2052	3.02	3.36
		0.70	1.02×10^{-7}		3100		3.54	
	1330	0.15	5.21×10^{-9}		19731*		0.92	
		0.19	8.22×10^{-9}		6768		0.71	
		0.30	1.60×10^{-8}		5101		1.15	
		0.45	6.39×10^{-8}		3032		2.72	
		0.70	1.33×10^{-7}		1678		2.92	
	1350	0.19	1.00×10^{-8}		5690		0.78	
		0.30	2.07×10^{-8}		8403		1.98	
		0.45	7.29×10^{-8}		2448		2.51	
		0.70	1.42×10^{-7}		1728		3.23	
	1380	0.15	1.54×10^{-7}		3446		0.93	
		0.30	3.54×10^{-8}		9289		3.01	
		0.45	1.02×10^{-7}		1881		2.79	

(continued)

Table 15.3 (continued)

Atmosphere	$T(°C)$	σ_a (GPa)	$\dot{\varepsilon}_m$ (s^{-1}) Hi-Nicalon	$\dot{\varepsilon}_{min}$ (s^{-1}) CGN	l_f (min) Hi-Nicalon	CGN	ε_f (%) Hi-Nicalon	CGN
	1400	0.70	2.56×10^{-7}		1193		4.18	
		0.30	9.15×10^{-8}		2616		1.96	
		0.45	2.21×10^{-7}		1851		4.49	
		0.70	4.60×10^{-7}		1520		8.41	
Argon	1200	0.45	2.11×10^{-9}		2992		0.59	
		0.70	2.76×10^{-8}		535		053	
	1250	0.45	6.56×10^{-9}		722		0.51	
		0.70	5.44×10^{-9}		441		0.61	
	1300	0.30	4.45×10^{-9b}		238[b]		0.38[b]	
		0.45	1.41×10^{-8b}		473[b]		0.47[b]	
		0.70	6.31×10^{-9b}		160[b]		0.44[b]	
	1400	0.15	1.11×10^{-9a}		5042		0.65	
		0.30	5.01×10^{-9}		4258		1.87	
		0.45	2.73×10^{-8b}		286[b]		0 57[b]	
			1.13×10^{-8}		3451		2.38	
			4.98×10^{-8b}		359[b]		0.59[b]	
	1500	0.70	4.18×10^{-7}		1288		14.09	
	1550	0.45	2.14×10^{-7}		748		5.05	

[a]The fiber did not fail at the end of the text
[b]Hi-Nicalon fibers prehent treated in argon

packets into a position more nearly aligned to the tensile axis, the sliding of these graphitic sheet-like structures, and the collapse of fiber pores.

15.4.4   Creep in SiC Composites

The most important SiC composite is the one strengthened by SiC fibers. Althouh SiC fibers, whiskers, and particulates are commonly used to strengthen metallic-based materials, known as "metal matrix composites" (MMC), the present focus is directed toward a SiC matrix with SiC fiber additives, belonging to the class of "ceramics matrix composites" (CMC).

Most ceramics are brittle, but often fiber additives induce certain amount of ductility in the composite, which is apparently also true in the case of SiC/SiC composites. This observation is related to a fiber-bridging effect, the consequence of which is the redistribution of stresses around strain-concentration sites, thus increasing toughness and reliability. As in previous chapters, the analysis of creep data is performed by the expression of the creep rate as given in Eqs. (13.4) and (15.10). Applying a constant tensile load produces an instantaneous strain, followed by a time-dependent creep strain, as seen in Fig. 15.31. As previously stated,

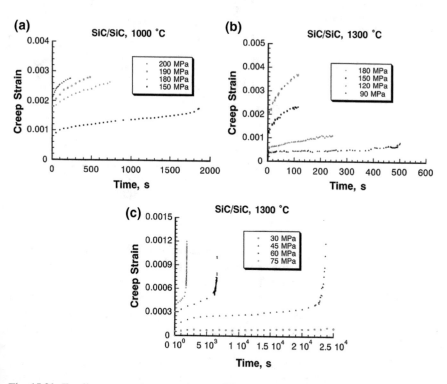

Fig. 15.31 Tensile creep strain versus time at different stresses in argon at 1000 and 1300 °C. **a** 1000 °C; **b** and **c** 1300 °C. Zhu et al. [9]. With kind permission of Elsevier

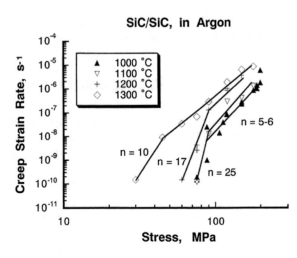

Fig. 15.32 Tensile minimum creep strain rate versus stress in argon at 1000, 1100, 1200 and 1300 °C. Zhu et al. [9]. With kind permission of Elsevier

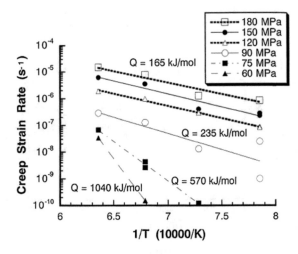

Fig. 15.33 Tensile minimum creep-strain rate as a function of absolute temperature. Zhu et al. [9]. With kind permission of Elsevier

whether all three stages of creep may be observed depends on stress and temperature. Creep rate, as a function of stress, is shown in Fig. 15.32. The stress exponents, in the 5–25 range, are indicated on the graph. Note that below an apparent threshold stress the creep rate decreases at a given temperature below a detectable level. From Eq. (15.10), an apparent activation energy may be calculated for creep. A plot of the creep-strain rate versus the inverse temperature is illustrated in Fig. 15.33.

As may be seen from Fig. 15.33, the apparent activation energy for creep decreases with increasing stress and a constant value of 165 is obtained at the high stress level of 120–180 MPa. In the lower stress range of 60–90 MPa, Q is in the range of 235–1040 kJ mol^{-1}. The apparent-stress exponent also decreases with increasing stress, as indicated in Fig. 15.32.

15.5   Creep Rupture in SiC

As indicated above, the use of SiC is generally as fiber-reinforced CMC and, therefore, most of the concern pivots around the SiC/SiC properties discussed above. However, it is no less important to discussion its creep properties and to know the work lifetime of such an CMC by evaluating its creep-rupture properties, in order to know its time-to-complete failure. This same SiC/SiC composite is now considered in regard to creep rupture. It is known that when applying a load to a specimen, an instantaneous strain is observed, followed by the strain rate. At low temperatures, this strain is elastic and recoverable but, at high temperatures, there is a nonrecoverable strain, creep strain, which is time- and stress-dependent, and may terminate in the failure of the specimen. Such failure is often referred to as "creep rupture". Whether all three stages of creep are observed or not depends on both the stress and the temperature, as previously stated. A creep rupture versus time plot is shown in Fig. 15.34. Note that creep is often thought to be composed of four stages, with "instantaneous strain" included.

The curves were fitted to the power-law relation in Eq. (13.31), now reproduced somewhat differently as:

$$t_r = B\sigma^n \tag{15.11}$$

Here, t_r is the time-to-rupture and n is the stress exponent for rupture. The derived values of n are 5.8 at 1000 °C, 4.1 at 1100 °C, 8.1 at 1200 °C, and 4.2 at 1300 °C. The steady-state creep-strain rates versus times-to-rupture are shown in

Fig. 15.34 Tensile stress versus time-to-rupture in argon at 1000, 1100, 1200 and 1300 °C. Zhu et al. [9]. With kind permission of Elsevier

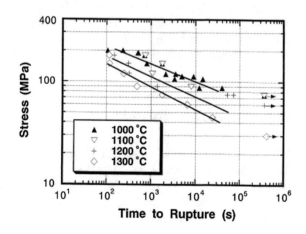

Fig. 15.35 Tensile minimum
creep-strain rate versus
time-to-rupture at different
stresses in argon at 1000,
1100, 1200 and 1300 °C.
Zhu et al. [9]. With kind
permission of Elsevier

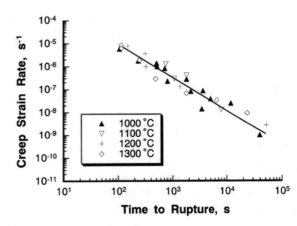

Fig. 15.35. The data are quite well-defined by a straight line, but with some scatter. The data fit the MGR, which was given above in Eq. (10.10) and reproduced here as Eq. (15.12):

$$C_{M-G} = t_r \dot{\varepsilon}^m \tag{15.12}$$

Here, C_{M-G} is the Monkman-Grant constant and m is the strain-rate exponent.

The MGR is a method for predicting the creep life of a material (in this case, the SiC/SiC composite). Clearly, knowing the creep rate enables the evaluation of the work lifetime, according to Eq. (15.12). Another empirical relation, useful mostly for metals, is the Larson-Miller Parameter (LMP). This is also valid for assessing lifetimes via Eq. (15.13), which is identical to Eq. (10.90):

$$P = (C + \log t_r) \tag{15.13}$$

A stress plot drawn according to the LMP is shown in Fig. 15.36. The value of the constant (20 in metals) is between 5–10 for the SiC/SiC composite at the temperatures shown in Fig. 15.36. After studying both Figs. 15.35 and 15.36, one cannot determine which of the two parametric techniques is preferable.

15.6 Superplasticity in SiC

In the tested nanocrystalline β-SiC, additional constituents were present, due to its fabrication and the ease of sintering. Thus, the material contained 3.5 wt% free C and amorphous 1% B (sintering aid). The sintering was performed at 980 MPa and 1600 °C for 1 h inside a glass-encapsulated vacuum, then followed by die pressing and cold isostatic pressing.

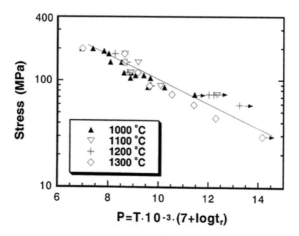

Fig. 15.36 Tensile stress versus Larson-Miller parameter at different stresses in argon at 1000, 1100, 1200 and 1300 °C. Zhu et al. [9]. With kind permission of Elsevier

Fig. 15.37 Hipped C, C-SiC specimen before and after tensile deformation. The tensile test was conducted at 1800 °C, and an initial strain rate of 3×10^{-5} s^{-1} in an argon atmosphere. The specimen deformed uniformly, and a superplastic elongation of 140% was achieved. Shinoda et al. [10]. With kind permission of John Wiley and Sons

Tensile tests were performed at 1600 °C in an argon atmosphere at an initial strain rate of 1×10^{-5}–1×10^{-4} s^{-1}. The sample preparation process resulted in 97.1% density and a 200 nm grain size. The HIPed B-, C-SiC specimens, before and after deformation, appear in Fig. 15.37. The stress-strain curves of the hot-pressed and HIPed specimens are shown in Fig. 15.38. In an additional example, a sample hot-pressed at 2000 °C for 1 h under a pressure of 30 MPa showed a superplastic elongation of >100%, whereas the hot-pressed specimen fractured without significant plastic deformation. Here, as in other materials, grain refinement is responsible for the superplastic deformation. The mechanism of superplastic behavior in this SiC is claimed to be associated with B segregation at grain boundaries promoting GBS.

However, an additional example showed the ability of a material to exhibit large strain during the deformation process. Superplasticity > 200% has been observed

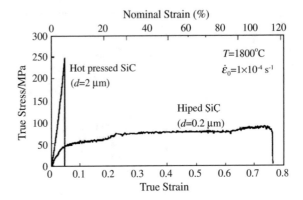

Fig. 15.38 Stress-strain curves of HIPped B, C-SiC and hot-pressed B, C-SiC. The tensile tests were conducted at 1800 °C and an initial strain rate of 1×10^{-4} s^{-1} in an argon atmosphere. HIPped, B, C-SiC exhibited superplastic elongation of 114%, because of grain refinement. On the other hand, hot-pressed B, C-SiC fractured without plastic deformation. Shinoda et al. [10]. With kind permission of John Wiley and Sons

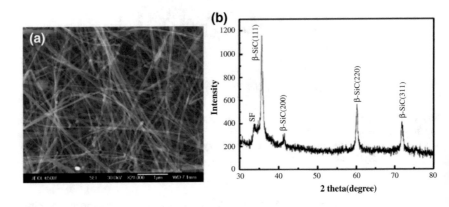

Fig. 15.39 a Typical morphologies of the SiC NWs from SEM observation; **b** XRD diffraction pattern indicating the SiC structure is cubic with stacking faults. Zhang et al. [6]. With kind permission of John Wiley and Sons

for the first time in nanowires (NW) at low temperatures ≤ 80 °C. The morphology of SiC nanowires is illustrated by SEM along with its X-ray diffraction (XRD) structure in Fig. 15.39. The XRD pattern indicates a well-developed cubic (3C) structure with stacking faults (SFs). The average NW length is several tenths of a micron and its diameter is up to 150 nm. TEM shows this general morphology in Figs. 15.40a, b in a magnified image, showing that the NW consists of two types of intergrowth segments, 1 and 2.

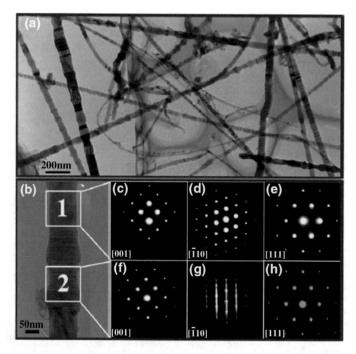

Fig. 15.40 a Typical morphologies of the SiC NWs under TEM observation; **b** in a selected piece of SiC NW showing the intergrowth feature with 3C and ODD/HD substructure segments; **c–e** are the [001], [−110] and [111] zone axes. EDPs taken from area 1 of **b** and **f–h** are the EDPs taken from area 2 of **b**. Zhang et al. [6]. With kind permission of John Wiley and Sons

Figure 15.40c–e show the corresponding electron-diffraction patterns (EDPs) along the zone axes [001], [−110], and [111] of area 1, respectively; and in Fig. 15.40f–h one sees the corresponding EDPs from area 2 along the same zone axes. HRTEM images taken along the [−110] zone axis, containing both the 3C and ODD/HD segments of the SiC nanowire, are found in Fig. 15.41. Segment 1 has an … ABCABC … stacking sequence, typical of a 3C structure, while segment 2 has a disordered structure in the [111] direction, namely, a SF sequence on a (111) plane, along the longitudinal growth axis.

Following the structural analysis, in situ tensile tests were performed in a nano tensile-testing stage. The setup is schematically shown in Fig. 15.42. Single SiC NWs were pulled by a bimetallic extensor and the process was recorded in situ by SEM imaging. All the NWs showed an extremely large tensile strain (with an average fracture strain ≥25%).

Figure 15.43a–g demonstrate the representative experiment in which a SiC NW is suspended and clamped between the two bimetallic actuating manipulators. The manipulator used to increase strain with increasing

Fig. 15.41 a is the HREM image of a SIC NW showing the intergrowth segments of 3C and ODD/HD structures at the atomic level. The FFT diffraction patterns for areas 1 and 2 are shown as insets in the *left bottom* corner and *right bottom* corner, respectively; **b** and **c** are the enlarged HREM images showing more clearly the atomic structure of 3C and ODD/HD structures. The atomic structural models and the simulated HREM images based on these models are overlaid in the corresponding figures. **d** are the EELS spectra for carbon K edge for the 3C and ODD/HD structure segments and; **e** shows the EDS spectra taken from the regions 1 and 2 of **a**. Zhang et al. [6]. With kind permission of John Wiley and Sons

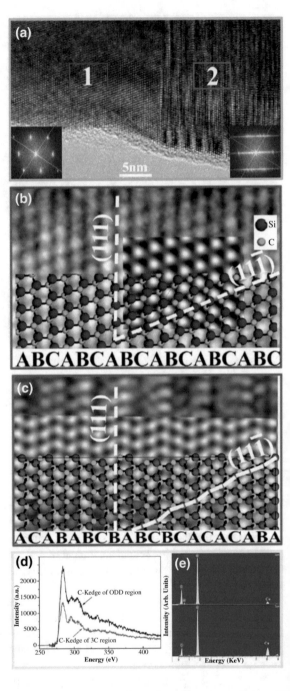

temperature from 30 to 80 °C. The strain rate was estimated as $\sim 5 \times 10^{-4}$. From a series of images, the total length of the wire before extension was

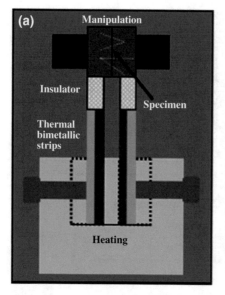
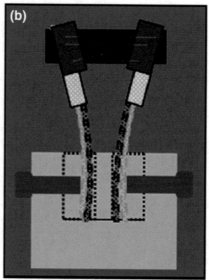

Fig. 15.42 a Schematic illustration of the tensile tool prior to extensile experiment with the SiC NWs scattered on the manipulator; **b** the conducting extensile experiment on the SiC NWs (*top view*). Zhang et al. [6]. With kind permission of John Wiley and Sons

$L = 25$ μm while, after the tension, the elongation was $T = 7$ μm. Thus, the average elongation is $T/L = 28\%$. However, the local elongation of the SiC *NW* is >200%. Looking at the white set of paired arrows, measured L_0 is 1.5 μm (Fig. 15.43b) and near the final length, l_f is 4.8 μm (Fig. 15.43e) and the local elongation is $(l_f - L_0)/L_0$ is ∼220.

A suggestion explaining the results of this superplastic deformation relates to an inhomogeneous structural feature in SiC NW. The TEM images shown in Fig. 15.44a) are typical TEM images. Figure 15.44b) provides a HREM image of a bi-structural model–one of a defect-free segment with a perfect cubic 3C structure, while the other is highly-defective, with a high density of SFs.

When putting a tensile load on a SiC NW specimen, the basic deformation is a shear one, thus a resolved-shear stress is acting on the active slip systems. For FCC SiC NW, the most favorable slip system is {111}/< 110 >, whereas for the ODD/HD parts of the SiC NW, the slip system lies on the (111) plane, in which the Scmidt factors are zero. Of the multiple slip systems present (see Zhang et al. [6]), the Schmidt factor on the (11−1) plane is 0.272, with respect to the loading axis along [111]. This means that slip was favorable on (11−1) in the structural 3C region when the load was applied along [111] or a [−1−1−1] orientation. The tensile process along this orientation would result in continuous dislocation nucleation and propagation in three sets of planes: (11−1), (1−11) and (1−1−1), with the favorable Schmidt factors. Such activity occurs within

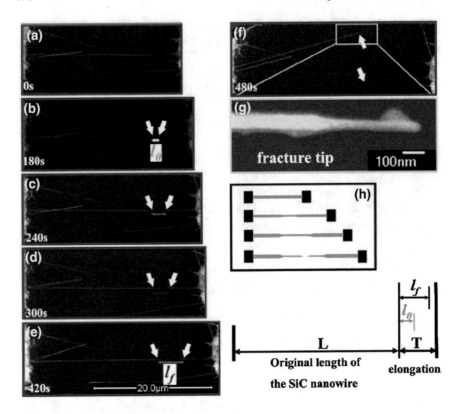

Fig. 15.43 a–f In situ sequential SEM images showing extensile experiments on a elongated SiC NW. **g** A high magnification back scattering electon (BSD) image showing a broken tip. **h** Schematic model of the tensile deformation process. Zhang et al. [6]. With kind permission of John Wiley and Sons

the region indicated by the yellow blockers shown in Fig. 15.44c. It is inferred that continuos plastic/supeplastic flow resulted in these cubic structural SiC NW regions to continuously transform into an amorphous state until fracture sets in.

The superplastic flow is confined to the 3C structural region, as illustrated in Figs. 15.46c, d between the two yellow blockers. In segments outside these yellow blockers, the ODD/HD regions, only elastic deformation occurs.

Thus, the deformation process in SiC NW consists of three stages: dislocation initiation, dislocation propagation, and amorphization with an additional super-plastic flow process in its wake. By combining dislocation initiation and propa-gation (as one activity, which includes the superplastic flow, one can generalize the plastic flow and fracture as:

Dislocation activity $\rightarrow$ amorphization $\rightarrow$ superplastic amorphous flow $\rightarrow$ fracture

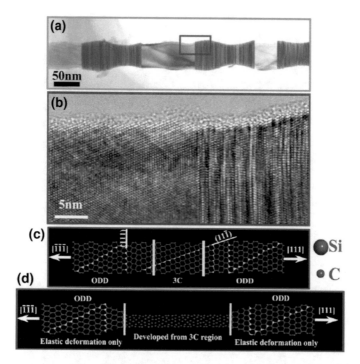

Fig. 15.44 a Atypical TEM image of a SiC NW; and **b** is the HREM image of the 3C and ODD structural segments; **c** is the two-dimensional atomic projection model of a SiC NW with two ODD/HD segments and one 3C segment in the middle; **d** illustration of the atomic model of plastic deformation process in 3C structural segment. The ODD/HD structural parts only conducted elastic deformation. Zhang et al. [6]. With kind permission of John Wiley and Sons

References

1. Atkinson A, Monty C (1989) Surfaces and Interfaces of Ceramic Materials. In: Dufour LC et al (ed), Kluwer Academic, Dordrecht, p 273
2. Carter CH, Jr Davis RF Bentley J (1984) *J Am Ceram Soc* 67[I]: 408
3. Carter CH, Jr Davis RF Bentley J (1984) *J Am Ceram Soc* 67[II]: 408
4. Weertman J (1957) J Appl Phys 28:1185
5. Bodet R, Bourrat X, Lamon J, Naslain R (1995) J Mater Sci 30:661
6. Zhang Y et al (2007) Adv Funct Mater 17:3435
7. Krishnamachari V, Notis MR (1977) Mater Sci Eng 27:83
8. Corman GS (1992) J Am Ceram Soc 75:3421
9. Zhu S, Mizuno M, Kagawa Y, Mutoh Y (1999) Comp. Sci. Tech. 59:833
10. Shinoda Y, Nagano T, Gu H, Wakai F (1999) J Am Ceram Soc 82:2916
11. Lane JE, Carter CH Jr, Davis RF (1988) J Am Ceram Soc 71:281
12. Farnsworth PL, Coble RL (1966) J Am Ceram Soc 49:264
13. Ashby MF (1972) Acta Metall 20:887
14. Newton RG, Spurrell DJ, Roy J (1967) Stat Soc Ser C: Appl Stat 16, 51
15. Niihara K (1979) J Less-Common Met 65:155

Chapter 16
Creep in Boron Carbide (B$_4$C)

Abstract Boron carbide is an excellent choice for high-temperature applications because of its properties. These are high hardness, high elastic modulus high thermal conductivity at room temperature, low thermal expansion, electrical conductivity, and its very high melting point (2447 °C). Moreover, it has a large neutron-capture cross section ($\sim$4000 barn), which makes B$_4$C a possible candidate for use in nuclear reactor components. Despite all these properties, it is puzzling that investigations of creep are lacking in the case of B$_4$C. A climb-glide power-law creep model is one concept regarding the mechanism. The density of dislocations and the presence of pileups support this creep model in B$_4$C. It was also suggested that vacancy diffusion model is operating during B$_4$C creep.

16.1 Introduction

Almost no creep data are available on the technologically important B$_4$C ceramic. This is surprising, due to its many much-sought properties, such as hardness ($\sim$30 Gpa), high elastic modulus ($\sim$450 Gpa), high thermal conductivity ($\sim$40 Wm^{-1} K^{-1} at room temperature), low thermal-expansion coefficient ($\sim$5 × 10^{-6} °C^{-1}), and good electrical conductivity ($\sim$3 Ω at room temperature). Moreover, it has a large neutron-capture cross section ($\sim$4000 barn), which makes B$_4$C a possible candidate for use in nuclear reactor components (properties given in Refs.: Thévenot, Domnich et al., Ashbee, Suri et al.,). All the above, taken together with its high melting point (2447 °C), makes this ceramic an excellent choice for many high-temperature applications, including protective armor (it also has the benefit of being light-weight). Despite all these properties, it is puzzling that investigations of creep (a high-temperature deformation) are lacking in the case of B$_4$C.

© Springer International Publishing AG 2017
J. Pelleg, *Creep in Ceramics*, Solid Mechanics and Its Applications 241,
DOI 10.1007/978-3-319-50826-9_16

16.2   Creep in B₄C

Compressive creep experiments were performed on dense, coarse-grained B_4C, in the stress and temperature ranges of 250–500 PMa and 1600–1800 °C, respectively. An argon atmosphere was used to prevent oxidation of the samples. Creep was analyzed by means of the usual relation, see, for example, Eq. (13.37), rewritten here once again as

$$\dot{\varepsilon} = A\sigma^n \exp\left(-\frac{Q}{RT}\right).$$ (16.1)

As is typically done in creep tests, the values of Q and n were determined for the activation energies from jumps at constant load and, for n, the jumps were performed at constant temperature. More experiments were done on specimens in which the temperature was kept constant (at 1650 °C or 1700 °C) and the stress varied from 250 to 500 MPa, or the stress was kept constant (at 255 MPa) and the temperature varied in the 1600–1800 °C range. The strain rate plotted against the total strain rate is shown in Fig. 16.1 for different stresses and temperature conditions.

At the end of the creep test, the total deformation was 13% and homogeneous. The specimen was intact and no transient creep was observed. The steady-state creep was achieved instantaneously following the (stress or temperature) jump tests, as indicated earlier. In Fig. 16.1, the activation energies are also indicated. The values of the activation energy and stress exponent from the single plot in Fig. 16.1 were determined to be $\sim$659 kJ mol^{-1} and $\sim$2.7, respectively. However, more accurate values may be determined from plots using constant stress and constant

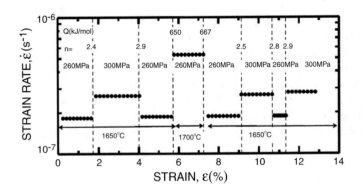

Fig. 16.1 Strain rate ($\dot{\varepsilon}$) versus total strain (ε) (%) data from a typical B_4C creep test at different stress (σ) and temperature (T) conditions. Values of the stress exponent (n) and the activation energy (Q) after stress or temperature jumps, respectively, are given at the top of the plot. Moshtaghioun et al. [4]. With kind permission of Elsevier

temperature tests. An Arrhenius-type plot of strain rate, $\dot{\varepsilon}$ versus the inverse temperature, is shown in Fig. 16.2a.

The activation energy and the stress are indicated in this figure (632 kJ mol^{-1} and 255 MPa). The total deformation at the end of the creep test is 18%. The values of n were determined from the plots in Fig. 16.2b, yielding $n = 3.1 \pm 0.3$ at 1650 °C and $n = 3.0 \pm 0.3$ at 1700 °C. In Fig. 16.2b, $\dot{\varepsilon}_{ss}$, namely, steady state creep rate, is plotted against stress. It is interesting to see the microstructural changes in the specimens before and after exposure to creep. The pre-creep structure is seen in the SEM micrographs in Fig. 16.3.

With a density of 99.2%, pores with a grain size of 17 ± 3 μm are visible in the microstructure. Twins also appear in Fig. 16.3a and b, by the surface offset resulting from twinning. The presence of pre-creep twinning may be observed in Fig. 16.3a and b, as confirmed by TEM bright-field images, shown in Fig. 16.4.

In Fig. 16.4a, straight-line dislocations with Burgers vector $b = \langle 1\,1\,0 \rangle$ gliding on $\{1\,1\,1\}$ planes are seen. A complicated substructure, located in areas with higher

Fig. 16.2 **a** Arrhenius $\dot{\varepsilon}_{ss}$ versus $1/T$ plot ($\sigma = 255$ MPa; T range 1600–1800 °C) yielding an activation energy of $Q \sim 632$ kJ/mol. Total deformation is ~18% at the end of the creep test. **b** Plot of $\dot{\varepsilon}_{ss}$ versus σ at 1650 and 1700 °C, yielding stress exponent values of $n \sim 3.1$ and $n \sim 2.9$, respectively. Moshtaghioun et al. [4]. With kind permission of Elsevier

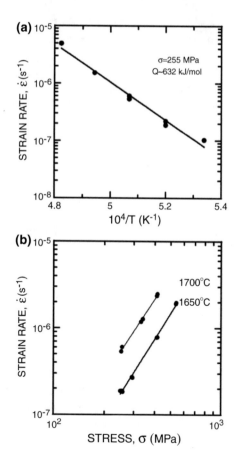

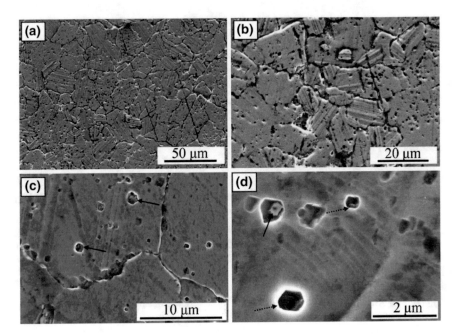

Fig. 16.3 SEM micrographs of as-sintered B₄C before creep testing at various magnifications: **a** low-magnification overall view, **b** moderate-magnification view, **c** high-magnification view showing pores (*dashed arrows*), and **d** high-magnification view showing the shapes of the pores (*dashed arrows*) and small grains within the pores (*solid arrow*). Moshtaghioun et al. [4]. With kind permission of Elsevier

dislocation densities, is visible in Fig. 16.4b and c, resulting from dislocation interaction from multiple slip planes and also from interaction with B₄C particles. A low density of dislocations is observed in some grains in Fig. 16.4c. The twins are seen in all the segments of Fig. 16.4. The twinning is along the {10$\bar{1}$1} rhombohedral planes. Since there are three variants of this plane, the angle between the rhombohedral planes is 65°36′. The specimens deformed to 13% did not show any change in grain size, as indicated in Fig. 16.5a. As mentioned above, the estimated grain size here is 17.6 ± 0.3 μm.

TEM images of B₄C are found in Fig. 16.6. Straight-line dislocations are shown in Fig. 16.6a, with Burgers vector **b** = ⟨110⟩ gliding on {111} planes. Stereographic images and a g · **b** analysis reveal the presence of long cross-slipping screw dislocation segments.

Again, the complicated substructure shown in Fig. 16.6b and c results from dislocation interaction from multiple slip planes and from interaction with B₄C particles, as evidenced in the areas having higher dislocation densities. The twins act as sub-boundaries and obstacles to dislocation glide. Consequently, pileups in appreciable numbers may be observed at the twin and grain boundaries shown in Fig. 16.6d–f. Various dislocation types are visible (partial, screw, and perfect),

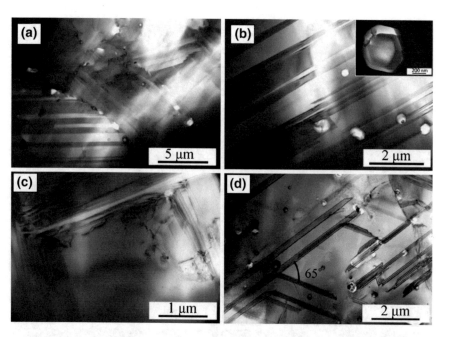

Fig. 16.4 Bright-field TEM images of B$_4$C before creep testing showing **a** twins and trapped small pores at grain boundaries, **b** small pores within larger grains and trapped small grains within coarse grains (*inset*), **c** dislocation debris in some grains, and **d** twins, with crystallographic orientation forming variants at $\sim 65°$ angle. Moshtaghioun et al. [4]. With kind permission of Elsevier

Fig. 16.5 SEM images of creep-tested B$_4$C specimen ($\sim 13\%$ total deformation): **a** low-magnification view and **b** higher magnification view of the same region showing a higher incidence of the surface offsets. Moshtaghioun et al. [4]. With kind permission of Elsevier

which is compatible with possible, simultaneous activity in several (111) slip planes. The estimated overall dislocation density of B$_4$C is 3×10^{12} m^{-2}.

The value of the stress exponent, $n \sim 3$, is an indication of climb-glide power-law creep, as are the additional observations of the density of dislocations, before and after creep. The presence of pileups also suggests the action of a

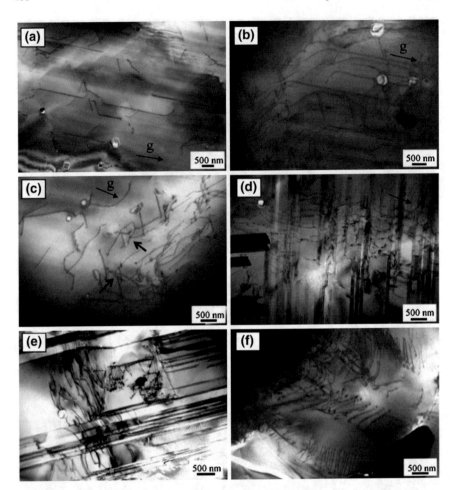

Fig. 16.6 TEM micrographs of creep-deformed B₄C specimens (∼13% total deformation) showing: **a** straight-line dislocations with Burgers vector $b = \langle 1\,1\,0 \rangle$ gliding on $\{1\,1\,1\}$ planes and long cross-slipping screw dislocation segments, (**b, c**) areas of higher dislocation densities and the dislocation debris, (**d, e**) interaction of dislocations and twins, and (**f**) dislocation pileups at the grain boundaries. *Arrows* indicate the presence of dislocation nodes. Moshtaghioun et al. [4]. With kind permission of Elsevier

power-law creep mechanism in B₄C. However, an additional work by Abzianidze et al. is worthy of consideration; it suggests that another creep mechanism may also be in play, evidenced by the much lower activation energy. As such, it is interesting to consider the creep evaluation by means of this new approach—providing a second observation of creep in B₄C.

Equation (16.1) is used, yet again, to analyze these new experimental creep results presented in Fig. 16.7. Here, creep is characterized by a short (10 min) transitional creep and low values (up to 0.3%) of instantaneous deformation.

Fig. 16.7 Boron carbide creep under 100 MPa; *1* 1500 °C, *2* 1650 °C, *3* 1800 °C. Abzianidze et al. [1]. With kind permission of Elsevier

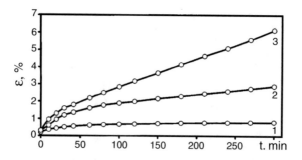

Fig. 16.8 Steady creep rate versus a stress: *1* boron carbide, at 1650 °C; *2* boron carbide, at 1800 °C; *3* Aluminum dodecaboride, density 2.37 g/cm^3, at 1500 °C; *4* aluminum dodecaboride, density 2.37 g/cm^3, at 1650 °C; *5* aluminum dodecaboride, density 2.22 g/cm^3, at 1650 °C; *5* aluminum dodecaboride, density 1.99 g/cm^3, at 1650 °C. Abzianidze et al. [1]. With kind permission of Elsevier

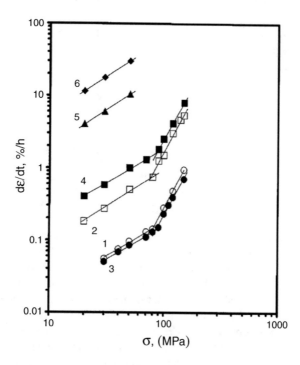

The creep rate versus stress relation is shown in Fig. 16.8. Note the change in the slope, indicating two different creep mechanisms operating in B$_4$C. The respective stress exponents are $n = 1$ and $n = 3$. The expected mechanisms are a vacancy-diffusive mechanism and a dislocation mechanism. The vacancy-diffusive mechanism, with exponent $n = 1$, occurs at stress values below 90 MPa. At stresses above 90 Mpa, the dominant mechanism is dislocation creep controlled, and has an exponent of $n = 3$. In Fig. 16.9, an Arrhenius-type plot is shown of the strain rate vs. the inverse temperature. The activation energies derived from the slopes appear in Fig. 16.9 and have the same value of 385 kJ mol^{-1}. Note that in Figs. 16.8 and 16.9 aluminum dodecaboride is included.

If the suggested vacancy diffusion model was operating during B$_4$C creep, under a stress of 90 MPa, then creep should be controlled by the slowest diffusing species.

Fig. 16.9 Steady creep rate versus a temperature: *1* boron carbide, under 50 MPa; *2* boron carbide, under 100 MPa; *3* aluminum dodecaboride, density 2.37 g/cm³, under 50 MPa; *4* aluminum dodecaboride, density 2.37 g/cm³, under 100 MPa; *5* aluminum dodecaboride, density 2.22 g/cm³, under 50 MPa; *6* aluminum dodecaboride, density 1.99 g/cm³, under 50 MPa. Abzianidze et al. [1]. With kind permission of Elsevier

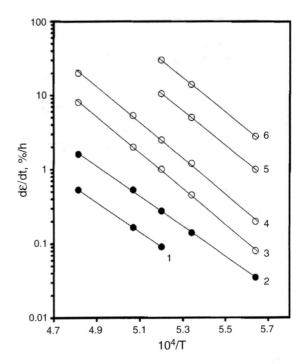

In B₄C, the carbon atoms are the slowest-moving atoms, based on the concept that the diffusion rate of B into carbon is higher than the diffusion rate of C into boron. This is supported by the fact that the self-diffusion of C is 382 kJ mol⁻¹, the same as the aforementioned 385 kJ mol⁻¹ for the creep-activation energy in B₄C.

At this stage, it is rather difficult to determine which of these two approaches to creep in B₄C is the more acceptable. Thus far, the author knows of no relevant third publication in the open literature that could enable him to make a scholarly judgment on the real creep phenomena occurring in B₄C. Such a determination must be postponed to such a time when sufficient research data will become available. This author hopes to encourage researchers to look into this matter regarding this very important ceramic–B₄C.

References

1. Abzianidze TG, Eristavi AM, Shalamberidze SO (2000) J Sol State Chem 154:191
2. Ashbee KHG (1971) Acta Metall 19:1079
3. Domnich V, Reynaud S, Haber RA, Chhowalla M (2011) J Am Ceram Soc 94:3605
4. Moshtaghioun BM, García DG, Rodríguez AD, Padture NP (2015) J Eur Ceram Soc 35:1423
5. Suri AK, Subramanian C, Sonber JK, Murthy TSRCh (2010) Int Mater Rev 55:4
6. Thévenot F (1990) J Eur Ceram Soc 6:205

Chapter 17
Creep in Silicon Nitride (Si₃N₄)

Abstract Si_3N_4 has a wide range of applications, such as in: automotive parts; mechanical bearings; high-temperature/thermal-shock-resistant ceramics; orthopedic solutions; metalworking tools; electronic insulators; and diffusion barriers in integrated circuits, to name just a few. The mechanical properties of Si_3N_4 play important roles in modern technology and industry. The evaluation of creep in Si_3N_4, is one of the subjects of this chapter performed by tensile and compressive creep tests in polycrystalline Si_3N_4. SiC-based composites were investigated by flexural and tensile loading of the specimens. A considerable discussion is devoted to cavitation in silicon nitride. Sections of superplasticity, nanosize silicon nitride and stress rupture are integral parts of this chapter. The final section deals with creep recovery in Si_3N_4.

17.1 Introduction

Si_3N_4 exists in three crystallographic structures: α-Si_3N_4 (trigonal), β-Si_3N_4 (hexagonal), and γ-Si_3N_4 (cubic). The most common allotropes are the α-Si_3N_4 and the β-Si_3N_4. They can be envisioned as consisting of layers … ABABAB … and … ABCDABCDABCD … for the β-Si_3N_4 and α-Si_3N_4 phases, respectively. The α-Si_3N_4 phase is also presented as a layered ABBA structure, emphasizing the mirror image of the AB layers, namely BA. Although the α-Si_3N_4 is harder, due to its longer stacking sequence, than the β-Si_3N_4, α-Si_3N_4 is not as chemically stable as the β-Si_3N_4. At high temperatures and in the presence of a liquid phase, α-Si_3N_4 transforms into β-Si_3N_4. Therefore, it is preferable to use the more stable β-Si_3N_4 in ceramic applications.

Si_3N_4 has a wide range of applications, such as in: automotive parts; mechanical bearings; high-temperature/thermal-shock-resistant ceramics; orthopedic purposes; metalworking tools; electronic insulators; and diffusion barriers in integrated circuits, to name just a few. The mechanical properties of Si_3N_4 play important roles in modern technology and industry. Thus, the following will deal with the evaluation of creep in Si_3N_4, assessing its impact on actual and potential applications.

© Springer International Publishing AG 2017   403
J. Pelleg, *Creep in Ceramics*, Solid Mechanics and Its Applications 241,
DOI 10.1007/978-3-319-50826-9_17

17.2   Creep in Polycrystalline Si$_3$N$_4$

The creep behavior of materials differs under different loading conditions. This section will deal with tensile and compressive creep behaviors in Si$_3$N$_4$ as affected by various loading conditions.

17.2.1   Tensile Creep in Si$_3$N$_4$

The Silicon nitride studied here was obtained by HIP and a densification aid of 4 wt% Y$_2$O$_3$ was added. This Si$_3$N$_4$ belongs to a class of silicon nitrides known as 'in situ reinforced composites,' because its microstructure consists of large acicular β-Si$_3$N$_4$ grains, whose interstices are filled with small equiaxed grains and the residual silicate densification aid. The second-phase glass in the interstitial region is almost completely crystallized, forming α-Y$_2$S$_2$O$_7$ and Y$_5$(SiO$_4$)$_3$N. Usually in silicon nitrides densified with the aid of a silicate glass, an amorphous grain-boundary film (about 1 nm thick) remains on some grain boundaries after heat treatment.

The specimens were tested until failure under a single stress and at a single temperature. In most cases, no tertiary stage was present, but a few specimens showed definite steady-state creep. The creep rate of most specimens continued to decrease until failure. The initial creep rate was ~ 4 times that of the creep rate at failure. The creep curves obtained are illustrated in the usual manner of ε versus time in Fig. 17.1 at three temperatures and under the indicated stresses. Clearly, higher stresses produced shorter creep strains to failure. As expected, tests at the same stress level provided the longest strain-to-failure at the lowest temperature. To obtain the stress exponents, a plot of strain rate $\dot{\varepsilon}$ versus stress was constructed, as shown in Fig. 17.2. This plot includes a compression graph for comparison. The stress exponents (even of a compression graph for comparison) are included from all tests.

Since it is known that compression tends to close pores and expend cracks, it is expected that compression creep will produce smaller creep strain than tensile creep. A comparison between tensile and compressive creep results may be seen in Fig. 17.3.

As seen, the creep rate under compression is much lower than under tension. At a strain of 0.005 (about the final strain in compression), the creep rate under tension is almost a 100 times greater.

17.2.2   Compressive Creep in Si$_3$N$_4$

Several silicon nitrides processed by different methods, together with SiALON, are compared below by means of compressive creep tests. The main methods

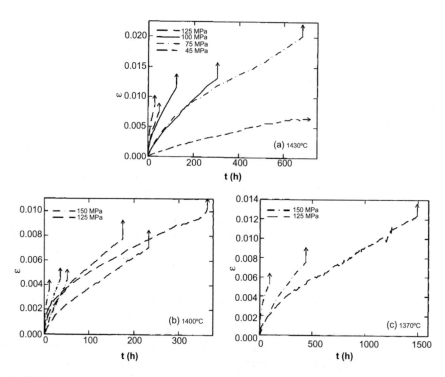

Fig. 17.1 Creep curves from all testing conditions: **a** 1430 °C, **b** 1400 °C, **c** 1370 °C. The figures omit some of the high-stress, short-time curves for clarity. *Upward arrows* denote failure; *horizontal arrows* denote interruption before failure. Note the different time and strain axes in each plot. Luecke et al. [1]. With kind permission of John Wiley and Sons

Fig. 17.2 Strain rate as a function of stress for tests conducted at 1430 °C. The extra data points for the low strains come from tests at low stress that were discontinued before failure. Luecke et al. [1]. With kind permission of John Wiley and Sons

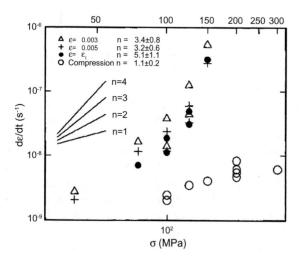

Fig. 17.3 Comparison of the creep curves in tension and compression for deformation at 125 MPa, 1430 °C. *Error bars* in the compression curve represent the standard deviation of the group of 10 measurements of the specimen length. To facilitate comparison, the sign of the compressive strain has been reversed. Luecke et al. [1]. With kind permission of John Wiley and Sons

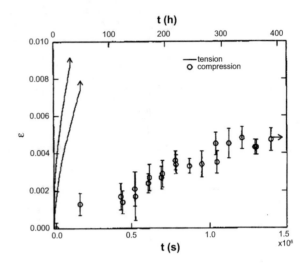

considered are the reaction bonded silicon nitride (RBSN) and the hot pressed silicon nitride (HPSN). In RBSN, a mixture of α-Si$_3$N$_4$ and β-Si$_3$N$_4$ with ~25% cavities is obtained, while HPSN yields a fullydense material when a proper additive is included. The additive used for HPSN-1 was 2% MgO, while in HPSN-2 it was ~5% MgO. The creep curves of these nitrides are compared in Fig. 17.4.

All the curves showed primary- and secondary-creep curves. If the creep test is allowed to continue, the steady-state creep eventually leads to accelerated failure along with consequent failure (such a case is not shown here). The creep rate versus stress is plotted in Fig. 17.5 and versus temperature in Fig. 17.6. The stress and temperature dependences of the steady-state creep rate were expressed by Eq. (13.4) above, reproduced here as:

$$\dot{\varepsilon} = A\sigma^n \exp\left(-\frac{Q_c}{RT}\right) \qquad (17.1)$$

Clearly here, A contains 1/kT of Eq. (13.4) and the conversion factor from R to k is also required. The stress exponents evaluated are 2.1–2.4, obtained from Fig. 17.5.

The activation energy for creep is ~650 kJ mol^{-1} for both RBSN and HPSN. One would expect to have the same creep properties under tension and compression–that at a given applied stress, comparable creep should result. However, in order to obtain a given creep rate under compression, the applied stress must be about an order of magnitude higher than during a tensile-creep test under the same conditions. The difference is claimed to be associated with crack formation at the grain boundaries during creep. The formation of grain-boundary cavities and cracks depends on the magnitude of the tensile stress that develops along the grain boundaries. In compression creep, the maximum tensile stress that develops across boundaries, parallel to the specimen axis, is about one-tenth of the applied

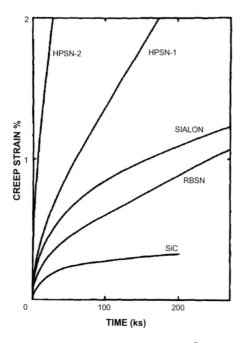

Fig. 17.4 Compression-creep curves recorded at 238 MN m⁻² and 1623 K for samples of reaction-bonded and hot-pressed silicon nitride, a sialon ($z = 1$) and sintered silicon carbide. The hot-pressed sample (designated HPSN-1) contained 2% MgO, compared with ∼5% MgO for the material labelled HPSN-2. Birch and Wilshire [2]. With kind permission of Springer

Fig. 17.5 Stress dependence of the secondary creep rate ($\dot{\varepsilon}_s$) for samples of reaction-bonded and hot-pressed silicon nitride, a sialon ($z = 1$) and sintered silicon carbide for compression-creep tests carried out at 1623 K. Birch and Wilshire [2]. With kind permission of Springer

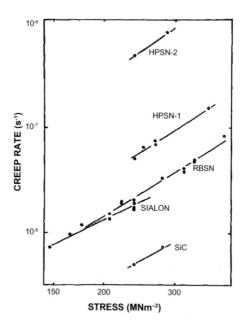

Fig. 17.6 Temperature
dependence of the
secondary-creep rate $(\dot{\varepsilon}_s)$ for
samples of reaction-bonded
and hot-pressed silicon
nitride, a sialon ($z = 1$) and
sintered silicon carbide for
compression-creep tests
carried out at 238 MN m^{-2}.
Birch and Wilshire [2]. With
kind permission of Springer

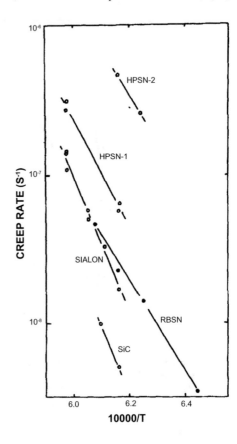

compression stress. The tenfold difference in creep strength between tension and compression may be related to the formation of microcracks that accommodate the relative movement of the crystals, rather than GBS itself. This is believed to be the rate-controlling mechanism in the silicon nitrides and explains differences in creep in nitrides of various sizes, porosity levels and impurity contents. Thus, creep resistance is influenced by the ease of GBS and crack formation, due to the presence of the aforementioned porosity and impurity levels, which contribute to the overall creep strain.

17.2.3   Cavitation in Si$_3$N$_4$

Cavity formation and expansion in HIPed silicon nitride is more critical than in compressive creep. Under tension, the volume fraction of the cavities usually increases linearly with strain, contributing to the strain. Under compression, there is a tendency for pores and cavities to close; therefore, the degree of cavitation is smaller when testing is performed under the same conditions (stress, temperature) as is the creep under tension. The material considered here has been indicated in Sect. 17.2.1—it contains 4 wt% Y_2O_3 for densification. The presence of the cavities is expressed in terms of the volume fraction in the gauge length at failure, as shown in Fig. 17.7. The volume fraction of the cavities is expressed as:

$$f_v = \frac{\rho_{grip} - \rho_{gage}}{\rho_{grip}} \tag{17.2}$$

where ρ_{grip} and ρ_{gauge} are the densities in the creep section and gauge section, respectively.

In Eq. 17.2, it is assumed that all the density changes in the gauge length result from cavitation close to the surface. A collection of TEM micrographs (in Fig. 17.8) shows cavitation in the crept specimens at the grain boundaries. Note that the cavities shown in Fig. 17.8 are of two types. Lens-shaped cavities are shown at two grain boundaries at lower temperatures and the less common, large irregularly shaped cavities appear at multigrain junctions in pockets of equiaxed, sub-micrometer-sized grains of Si_3N_4 and crystalline silicate. The size distribution of the cavities from both sections, the grip and the gauge, is illustrated in Fig. 17.9. The volume fraction of the cavities in the specimens tested at 1400 °C and 125 MPa, as a function of strain, is shown in Fig. 17.10. The testing of some of the specimens was interrupted before failure, while others crept to failure. The scatter in the specimens tested to failure is quite large. However, the interrupted creep test results show a reasonable, linear increase in the cavity-volume fraction under strain. However, this plot does not distinguish between cavity growth and additional cavity formation during creep. Under the same test conditions (stress, temperature), specimens creep under a tensile load 100 times faster than under compression. This is also reflected by the values of the stress exponent in the creep rate relation. Recalling that the creep rate is proportional to the stress, $\dot{\varepsilon} \propto \sigma^n$, one expects a larger creep rate under tension, since $2 < n < 7$, while under compression $n \cong 1$.

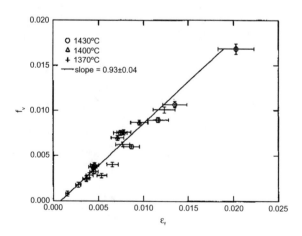

Fig. 17.7 Total volume fraction of cavities in the gauge length for all specimens tested to failure. Luecke et al. [1]. With kind permission of John Wiley and Sons

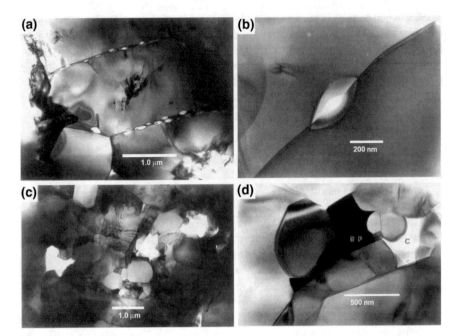

Fig. 17.8 Transmission electron micrographs showing both small, lens-shaped cavities, as well as the larger, more irregularly shaped cavities. **a** and **b** show the lens-shaped cavities that formed in a specimen crept at 1370 °C under 125 MPa for 1521 h, while **c** and **d** show the irregularly shaped, interstitial cavities that formed in the silicate phase in a specimen deformed at 1430 °C under 125 MPa for 30 h. In both micrographs, the tensile axis is vertical. In **d** "C" is a cavity and "SP" is the silicate phase. Luecke et al. [1]. With kind permission of John Wiley and Sons

Fig. 17.9 Derived cavity volume fraction size distributions for a specimen tested to failure at 1430 °C and 100 MPa. The difference between the curves is the cavity volume fraction distribution for the specimen. Luecke et al. [1]. With kind permission of John Wiley and Sons

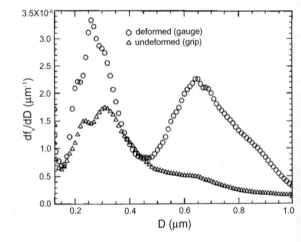

Fig. 17.10 Volume fraction of cavities as a function of strain for specimens crept at 1400 °C and 125 MPa, along with the best-fit line from Fig. 17.7. Luecke et al. [1]. With kind permission of John Wiley and Sons

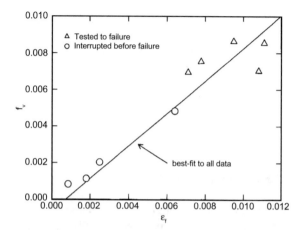

This is significant for cavitation, since there is a linear increase in the cavity content with creep strain. It is possible to conclude this section on cavity formation and growth with the following information: (a) the volume fraction of cavities increase linearly with strain, which accounts for a substantial contribution to overall strain; (b) there is a linear dependence of creep rate on stress; (c) cavities are formed throughout the life of creep, but the growth of cavities contribute significantly to the total volume of cavities at large strains; (d) cavities are more readily formed during tensional creep than during compressional creep; (e) with increasing strain, the number of large, multigrain-junction cavities increases; (f) furthermore, in tensile creep, the silicon nitride grains remain essentially un-deformed and the silicate-filled interstitial volume of the microstructure increases in volume during deformation, leading to cavitation in the weaker silicate phase at multigrain junctions.

The sliding of the silicon-nitride grains accommodated this expansion. Under compressional creep, cavitation is barely possible, ruling out the possibility of the existence of a sliding mechanism. In its stead, diffusional creep produces the observed creep strain.

17.3   Creep in Composite Silicon Nitride

17.3.1   Introduction

As stated above, silicon-nitride-based ceramics are of great technological interest, because they possess excellent high-temperature strength, oxidation resistance, have low density, and a low coefficient of thermal expansion. Again, these properties are of major importance for high-temperature applications in the aerospace industry and in turbine engines, to name just a couple of examples. However, their inherently brittle nature and low creep resistance at high temperatures must be resolved in order to make Si₃N₄ a plausible high-temperature structural material. Effective improvement of Si₃N₄ is achieved by reinforcing it with an appropriate, continuous fiber. Nitride-, boride- and carbide-based fibers were tested as possible reinforcing agents. One of the most commonly used fibers for Si₃N₄ reinforcement is SiC.

17.3.2 Si₃N₄-Based SiC Composite

17.3.2.1 Flexural Test

The addition of SiC reinforcing fibers into a Si₃N₄ matrix can result in a stronger and tougher material than monolithic Si₃N₄, as indicated by the following experiments. Four-point flexural tests were performed at temperatures of 1200–1450 °C and at stress levels of 250–350 MPa. The tested fiber is designated as SCS-6. In order to consolidate the composite and to facilitate its densification, sintering additives were required; in the present case, 5 wt% Y₂O₃ and 1.25 wt% MgO were

Fig. 17.11 Creep response for monolithic Si₃N₄ and the 30 vol% composite at 1200 °C. Thayer and Yang [3]. With kind permission of Elsevier

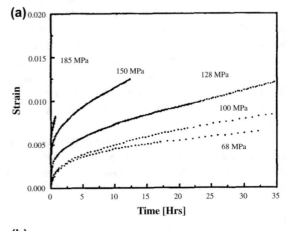

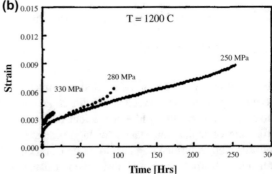

added to the Si_3N_4 powder. The composite preforms were consolidated by uniaxial hot-pressing at 1700 °C and 70 MPa for 1 h in a N_2 atmosphere. (Experimental details are found in Thayer and Yang). Creep curves, in the usual presentation of strain vs. time, are shown for the 30 vol% SiC composite in Fig. 17.11 and compared with the monolithic Si_3N_4.

All the curves show primary creep and steady-state creep, and some exhibit tertiary creep leading to rupture. The results are clearly dependent on the test conditions. The strain-rate data versus applied stress are plotted on a logarithmic scale in Fig. 17.12.

The apparent activation energy was derived from the strain rate versus inverse temperature plots shown in Fig. 17.13. Linear regression was applied and the values of the activation energies are listed on the graph. For the monolithic matrix, an activation energy of 532 kJ mol^{-1} was obtained in the temperature range of 1200–1350 °C. The stress applied was 100 MPa. In the case of the 30 vol% Si_3N_4-SiC composite, the applied stress was 250 MPa and the temperature range, 1200–1450 °C. The activation energy of the 20 vol% SiC was evaluated at 250 MPa and a temperature range of 1200–1350 °C.

Reinforcement of the Si_3N_4 by the SCS-6 fiber resulted in a steady-state strain rate reduction by about three orders, relative to the monolithic Si_3N_4. The tertiary creep in the composite resulted from the rapid growth of microcracks, initiated from the fiber-rupture sites. A repetitive scheme of matrix-stress relaxation/fiber rupture/load transfer occurs in the composite by means of a synergistic effect, which is supposed to be the basis of the improvement. The interfacial property, between the fiber and the matrix, controls the mechanical performance of the composite.

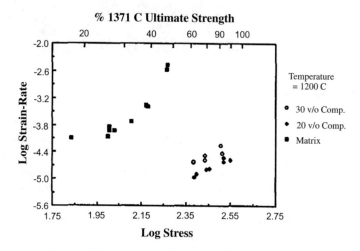

Fig. 17.12 Steady-state strain rates for all materials plotted against the applied stress on the lower scale and against the percent of the ultimate flexural strength of the 30 vol% composite at 1371 °C (390 MPa). Thayer and Yang [3]. With kind permission of Elsevier

Fig. 17.13 Arrhenius plot of the strain-rates for all materials. The matrix stress level was 100 MPa and that of the composites was 250 MPa. Thayer and Yang [3]. With kind permission of Elsevier

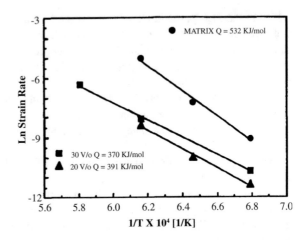

Fig. 17.14 SEM micrograph showing the typical fiber distribution in the hot-pressed SiC-Si$_3$N$_4$ billet studied. Holmes [4]. With kind permission of Springer

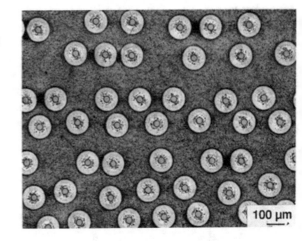

17.3.2.2   Tensile Test

For the sake of comparison, here are the results of a tensile-creep test done on a similar composite containing of 30 vol% SCS-6 fiber. This unidirectional composite was tested in air at stress levels of 70–190 MPa. Steady-state creep was seen, except at 190 MPa, where tertiary creep was observed. This composite was prepared by hot pressing. A micrograph showing the fiber distribution in the hot-pressed composite is shown in Fig. 17.14

The creep behavior of the composite at 1350 °C is shown in Fig. 17.15 for stresses of 70, 110, and 150 MPa. The stress dependence of the apparent steady-state creep is seen in Fig. 17.16. Fitting the steady-state creep rate to the power law, namely $\dot{\varepsilon} \propto \sigma^n$, provides a somewhat high stress exponent, $n \sim 7$, for the HP-SiC$_f$-Si$_3$N$_4$ composite.

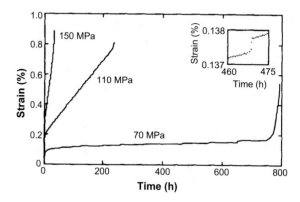

Fig. 17.15 Tensile creep behaviour of HP-SiC$_f$-Si$_3$N$_4$ at 1350 °C. Representative creep curves for stress levels of 70, 110 and 150 MPa are shown (at 190 MPa, only tertiary creep was observed, with failure occurring in under 5 min). The *Inset* shows a typical strain jump observed during creep testing at 70 MPa. Holmes [4]. With kind permission of Springer

Fig. 17.16 Steady-state tensile creep rate plotted against stress for HP-SiC$_f$-Si$_3$N$_4$ at 1350°C (●), monolithic HP-Si$_3$N$_4$ at 1315 °C (○) and monolithic HIP-Si$_3$N$_4$ at 1370 °C (Δ). Holmes [4]. With kind permission of Springer

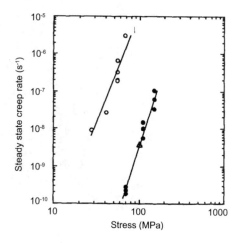

Creep failure was accompanied by fiber pullout. The pullout of fibers is shown in Fig. 17.17. In Fig. 17.18, the fiber debonding along the fiber-matrix is illustrated. One can thus conclude that in this composite, under the conditions of tensile creep, failure was accompanied by extensive fiber pullout and debonding along the fiber-matrix interface.

17.4 Creep Rupture in Si$_3$N$_4$

Creep tests are performed to establish the lifetime of Si$_3$N$_4$ before failure. The fracture or failure of the test specimens are usually referred to as 'creep rupture.' The specimens used for the creep tests and the creep-rupture evaluation were

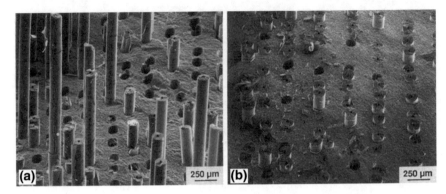

Fig. 17.17 SEM micrographs comparing the extent of fiber pullout observed after creep failure at **a** 70 MPa, t_f = 794 h and **b** 150 MPa, t_f = 34 h. Holmes [4]. With kind permission of Springer

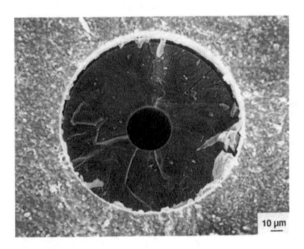

Fig. 17.18 SEM micrograph showing debonding along the fiber-matrix interface observed after creep failure at 70 MPa (t_f = 794 h). Holmes [4]. With kind permission of Springer

prepared by HIP. The creep and the resulting creep rupture data are presented in Table 17.1 and the creep curves are shown in Fig. 17.20.

Based on these creep curves, a transition region may be drawn, as shown in Fig. 17.21. In the region above the transition stress, rupture occurs. Hence, the creep rate is high and the lifetime is short. When the creep rupture occurs below the transition region, the creep rate may be very low, depending on the stress level and the longevity of the creep. The models used to analyze creep rupture are: (a) the LMP; (b) the minimum-commitment method (MCM); and (c) the MGR. Table 7.1 was constructed based on all three of these models (Fig. 17.19).

The analysis of the creep data was done using Norton's relation, rewritten here as:

Fig. 17.19 SEM micrograph
showing the two types of fiber
fractures observed after creep
failure at 70 MPa. The rough
(porous) appearance of the
two fibers on the right-hand
side of the micrograph is
attributed to the localized
reaction of fibers with
segregated sintering oxides.
Approximately 15% of all the
fibers had a rough fracture
surface; all other fibers
showed a brittle failure mode
(e.g., fiber on left-hand side of
micrograph and in
Fig. 17.18). Holmes [4]. With
kind permission of Springer

$$\dot{\varepsilon} = A\sigma^n \exp\left(-\frac{Q}{RT}\right) \tag{17.3}$$

(a) The creep-rupture analysis, in accordance with the LMP is:

$$\log t_r = B_0 + \frac{B_1}{T} + \frac{B_2}{T}\log\sigma \tag{17.4}$$

B_0, B_1, and B_2 are constants. A multivariate regression analysis was performed on the non-arrowed data in Fig. 17.22, providing values for the constants: $B_0 = -58.28$; $B_1 = 136{,}600$; and $B_2 = -20510$.

(b) The MCM has the following form:

$$\log t_r + \left[R_1(T - T_m) + R_2\left(\frac{1}{T} - \frac{1}{T_m}\right)\right] = B + C\log\sigma + D_\sigma + E\sigma^2 \tag{17.5}$$

The MCM is included in Fig. 17.22. T_m represents the middle temperature (1498 K) of the temperature range used. A multivariate regression analysis provided: $R_1 = 0.1731$; $R_2 = 303700$; $B = 87.24$; $C = -45.69$; $D = 0.1156$;   and

Table 17.1 Matrix for creep tests of GN-10 Si$_3$N$_4^a$. Ding et al. [5]. With kind permission of John Wiley and Sons

Stress (MPa)	1150 °C	1200 °C	I250 °C	1300 °C
75			$D(>2238$ h)^d	D (>1125 h)^e
100			$D(>1030$ h)^d	$X(1721$ h)
125		$D(>1031$ h)^b	$X(2996$ h)	$X(15.2$ h)
150		$X(1204$ h)	$X(135.9$ h)	$X(0.2$ h)^f
175		$D(>3405$ h)^c	$X(25.5$ h)	
200		$X(203.1$ h)		
225		$X(96.3$ h)		
250	$X(733.8$ h)	$X(7.5$ h)		
275	No test			
300		$X(365.4$ h)		

[a]The numbers in parentheses indicate the duration of the lest, which is denoted either as completed (X) or as disrupted (D)
[b]Stress increased to 225 MPa after 1031 h of testing
[c]Fractured at specimen buttonhead due to a power outage
[d]Fractured at specimen shank
[e]Stress increased to 100 MPa after 1125 h of testing
[f]No meaningful creep rate available due to fast fracture

$E = -0.9733 \times 10^{-4}$. Both the experimental data and the predictions appear in Fig. 17.22. The LMP describes a linear relation between log (stress) and log (rupture time), whereas the MCM is not linear (see Fig. 17.22).

(c) The MGR uses a power-law relation to describe the correspondence between creep, rupture time, and the minimum creep rate, given as:

$$t_r = A\dot{\varepsilon}^{-p} \qquad (17.6)$$

The constants A and p are 0.0021 and 0.91, respectively, and were determined from Fig. 17.23. The stress exponent and the activation energy of Norton's power law were evaluated giving 12.6 and 1645 kJ mol^{-1}, respectively. The MGR curve yielded a p value of 0.91.

Based on the high values derived from the data analyses for both the stress exponents, m and n, the growth of macrocracks is probably the most dominant mechanism of creep fracture. Recall that m is the slope of the LMP curve, with values between 13 and 14.4. In an additional creep-rupture evaluation, a S$_3$N$_4$ composite containing 4 wt% yttria underwent a uniaxial tensile-creep test, performed in the 1422–1673 K temperature range. In Fig. 17.24, a MGR-type relation is shown, indicating rupture-life dependence on the secondary-creep rate. The slopes of the lines have approximately the same values. The strain at fracture is seen in Fig. 17.25 versus the rupture time in 17.25a and versus the stress in 17.25b. In this figure, (a) is the total accumulated strain (excluding elastic strain) at fracture, seen as a function of failure time; while in (b) the failure strain is shown as a

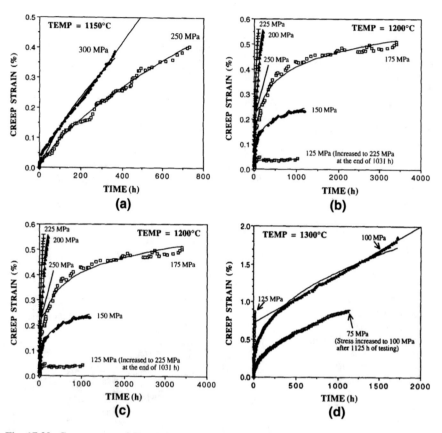

Fig. 17.20   Creep curves of GN-IO Si₃N₄ tested at **a** 1150 °C; **b** 1200 °C; **c** 1250 °C; **d** 1300 °C. *Symbols* are experimental data; *solid lines* are curve fitting. Ding et al. [5]. With kind permission of John Wiley and Sons

decreasing function of stress. Typical fractures, showing the crack-growth regimes in internal and surface initiations, are shown in Fig. 17.26a, b, respectively.

This section on creep rupture ends with the observations that cavitation occurs at two grain boundaries and preferentially at triple junctions. As such, cavitation is an inherent part of the creep process. Failure occurs by crack initiation and propagation. And finally, the MGR can be applied to rupture-life prediction.

17.5   Superplasticity in Si₃N₄

Fine-grained material is essential, but not sufficient, for superplastic behavior. Superplasticity has been observed in various silicon nitrides. The following considers the deformation behavior of hot-pressed, fine-grained β-Si₃N₄ in the

Fig. 17.21 In the stress-rupture region, the creep rate is high and the time-to-rupture is short. As temperature and stress decrease below the approximate transition region (*shaded area*), the behavior features are reversed. Ding et al. [5]. With kind permission of John Wiley and Sons

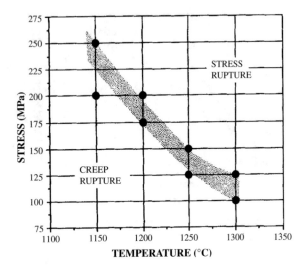

Fig. 17.22 Comparison of experimental data and predictions of the Larson-Miller and minimum commitment models. *Arrowed* data were not included in the regression analysis. Ding et al. [5]. With kind permission of John Wiley and Sons

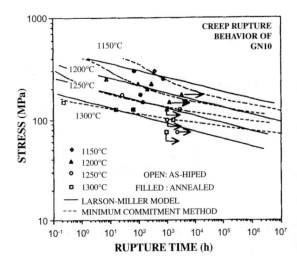

1450–1650 °C temperature range. This deformation test is performed under compression and the flow stress is expressed as a function of strain rate and temperature. A typical curve is seen in Fig. 17.27. In the figure, a corrected curve is shown. Corrections were required, because the experiments were performed at a constant displacement, rather than a CSR, and because large deformations were involved. The corrected flow stress is given as:

$$\sigma_c = \sigma_0[\exp(\varepsilon)]^{1/n} \tag{17.7}$$

Also note that the n values barely change with temperature. The stress exponent is about 1 in the lower temperature region and somewhat larger in

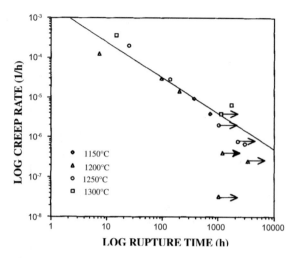

Fig. 17.23 Comparison of experimental data and the Monkman-Grant relation represented by a straight line. *Arrowed* data were not included in the regression analysis. Ding et al. [5]. With kind permission of John Wiley and Sons

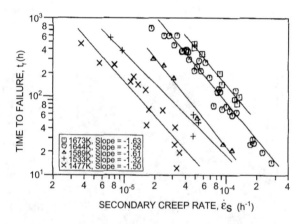

Fig. 17.24 Dependence of rupture life on secondary-creep rate. Note the stratification of the Monkman-Grant lines with respect to temperature. Menon et al. [6]. With kind permission of John Wiley and Sons

temperatures >1550 °C. For the activation-energy evaluation, the expression indicated several times in earlier sections and reproduced here is:

$$\dot{\varepsilon} = \frac{A}{d^p} \sigma^n \exp\left(-\frac{Q}{RT}\right) \qquad (17.8)$$

The temperature dependence of the strain rate is plotted as an Arrhenius curve in Fig. 17.30. σ_0 stands for the corrected flow stress. The stress exponent, n, is expressed by $\dot{\varepsilon} = A\sigma^n$. A typical set of stress–strain curves, for various strain rates, at 1550 °C under compression is shown in Fig. 17.28. All the data shown here and below are for true stress, true strain, and true strain rate (corrected data). After the initial transient state, a steady state is reached for all the strain rates. The initial, transient state reflects some elastic deformation of the sample. Note that no strain hardening is observed in the lines of Fig. 17.28 and a large deformation strain, up

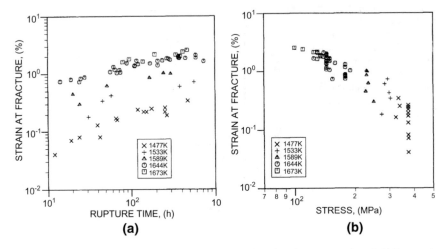

Fig. 17.25 a Failure strain in creep-tested specimens plotted against rupture time. **b** Failure strain in creep-tested specimens plotted against applied stress. Menon et al. [6]. With kind permission of John Wiley and Sons

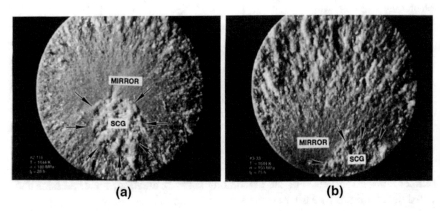

Fig. 17.26 Optical photographs of fracture surfaces from **a** an internal initiation at 1644 K/180 MPa/28 h, and **b** a surface initiation at 1644 K/150 MPa/75 h. Note the subcritical crack growth (SCG) region and the mirror region that follows SCG zone. Menon et al. [6]. With kind permission of John Wiley and Sons

to ~55%, is obtained. Quasi-steady-state flow stresses may be determined beyond the initial regions. The strain rate, as a function of flow stress at various temperatures, is plotted on a logarithmic scale in Fig. 17.29.

The activation energies calculated from the slopes of the lines in Fig. 17.30 are 344 ± 26 kJ mol^{-1} at 20 MPa and 410 ± 46 kJ mol^{-1} at 100 MPa. Un-deformed and deformed specimens showed (imaging analysis) that the average grain diameter and aspect ratio changed very little after superplastic deformation. Figure 17.31

Fig. 17.27 Typical
correction curve for a
compression test at 1550 °C
and an initial strain rate of
3 × 10^{-4} s^{-1}, in the
as-hot-pressed β-Si$_3$N$_4$
material. Zhan et al. [7]. With
kind permission of John
Wiley and Sons

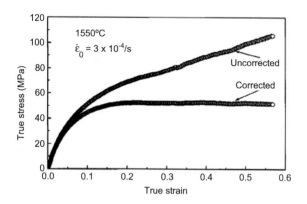

Fig. 17.28 Compressive
stress–strain curves for
various strain rates of the
as-hot-pressed material at
1550 °C. Zhan et al. [7]. With
kind permission of John
Wiley and Sons

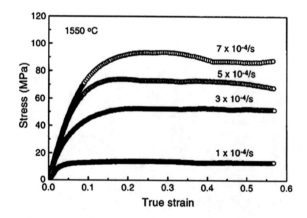

Fig. 17.29 Strain rate versus
stress at various temperatures,
under compression
(*n* = slope), in the
as-hot-pressed β-Si$_3$N$_4$
material. Zhan et al. [7]. With
kind permission of John
Wiley and Sons

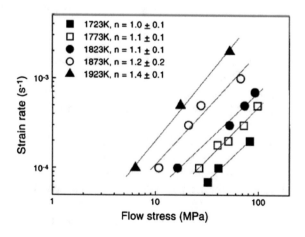

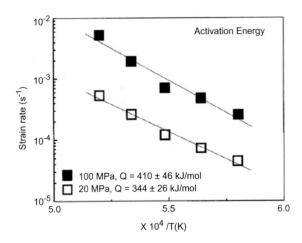

Fig. 17.30 Determination of activation energy for flow equation in the as-hot-pressed β-Si$_3$N$_4$ material. Zhan et al. [7]. With kind permission of John Wiley and Sons

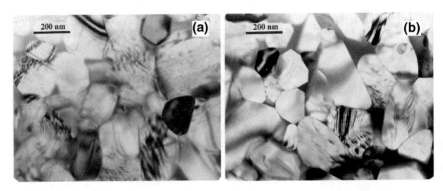

Fig. 17.31 TEM photographs of **a** an un-deformed sample and **b** a deformed sample at 1600 °C, with a true strain of −1.1, showing no dynamic grain growth. Zhan et al. [7]. With kind permission of John Wiley and Sons

shows TEM photographs of (a) an un-deformed sample and (b) a sample deformed at 1600 °C and a large true strain of −1.1.

RTEM photographs are shown in Fig. 17.32. Boundaries, oriented parallel or perpendicular to the applied-load direction in a superplastically deformed sample, are seen.

Contrary to some silicon nitrides, no strain hardening is observed in the β-Si$_3$N$_4$ and no shape change occurred because of the uniform particle-size distribution of the original powder, and because no α-to-β phase transformation occurred. Apparently, the grains remained equiaxed, with no grain growth, although a large deformation was involved. The strain hardening in other silicon nitrides is attributed to microstructural changes during deformation, such as dynamic grain growth and

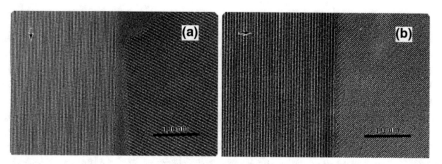

Fig. 17.32 Representative HRTEM photographs of boundaries oriented **a** parallel and **b** perpendicular to the applied-load direction, indicating that the grain-boundary film thickness decreased after superplastic deformation, under compression [(→) applied stress direction during deformation]. Zhan et al. [7]. With kind permission of John Wiley and Sons

α-to-β phase transformation. As mentioned above, a fine grain size is necessary, but not sufficient alone, to cause superplasticity. Also, it is important to ensure the stability of the fine-grained microstructure during superplastic deformation, thus preventing dynamic grain growth and keeping the cohesive strength of the grain boundaries high enough to sustain high ductility without fracture. The β-Si₃N₄ used in these experiments has great microstructural stability and can stave off static or dynamic grain growth. It is believed that the equiaxed grain shape remaining after a large deformation, in the absence of dislocation activity, suggests that GBS and grain rotation, accommodated by viscous flow, may form the mechanism of superplasticity in β-Si₃N₄. This is also supported by the stress exponent of ∼ 1, which is a Newtonian viscous flow, meaning that the strain rate is directly proportional to the stress, $\dot{\varepsilon} = A\sigma$.

17.6 Creep in Nano-Si₃N₄

Limited data, if any, are available on creep in pure nano-Si₃N₄, despite the extensive research devoted to evaluating properties of Si₃N₄ itself. However, more information is available on the nano-Si₃N₄ composite. A frequently discussed nanocomposite is the Si₃N₄/SiC system, the topic of this section. This system has a high creep resistance at elevated temperatures, although a high-density material is required for this purpose. Both of the components comprising this system are covalent ceramics and, for high-density achievement, liquid-phase sintering is required. Additives, such as Y₂O₃, Al₂O₃, MgO, etc., serve as sintering aids. At the sintering temperature, the additive reacts with the silicon-oxide layer (which is always present at the surface of the silicon nitride particles) forming a liquid phase and, thus, promoting sintering. Upon cooling, a thin (0.5–2 nm) liquid, glassy phase forms at the grain boundaries. This glassy phase determines the creep behavior of the silicon-nitride composite.

To examine the creep behavior in this nano-nanocomposite and to avoid problems, which may arise from cavitation (generally occurring during tensile creep), compression tests were performed. The objective of this investigation was to determine the steady-state creep rate at various temperatures and stresses. Creep strain–time curves for one of the nano-nanocomposites (sintered at 1600 °C/10 min with 1 wt% yttria additive) is illustrated in Fig. 17.33.

The compression-creep tests were conducted in air in the temperature range of 1350–1450 °C applying a uniaxial constant stress, using a 'step-stress' technique. The specimen was first subjected to a 50 MPa stress and, when the creep deformation reached steady state, the stress was increased at 50 MPa steps and the secondary creep rate was measured at each new stress level. TEM results are shown in Fig. 17.34. The analysis reveals that, with decreasing additive, the grain size decreases and there is a transition from a micro-nano to a nano–nano structure. In Fig. 17.34a for 8 wt% Y₂O₃ additive, the microstructure is composed of a submicron matrix of Si₃N₄/SiC matrix having a mean grain size of 180 and 10–30 nm SiC particles are present as inclusion in most Si₃N₄ grains. This microstructure also remained with 5 wt% Y₂O₃ additive, but the grain size had decreased to 130 nm. When the yttria content is further decreased to a level of 3 wt%, both components in this system are nano-sized and the grain sizes of the silicon nitride and the SiC are, respectively 62 and 35 nm for the 3 and 1 wt% Y₂O₃ (Fig. 17.34b, c). The grain size further decreases when no Y₂O₃ is added. This is shown at Fig. 17.34d, where the grain size is 27 nm for 10 min sintering at 1600 °C and 40 nm with 30 min sintering. An elemental distribution of the nano-nanocomposite was performed by EELS illustrated in Fig. 17.35.

The two constituent phases of the composite are randomly mixed with about equal grain sizes. The important information received from the EELS analysis is about the presence of oxygen, which is responsible for liquid-phase formation at the boundaries and pockets, despite the absence of metal-oxide sintering phases. This is an indication that O diffused during sintering from the particle surfaces of the

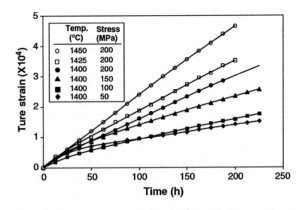

Fig. 17.33 Compression-creep strain–time curves for one of the nano-nano composites (1 wt% Y₂O₃, 1600 °C/10 min sintered). wan et al. [8]. With kind permission of John Wiley and Sons

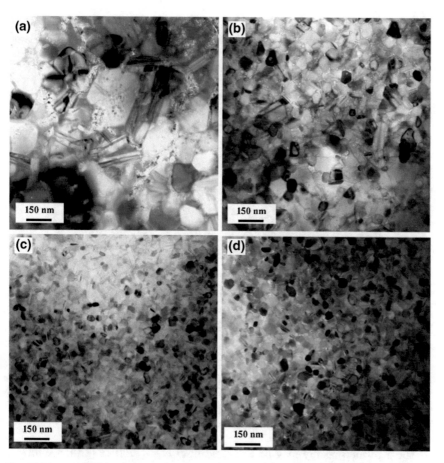

Fig. 17.34 Transmission electron microscopy (TEM) observations of nanocomposites of Si_3N_4-SiC. **a** Sintered with 8 wt% Y_2O_3 at 1600 °C for 10 min, micro-nano structure, **b** sintered with 3 wt% Y_2O_3 at 1600 °C for 10 min, nano-nano structure, **c** sintered with 1 wt% Y_2O_3 at 1600 °C for 10 min, nano-nano structure, **d** sintered without additive at 1600 °C for 30 min, nano-nano structure. wan et al. [8]. With kind permission of John Wiley and Sons

original, amorphous powder (silica or oxynitride) into the interior of the particles and is distributed along the grain boundaries. In Fig. 17.36, one sees HRTEM of the grain-boundary region. It is important to note that a large number of grain boundaries exist without the amorphous layer (Fig. 17.36a), although EELS revealed the existence of oxygen. This is probably because the amount of segregated O was insufficient to form a glassy phase. However, some bi-grain junction, glassy layers are visible (1 nm thickness) in Fig. 17.36b. Most of the glassy grain-boundary phase exists at multigrain junctions, as indicated in 17.36c. The creep results from this research are compared with those from other microcrystalline silicon-nitride ceramics in Fig. 17.37.

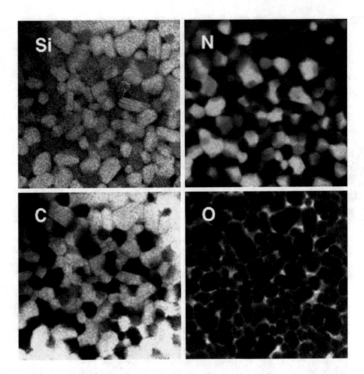

Fig. 17.35 Electron energy loss spectroscopy (EELS) analysis of the component elements in the Si₃N₄-SiC nanocomposite sintered at 1600 °C for 30 min without additive. wan et al. [8]. With kind permission of John Wiley and Sons

The activation energy may be evaluated from Eqs. (17.8) or (8.9), resulting in 205 kJ/mol for the 1 wt% Y_2O_3, which is significantly lower than that of the microcrystalline Si₃N₄. Experimental creep data are usually compared with the established theoretical models for their n, p, and Q values (Eq. 17.8), in order to determine the creep-deformation mechanism. The different n values in the various silicon nitrides in the range <1 and >3 make it somewhat difficult to compare nano-Si₃N₄ and micro- or macro-sized silicon nitrides. Moreover, the activation energy varies in the 300–1300 kJ mol^{-1} range and, thus again, no definitive, universal-creep mechanism, applicable to all types of silicon nitrides, is plausible, due to the various values that complicate such a formulation.

The further cavitation associated with tensile creep makes the above task even more difficult. Nevertheless, and despite these discrepancies, it is generally agreed that the steady-state creep for silicon nitride with a glassy phase proceeds by a solution-precipitation mechanism through the amorphous grain-boundary phase (glassy phase).

The extraordinarily high creep resistance found in the nanocomposites (as mentioned above) strongly suggests a fundamental change in creep mechanism. Another indication that the creep mechanism may be different in nano-nanocomposites is the

Fig. 17.36 High-resolution transmission electron microscopy (HRTEM) analysis of the grain boundary of the nano-nano composite (no additive, 1600 °C/30 min sintered). **a** Glass-free grain boundary, **b** Grain boundary containing glassy layer, **c** triple junction. wan et al. [8]. With kind permission of John Wiley and Sons

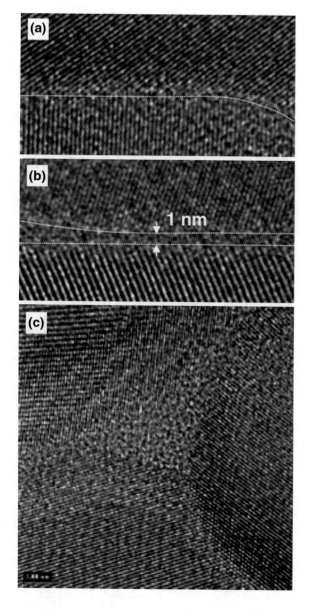

low activation energy found in these materials, only 205 kJ mol^{-1}. A dislocation-based mechanism for creep in silicon nitride is not a possibility, because of the strong covalent bonds in both the Si$_3$N$_4$ and the SiC, and due to the high Peierls force. Also, alternative mechanisms, based on solid-state diffusion in lattices (Nabarro-Herring) and grain boundaries (Coble creep) are unlikely, as indicated by the diffusion data in Table 17.2.

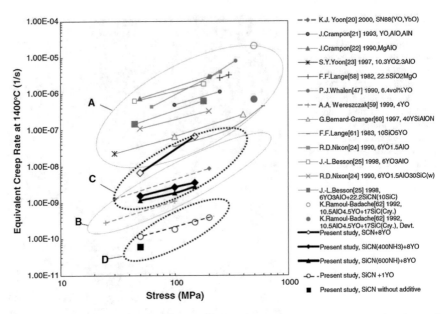

Fig. 17.37 Comparison of the compression-creep property of nanocomposites with those of existing silicon-nitride ceramics (additive in weight percentage unless specified, molecular formula simplified for clarity. For instance, "6YO" in figure legend stands for "6 wt% Y₂O₃"). wan et al. [8]. With kind permission of John Wiley and Sons. Crampon et al. [9], Yoon et al. [10]

Table 17.2 Activation energy for diffusion processes in the Si₃N₄/SiC system. wan et al. [8]. With kind permission of John Wiley and Sons. Crampon et al. [9], Yoon et al. [10]

Medium	Diffusing particle	Temperature range (°C)	Activation energy (kJ/mol)	Note	References
α-Si₃N₄	Si	1400–1600	199	Self diffusion	Kunz et al.[53]
	N	1200–1410	233	Lattice diffusion	Kijima and Shirasaki[54]
	Si	NA	NA	Grain-boundary diffusion	NA
	N	NA	NA	Grain-boundary diffusion	NA
β-Si₃N₄	Si	1490–1750	390	Lattice diffusion (β with some α)	Batha and Whitney[55]
	N	1200–1410	777	Lattice diffusion	Kijima and Shirasaki[54]
	Si	NA	NA	Grain-boundary diffusion	NA
	N	NA	NA	Grain-boundary diffusion	NA
β-SiC	Si	1960–2260	911	Lattice diffusion	Ziegler et al.[2]
	Si	2010–2270	612	Grain-boundary diffusion	Hon et al.[56]
	C	1860–2230	841	Lattice diffusion	Ziegler et al.[2]
	C	1855—2100	564	Grain-boundary diffusion	Hon and Davis[57]
In GB of HPSN	Si/N	1450–1550	448	Grain-boundary diffusion	Ziegler et al.[2]
(10 wt% Y₂O₃)	Si/N	1550–1760	695	Grain-boundary diffusion	Ziegler et al.[2]

In all the cases indicated in Table 17.2, their activation energies are much higher than in the nanocomposites considered here, with the low value of 205 kJ mol^{-1}. This low activation energy implies that, in the nano-nanocomposites, creep is not controlled by either diffusion in silicon carbide or by the lattice diffusion in β-Si$_3$N$_4$.

Summing up this section, creep deformation in covalent ceramics, such as silicon nitride and silicon carbide, is dominated by a solution-precipitation process via a glassy phase at the grain-boundary regions.

17.7 Creep Recovery in Si$_3$N4

The definition of 'creep recovery' is the time-dependent portion of the strain in materials following the removal of the stress that was responsible for that strain. As such, to get information on recoverable strain, often cyclic creep tests are performed, involving the removal of stress and the observation of the recovered strain. The time-dependent portion of the decrease in strain in a material, following the removal of a stress that has deformed it, is shown in a schematic illustration of recovery in cyclic-creep experiments in Fig. 17.38.

The definition of 'total strain' is shown in Fig. 17.39.

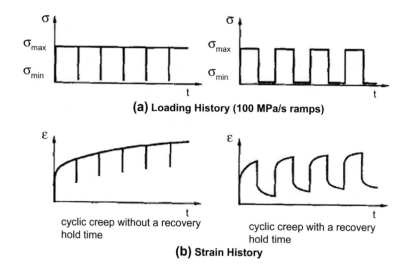

Fig. 17.38 Cyclic creep experiments. **a** Schematic representation of the cyclic loading histories examined and **b** idealized strain response. For all cyclic creep experiments, the creep stress (σ_{max}) was fixed at 200 MPa and the recovery stress (σ_{min}) at 2 MPa. The loading and unloading ramps were performed at 100 MPa/s. Holmes et al. [11]. With kind permission of John Wiley and Sons

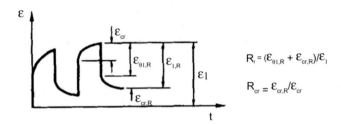

Fig. 17.39 Definition of the total-strain recovery ratio (R_t) and creep strain recovery ratio (R_{cr}) used to quantify the amount of strain recovery during the cyclic creep experiments. Holmes et al. [11]. With kind permission of John Wiley and Sons

The two types of ratios used to quantify the amount of strain recovered after unloading are: (1) total-strain recovery ratio and (2) creep strain recovery ratio. They are both shown in Fig. 17.39 and may be expressed as:

$$R_t = \frac{\left(\varepsilon_{el,R} + \varepsilon_{cr,R}\right)}{\varepsilon_t} \qquad (17.9)$$

and:

$$R_{cr} = \frac{\varepsilon_{cr,R}}{\varepsilon_{cr}} \qquad (17.10)$$

ε_t is the total accumulated strain and ε_{cr}, R is the creep strain recovered in a particular cycle. Following the loading cycle, the creep behavior in Si$_3$N$_4$ reinforced with SiC fibers is presented in Fig. 17.40.

One may infer from the Fig. 17.40a that, in the 50 h creep/50 h recovery cycle, transient creep is characterized by rapid strain accumulation and a decreasing creep rate for the first 50 h of creep. On subsequent cycles, transient creep decreased to approximately 25 h per cycle. Rapid, repeated unloading, and reloading cycles every 50 h (no recovery hold period) did not introduce transient creep and the prior creep rate continued, namely was immediately reestablished (Fig. 17.40b). A significant change is observed in the shorter 300-s creep/300 s recovery cycles, as seen in Fig. 17.40c. In this case, the transient-creep duration was significantly reduced (less than 20 h, compared to 70 h for sustained creep at 200 MPa). On reloading, every indication for transient creep is seen. The insets in Fig. 17.40c show the creep behavior at selected times. The accumulated creep strain was much less than in the sustained creep loading at 200 MPa.

There was not much difference between the sustained and the accumulated creep, once the recovery period was eliminated by rapid unloading and reloading (every 300 s). The reduction in the accumulated strain was the result of strain recovery (that occurs during 2 s unloading and 2 s reloading at a rate of 100 MPa s^{-1}). In Figs. 17.41 and 17.42, the cyclic and sustained creeps were loaded at 200 MPa.

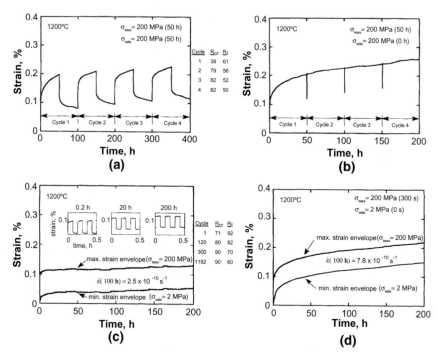

Fig. 17.40 Isothermal (1200 °C) cyclic creep behavior of 0° SCS-6 SiC/Si₃N₄. Specimens were cycled between stress limits of 200 and 2 MPa. For loading histories with a finite recovery hold time, the total-strain recovery and creep strain recovery ratios ($R_t = (\varepsilon_{cr,R} + \varepsilon_{xp,P})/\varepsilon_t$, and $R_{cr} = \varepsilon_{cr,}$ $/\varepsilon_{cr}$, respectively) are shown adjacent to the creep curves. **a** 50-h creep/50-h recovery. **b** 50-h creep/0-s recovery. **c** 300-s creep/300-s recovery. **d** 300-s creep/0-s recovery. Holmes et al. [11]. With kind permission of John Wiley and Sons

The cycle period has a large impact on creep behavior. In the case of the 50-h creep/50-h recovery cycles, only a moderate reduction in accumulated creep strain was observed (Fig. 17.41). For an equivalent time at 200 MPa, the higher frequency 300-s creep/300-s recovery cycles showed great reductions in accumulated creep strain and strain rate (e.g., at 100 h, the accumulated strain was 60% lower and the creep rate was 43% lower than that found for sustained loading at 200 MPa, see Fig. 17.42). The reduction in accumulated creep strain found for short-duration cyclic loading is a consequence of the reduced duration of transient creep and the reduction in the overall creep rate.

Significant strain recovery was observed for the loading histories with a finite recovery hold time. For single-cycle 200-h creep/25-h recovery experiments (see Fig. 17.43), the creep strain recovery ratio (R_{cr}) was roughly 50% after 2.5 h of recovery; the total-strain recovery ratio was approximately 45–46%. A larger recovery in creep strain has occurred during the multi-cyclic creep experiments. Creep strain recovery ratios could reach 82% in a (four-cycle, 50-h creep/50-h) recovery experiment.

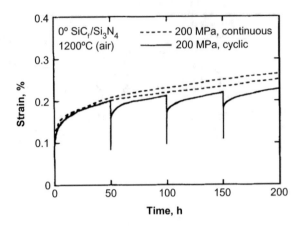

Fig. 17.41 Comparison of the accumulated creep strain and tensile-creep rate for sustained loading at 200 MPa and long-duration cyclic loading (50-h creep/50-h recovery) between stress limits of 200 and 2 MPa. Only the loading portions of the cyclic-creep curve are shown (the recovery segments were deleted, and the resulting curves were shifted to the left to allow a comparison of creep strain accumulation to be made for an equivalent time at the creep stress of 200 MPa). Holmes et al. [11]. With kind permission of John Wiley and Sons

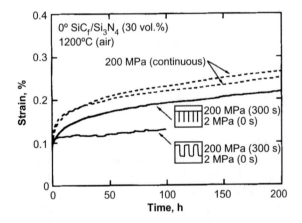

Fig. 17.42 Comparison of the accumulated creep strain and creep rate for sustained loading at 200 MPa and for short-duration cyclic loading (300-s creep/300-s recovery and 300-s creep/0-s recovery). For the cyclic creep experiments, only the traces of the strain versus time curves obtained at the creep stress of 200 MPa are shown. As in Fig. 17.41, the recovery segments of each cycle have been removed to allow a comparison of accumulated creep strain to be made for an equivalent time at the creep stress of 200 MPa

Strain recovery provides a powerful mechanism for the reduction of creep strain during cyclic-creep loading. In the absence of cyclic-crack growth, the strain recovery is expected to significantly increases the life of the

Fig. 17.43 Creep-recovery behavior of specimens subjected to sustained tensile creep for 200 h at 200 MPa, followed by 25 h of recovery at 2 MPa. After 25 h of recovery, the total-strain recovery ratio ($R_t = \varepsilon_{el,R} + \varepsilon_{cr,R})/\varepsilon_{el}$) was approximately 50%. Holmes et al. [11]. With kind permission of John Wiley and Sons

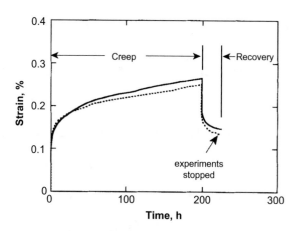

cyclically loaded structures. The main findings of these experiments may be summarized as:

(a) The basic damage in fiber-reinforced composites is the periodic fiber fracture that occurs during long-duration tensile creep. This is expected to occur in composites in which the creep rate of the matrix exceeds that of the fiber.

(b) Primary creep is significantly reduced during cyclic loading with a finite recovery hold time. Thus, under sustained loading, primary creep persists for ∼70 h, under the experimental conditions, while it can be reduced during cyclic loading to ∼20 h (by means of a 300 s hold time at 200 MPa, followed by a 300 s recovery per cycle).

(c) Knowledge of creep strain recovery behavior is essential, if the lifetime of the composite is to be increased under sustained or cyclic loading. Significant residual tensile stresses may develop in the component of a composite that has a lower creep rate. A reduction in residual stresses may be practically achieved, if a specific component of interest (with a lower creep rate) is periodically removed from service, isothermally annealed (under zero load), in order to remove the residual tensile stresses and accumulated creep strain.

One cannot finish this section without considering recovery in hot-pressed, pure silicon nitride. This creep test may be characterized by (a) persistent non-recoverable plastic deformation and (b) a transient recoverable (viscoelastic) deformation. There is a power-law stress exponent of $n = 4$ and an activation energy of 848 kJ mol^{-1}. The persistent creep component is time-dependent and is described by a parabolic law, while the recoverable (viscoelastic) component is independent of the total strain. Various creep data must be accumulated in order to amass the relevant data from such creep experiments.

At this point, a review of some important relations, before considering the actual recovery information, is in order.

The power law:

$$\varepsilon = at^b \tag{17.11}$$

where b is a least-square slope of a log–log plot and a is the value of strain at log $t^b = 0$. At constant stress, a and b are constants. From Eq. (17.11), one gets:

$$\dot{\varepsilon} = abt^{b-1} = c\varepsilon^d \tag{17.12}$$

As b approaches unity, or as t gets longer, $d\dot{\varepsilon}/dt$ is given as:

$$\ddot{\varepsilon} = ab(b-1)t^{b-2} \tag{17.13}$$

Another commonly used empirical expression is:

$$\varepsilon = \left(\frac{\sigma}{k_2}\right)\left(1 - \varepsilon^{-k_1 k_2 t}\right) + k_3(\sigma - \sigma_0)t \tag{17.14}$$

where $k_{1,2,3}$ are constants at constant temperature, σ is the applied stress, and $\sigma_0 \geq 0$ is an apparent yield stress. Equation (17.14) does not apply to the observed creep curves considered here, because no steady-state creep was observed. The activation energy is determined from the known expression, given earlier, but reproduced here in a different form as:

$$\dot{\varepsilon} = Af(s)\sigma^n \exp\left(-\frac{Q}{RT}\right) \tag{17.15}$$

where A is a constant, $f(s)$ is some function of creep structure and, σ is the applied stress. Q may also be evaluated experimentally from data at two temperatures by:

$$Q = \frac{R \ln\left(\frac{\dot{\varepsilon}_1}{\dot{\varepsilon}_2}\right)}{\left(\frac{1}{T_2}\right) - \frac{1}{T_1}} \tag{17.16}$$

Those expressions previously used for the analysis of creep in pure Si$_3$N$_4$ are not now reproduced here again.

Finally, creep strain recovery in Si$_3$N$_4$ is characterized by very high initial rates, which rapidly decreased over time, as illustrated in Fig. 17.44.

The total recovered strain is $\sim 0.1\%$ absolute strain, or roughly 5–10% of the previous creep strain, and recovery lasted up to 30 h. When the specimen is reloaded, the curve displays an inverse of the recovery in parallel, along with the continued accumulation of non-recoverable plastic deformation. The results of the strain-recovery experiments are presented in Table 17.3.

A plot of the normalized, corrected recovery rate versus the normalized, prior creep stress at constant strain and temperature is available in Fig. 17.45. The linear dependence, with a slope of 1.1 ± 0.2, demonstrates that the recovery phenomenon is linear viscoelastic. The most general, and simplest, linear viscoelastic analogue

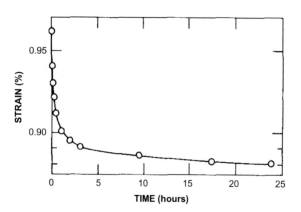

Fig. 17.44 Strain versus time curve for recovery at 1204 ° C; prior stress 103.3 MN m⁻². Arons and Tien [12]. With kind permission of Springer

model, which predicts a creep transient and subsequent recovery, is the Kelvin–Voight model, consisting of a spring and dashpot connected in parallel. Under a step-loading function, $\sigma = \sigma_0 H(t)$, where σ_0 is the applied stress and $H(t)$ is a unit-step function; the resultant strain is:

$$\varepsilon(t) = \left[1 - \exp\left(-\frac{t}{\theta}\right)\right]\sigma_0 k \qquad (17.17)$$

where θ, the retardation time, equals n/k, the dashpot viscosity divided by the spring constant. The quantity $[1 - \exp(-t/O)]/k$ is usually denoted by $C(t)$, and is called the 'creep compliance function.' Thus, Eq. (17.17) becomes:

$$\varepsilon(t) = C(t)\sigma_0 \qquad (17.18)$$

On removal of the stress, $t = t'$. The superposition principal demands that for $t > t'$:

$$\varepsilon(t) = \varepsilon(t') - C(t - t')\sigma_0 \qquad (17.19)$$

Accordingly, it is predicted that a plot of the $\ln\{[\varepsilon(\infty) - \varepsilon(t)]/\varepsilon(\infty)\}$ versus t should be a straight line of slope = $\pm\theta$. Such a plot is found in Fig. 17.46.

The activation energy for the viscoelastic process is obtained through the temperature dependence of the recovery rate. The activation energy for the viscoelastic mechanism may be obtained by allowing:

$$Q_v = -\left\{[\partial]n\dot{C}(t,T)]/\left[\partial\left(\frac{1}{RT}\right)\right]_{t=\text{const}} = -[\partial]n\dot{\varepsilon}(t.T)]/\left[\partial\left(\frac{1}{RT}\right)\right]_{t=\text{const.}}\right\} \qquad (17.20)$$

The natural logarithm of the recovery-strain rate vs. the reciprocal of the temperature is plotted in Fig. 17.47. The resultant activation energy is $Q_v = 722 \pm 25$ kJ mol⁻¹ at 4 h. Figure 17.47 was taken 4 h after the load removal.

Table 17.3 Recovery data. Arons and Tien [12]. With kind permission of Springer

Specimen	Temperature (°C)	Stress (MN m^{-1})	Accumulated strain (% ± 0.05)	Recovery rate[a] at $t = 4$ h (sec^{-1})	Recovered strain (%)	Duration (h)	Corrected recovery rate (sec^{-1})
7	1204	68.9	0.50	-1.7×10^{-9}	0.020	4	-2.4×10^{-8}
9	1204	86.1	0.50	-5.0×10^{-9}	0.062	19	-2.7×10^{-8}
2	1204	103.3	0.25	-1.5×10^{-8}	0.178	118	-3.7×10^{-8}
6	1204	103.3	0.50	-1.2×10^{-8}	0.090	18	-3.4×10^{-8}
6	1204	103.3	1.00	-1.0×10^{-8}	0.080	24	-3.2×10^{-8}
6	1204	103.3	1.50	-1.5×10^{-8}	0.085	10	-3.7×10^{-8}
8	1177	103.3	0.50	-4.1×10^{-9}	0.058	22	-1.0×10^{-8}
3	1233	103.3	0.50	-5.9×10^{-9}	0.064	8	-8.5×10^{-8}
3	1233	103.3	1.00	-6.3×10^{-9}	0.132	32	-8.5×10^{-8}
5	1260	103.3	0.50	$+1.0 \times 10^{-8}$	0.050[b]	1[b]	-2.6×10^{-7}

[a]Recovery rate taken at 4 h after load removal
[b]Before turn about in curve (see text)

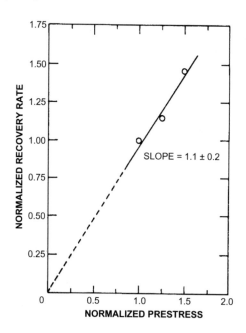

Fig. 17.45 Dependence of corrected recovery rate upon prior stress. Curved slope of ∼1 is indicative of linear viscoelasticity. Arons and Tien [12]. With kind permission of Springer

Fig. 17.46 Plot of natural log of the fraction of recovered strain versus time. Lack of straight line fit indicates that recovery is characterized by a spectrum of retardation times. Arons and Tien [12]. With kind permission of Springer

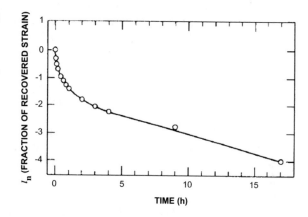

When a glass phase is present in the grain boundaries, it is assumed that GBS and grain-boundary fluid motion vary with strain and the creep rate of the entire composite is given by:

$$\dot{\varepsilon} = \text{constant} \cdot \sigma \frac{x_0}{x_0 - yx(\sigma)} \tag{17.21}$$

Here, according to the model, any given volume of the aggregate will contain x_0 number of glassy areas of which $x(\sigma)$ contain voids under an applied stress, σ.

Fig. 17.47 Temperature
dependence of corrected
recovery rate. Arons and Tien
[12]. With kind permission of
Springer

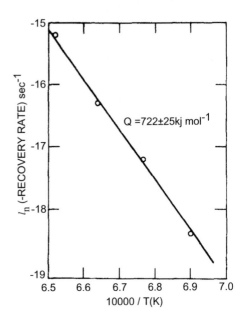

Fig. 17.47 Temperature dependence of corrected recovery rate. Arons and Tien [12]. With kind permission of Springer

If each void containing glassy regions affects y nearest-neighbor regions, such that they have negligible resistance to deformation (compared to voidless regions), then the voidless region must support a greater stress by a factor of $x/x_0 - yx(\sigma)$. Under the assumption that GBS and grain-boundary fluid motion vary with stress, the creep rate of the composite is given by Eq. (17.21). The stress exponent is defined by $n = \ln \dot{\varepsilon}/d\ln\sigma$, then given by:

$$n = 1 + \frac{y}{x_0 - yx(\sigma)} \frac{\partial x(\sigma)}{\partial \sigma} \qquad (17.22)$$

If the hot-pressed silicon nitride is again step-loaded after the loading-unloading cycle has been completed, once again, the elastic strains will accumulate and the grains will continue to slide and rearrange as before. The persistent creep deformation will simply continue from where it was interrupted by the unloading sequence.

According to the creep results (the creep part is not shown here, as was indicated), it is proposed that creep deformation in hot-pressed silicon nitride is due to relative grain motion, accommodated by grain-boundary phase flow and cavitation. (Note that the concepts of viscous material and fluid are used in this work, which were once considered to be characteristics of grain boundaries). The reason for using such early terminology for creep is a consequence of the discussion about recovery in pure Si$_3$N$_4$, which focused on recovery occurring during cyclic creep, which is of practical importance for periodic loading and the examination of test specimens and, perhaps, while heat-treating and restoring whatever strain can be restored, before reloading them for further use.

Thus, creep deformation is explained by GBS which is rate-controlled and accommodated by grain-boundary phase percolation, cavitation, and void and wedge opening.

References

1. Luecke WE, Wiederhorn SM, Hockey BJ, Krause RE Jr, Long GG (1995) J Am Ceram Soc 78:2085
2. Birch JM, Wilshire B (1978) J Mater Sci 13:2627
3. Thayer RB, Yang J-M (1993) Mater Sci Eng A 160:169
4. Holmes JW (1991) J Mater Sci 26:1808
5. Ding J-L, Liu KC, More KL, Brinkman CR (1994) J Am Ceram Soc 77:867
6. Menon MN, Fang HT, Wu DC (1994) J Am Ceram Soc 77:1217
7. Zhan G-D, Mitomo N, Nishimura T, Xie R-J, Sakuma T, Ikuhara Y (2000) J Am Ceram Soc 83:841
8. Wan J, Duan R-G, Gasch MJ, Mukherjee AK (2006) J Am Ceram Soc 89:274
9. Crampon J, Duclos R, Rakotoharisoa N (1993) J Mater Sci 28:909
10. Yoon KJ, Wiederhorn SM, Luecke WE (2000) J Am Ceram Soc 83:2017
11. Holmes JW, Park YH, Jones JW (1993) J Am Ceram Soc 76:1281
12. Arons RM, Tien JK (1980) J Mater Sci 15:2046

Index

© Springer International Publishing AG 2017
J. Pelleg, *Creep in Ceramics*, Solid Mechanics and Its Applications 241,
DOI 10.1007/978-3-319-50826-9

Printed in the United States
By Bookmasters